THE PACIFIC CAMPAIGN

THE PACIFIC CAMPAIGN

THE SECOND WORLD WAR
THE US–JAPANESE NAVAL WAR (1941–1945)

DAN VAN DER VAT

Birlinn

This edition published in 2001 by
Birlinn Limited
West Newington House
10 Newington Road
Edinburgh
EH9 1QS

www.birlinn.co.uk

First published in Great Britain in 1992 by
Hodder and Stoughton Ltd

ISBN 1 84158 123 2

British Library Cataloguing-in-Publication Data
A catalogue record for this book is available from the British Library

Printed and bound by
Creative Print and Design, Ebbw Vale

Contents

List of Maps   8
Preface: The Rise of Japan   13

Introduction   19
    US errs – Japan strikes – Pearl Harbor –
    Philippines – Guam – Wake Island – British
    battleships sunk

PART I
COLLISION COURSE

  1 The View from the East   43
    Japan expands – German alliance – Japanese
    militarism – Hirohito – the Navy – Manchuria –
    the Army revolts – 'China Incident' – clashes with
    Russia – autarky and reform – US reacts – Japan
    takes the southern option – Yamamoto – US
    overreacts – the junta goes to war
  2 The View from the West   91
    Origins of the Second World War – US neutrality
    – Washington Conference – US isolationism –
    Roosevelt – US intelligence – 'Purple' – hunt for
    scapegoats – 'Magic' and 'Ultra' – who knew?
    – the Maryknoll mission – last talks – US war
    warning – war

PART II
JAPAN RAMPANT

  3 Japan Attacks   133
    Churchill on the rack – British trounced in
    Malaya – Singapore falls – US Navy stretched
    – uncommon front from Burma to Australia –
    Japan's strategy – the 'Zero' fighter – ABDA

command – Dutch East Indies – Indian Ocean –
Bataan and the 'Death March'

4 The Giant Awakes   167
US admirals Kimmel – King – Nimitz – Halsey –
US Navy hits back – Doolittle bombs Tokyo – US
submarines – and dud torpedoes – Japan goes too
far – General MacArthur, USA

5 The Turn of the Tide   204
Junta dazed by its own success – what next?
– Battle of the Coral Sea – Japan halted –
Rochefort's intelligence – Battle of Midway
– Admiral Spruance – Japan thrashed – US
Aleutians seized

6 Papua and Guadalcanal   239
US goes for Rabaul – gallant little Australia
– General Sir Thomas Blamey – MacArthur
takes over – Papua campaign – target southern
Solomons – Marines land – Japanese counter –
US disaster at Savo Island – US raid on Makin –
'Tokyo Express' – Henderson Field – carriers clash
– Japan reinforces

7 At Sea in the Solomons   272
US reinforces – submarine *I19* – 'Bloody Ridge' –
battles in the Slot – Halsey takes over – Sergeant
Basilone, USMC – US carriers hit – US admirals
and cruisers lost – 'L' for language – 'Tenacious
Tanaka' – know thine enemy – Japan withdraws
from Guadalcanal and Papua – US torpedo
debacle

PART III
AMERICA RESURGENT

8 Intelligence Applied   307
Casablanca – Operation Cartwheel – crisis in
Japan – Bismark Sea battle – blood in the water
– Yamamoto's last flight – code-breaking chaos –
Japanese intelligence – a German view – Japan's
substandard subs

9 Birth of the Leapfrog   331
Aleutians regained – death of a pilot – Lae
and Salamaua, New Guinea – up the Solomons
ladder – bypass and leapfrog – Bougainville –
New Britain – Japan trims – soldiers' gripes –
conferences

10 Hitting the Beaches   359
The Gilberts – shock of Tarawa – divided
US command – the Marshalls invaded – Truk
battered – Rabaul bypassed – sitting out the war
– MacArthur leapfrogs – western New Guinea –
Morotai

11 The Marianas and the Great Turkey Shoot   382
US prepares – the B-29 bomber – Saipan –
Philippine Sea battle – 'Turkey Shoot' – Spruance
blamed for winning – US Marines versus US Army
– Nagumo's last stand – human lemmings – Tinian
– Guam recovered – Tojo *à gogo* – US submarine
heyday – wolfpacks and convoys – Japanese
interservice rivalry – MacArthur shall return

12 The Philippines and Leyte Gulf   416
British return – Mindanao bypassed – 'hitting
'em where they ain't' – more Morotai – Palau –
Formosa – Leyte invaded – Mac back – kamikaze
– Battle of Leyte Gulf – 'The world wonders' –
struggles ashore – Halsey's typhoon – Luzon

13 '. . . Not Necessarily to Japan's Advantage . . .'   450
Bombing Japan – junta blames people – subs
strangle Japan – Iwo Jima – flag up (twice) –
Okinawa – supership *Yamato* sunk – US mighty,
slow – Prime Minister Suzuki – death of FDR
– Truman – the Bomb – USS *Indianapolis* –
Potsdam Declaration – Hiroshima – enter the
USSR – Nagasaki – surrender – the final score –
MacArthur's last word

Acknowledgments and Sources   486

Indexes   493

Maps

Pacific Campaign   10–11

The Japanese Empire   44–5

The Philippines, Singapore
   and the Dutch East Indies   135

New Guinea and Papua   240

Guadalcanal and Southern Solomons   254

Northern and Central Solomons   273

Mariana Islands   383

Philippine Islands   417

A military man can scarcely pride himself on having 'smitten a sleeping enemy'; it is more a matter of shame, simply, for the one smitten.

Admiral Yamamoto Isoroku, C-in-C, Combined Fleet
(letter to a Japanese citizen, early 1942)

We must not again underestimate the Japanese.

Admiral Chester W. Nimitz, C-in-C, Pacific Fleet
(entry in Command Summary, 10 December 1941)

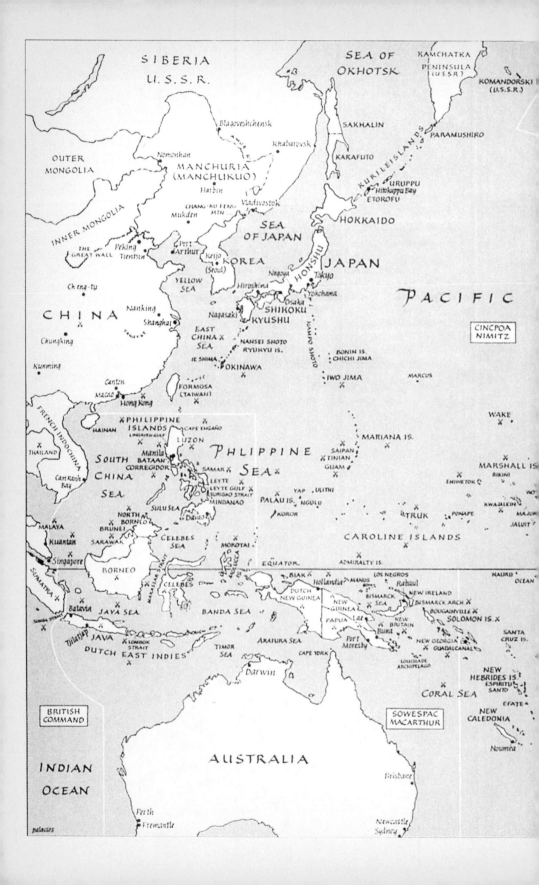

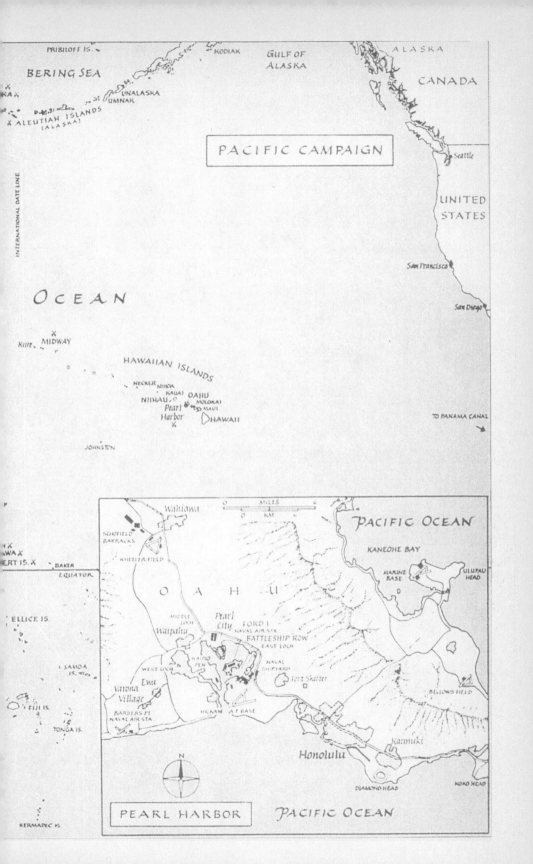

Preface: The Rise of Japan

On August 9, 1945, the centre of Nagasaki was destroyed by the second of two American atomic bombs, which shortened the Second World War by helping to force Japan to surrender before it was invaded. On January 18, 1990, the mayor of Nagasaki was shot in the back at point-blank range by a fanatic for daring to say, when Emperor Hirohito lay dying in 1988, that the wartime monarch was partly to blame for the conflict (and thus, by inference, for the way it ended). Nagasaki was rebuilt and is thriving; Mr. Hitoshi Motoshima, its mayor since 1979, was saved by two hours of surgery. The forty-four and a half years between these two events in the same city, during which Japan recovered from its crushing defeat to become the world's strongest economy, were manifestly not enough to enable the country to come to terms with its past. Yet the war whose strategically decisive campaign this book sets out to re-examine grew directly and organically out of Japan's history in the modern era, from when the United States forced it to emerge from its voluntary isolation. The two nations went to war in 1941; but the conflict, which still casts its shadow upon the present, went back to 1853.

The outbreak of general war in the Pacific in 1941 cannot be properly understood without reference to the general historical background, but Japan's early history need not detain us here. To cut a very long story short, the Japanese are descended from aboriginal Ainu people, whose strongest traces are to be found on the northern island of Hokkaido, and immigrants from China and Korea. This migration ended in about A.D. 600, and for a thousand years, while links were maintained with the Chinese mother-culture (the influence of Chinese writing, Buddhism and Confucian ideas was particularly strong), hardly any outsiders came. Between the twelfth and the nineteenth centuries, self-isolated Japan went through a feudal phase similar to Europe's in the ninth to fifteenth. Two attempts at invasion by the Mongols in the thirteenth century were driven off by 'divine wind' – kamikaze. Warfare between powerful barons, nominally the vassals of an imperial monarchy said to have been founded in 660 B.C., was endemic for long periods. The warlords were served by a knightly class called 'samurai'; the emperors continued to reign after they were supplanted as rulers by their marshals, the shoguns. The latter also remained in place as hereditary prosovereigns (the Tokugawa dynasty, from 1603 to 1868) after real power passed to a shifting group of the most powerful lords, the daimyo, reflecting a still-prevalent Japanese preference for group leadership. Then the Dutch came calling from their new colonies in the East Indies in the seventeenth century. From 1638 onward, only they and the Chinese were tolerated as visitors and trading partners on a small scale, and then only at the port of Nagasaki – until 1853.

By that time Japan's rulers had two worries on their minds. There was the 'domestic anxiety' in the shape of widespread social unrest caused by famine; and the 'foreign anxiety' caused by Western intervention in China, which was seen as a threat to Japan. Internal reforms could deal only with the former; there was no homemade answer to the age of steam. That technology enabled the Westerners to reach China in increasing numbers – and drew their attention to Japan as a way station on the new trans-Pacific route for the mechanically temperamental, short-ranged steamships of the period. The Americans, drawn across the Pacific by the same 'manifest destiny' which had brought them to the West coast of the United States, were annoyed by the Japanese practice of interning shipwrecked seamen and retaliated in kind; an American warship entered Nagasaki in 1849 and rescued twelve of them. When the Dutch tried to warn the Japanese of Western expansionism and urged them to make an orderly accommodation with it, the Japanese snubbed them and absolutely refused.

On the orders of President Millard Fillmore, Commodore Matthew C. Perry, United States Navy, began assembling his new East India Squadron in 1852. Its stated purpose was to open up Japan to limited foreign trade and to shipping, and to make consular arrangements for stranded seamen. He anchored in Tokyo Bay on 8 July 1853, with two armed steam frigates and two supporting ships, and insisted on waiting until the shogun agreed to forward the letter he had brought from the president to the emperor, containing these demands. The missive was accepted after a tense confrontation on the beach between 250 US Marines and bluejackets armed with rifles and ten thousand Japanese armed with bows, pikes and muskets. Perry said he would be back in due course and returned with eight ships in February 1854. His ultimate argument was his guns, vastly more powerful than the backward cannon still in use in Japan, but he was also a skilled diplomat. Negotiations through a Dutch interpreter took place at Yokohama, leading to the Treaty of Kanagawa, signed on 31 March. It provided for a consulate, supply facilities for shipping and the reciprocal return of stranded seamen. Lavish and bibulous exchanges of banquets ensued to celebrate Japan's entry into the modern world. Japan began to imitate Western industries almost at once, and the ports were thrown open in 1858.

The shoguns were thrown out ten years later, on the restoration of imperial rule under the Emperor Meiji. This was actually a revolution by the more progressive lords with the real power, who decided that, if Japan had to come to terms with the outside world, it should do so thoroughly, with the adoption not only of Western scientific and technical ideas, foreign trade and modern transport but also of Western institutions. Feudalism was abolished as the ruling oligarchy, the genro, copied the leading Western powers and set up a two-chamber parliament, an independent judiciary, a modern army and navy and very limited political rights (only 1 percent had the vote at first).

A society that had invented itself in all but total isolation over a millennium now proved astoundingly adaptable, yet stable enough to absorb overseas ideas without surrendering any important aspect of national identity. The 1889 constitutional reforms set the seal on the transformation as Japan decided to become a world power of the first rank. In 1895, when the

Emperor Taisho succeeded to the throne, Japan fought a short, sharp war with the Chinese in which it won control of Korea. In 1904–5 it fought another, against imperial Russia, taking over Russian concessions and spheres of influence in Manchuria and China proper. It was the first time that a 'non-white' nation had trounced a white one. Both wars were begun by a Japanese surprise attack. Successful involvement in war hugely enhanced the prestige and power of the general staff, which did not have an autocratic ruler like the German Kaiser but only a constitutional monarch to contend with. The generals thus began to supplant the genro as Japan's most important ruling group, far surpassing even its role model, the old Prussian general staff, in influence. And as they did so, Japan's ambition soared: now nothing less than world domination would do. Recent events suggest that the end is not far-fetched; the events described in this book merely show that the military means to that end were ill-chosen.

Japan won a remarkable tribute and certificate of respectability in 1902, when the British sought an alliance. The gain for Japan was recognition and influence; the British acquired protection for their interests in the Far East as they concentrated on their life-and-death, naval arms race with Germany closer to home. In the First World War, from 1914 to 1918, Japan honored the agreement and was handsomely rewarded with industrial expansion, the former German island possessions in the western Pacific, and more influence in China. Imperial ambition was strongly encouraged when Japan took its seat at the world's top table, the Versailles peace conference, as one of the 'big five' victorious Allied and Associated Powers. The Western refusal to accept the Asian upstart as a true equal only fueled Japanese expansionist ambition even further.

So did the great economic recession, which began to grip the world a decade after the Great War. Unable, in marked contrast to the United States, to find a one-country solution because of its almost total lack of material resources, and unable also to imitate the British Commonwealth-preference policy, Japan turned an acquisitive eye on China, provoking a war in Manchuria in 1931 and another in China proper in 1937. In so doing it clashed directly with the widespread and strong, if largely sentimental, American interest in that crumbling empire.

Even though it was doing a much more lucrative trade with Japan, which was helplessly dependent on American strategic supplies, the United States still focused its commercial and expansionist ambition on China, where it pursued what it termed an 'Open Door' policy of free trade, free access for missionaries, and special privileges for Westerners. Washington did not shrink from sharing in the spoils opened up to the West by that most disgraceful manifestation of British imperialism, the Opium War of 1839–42, which contributed so much to rendering China defenseless against Western exploitation. But when Japan sought to monopolize foreign influence in China from 1921 onward, American mistrust, jealousy, and finally active hostility, were aroused.

When Nazi Germany went on the rampage in Europe in 1939–40, Japan perceived an irresistible opportunity to take over the Asian possessions of embattled Britain, defeated France, and (especially) the overrun Netherlands, now prevented by Hitler from sending reinforcements to the fabulously wealthy Dutch East Indies. Here, it seemed, was Japan's chance to break out of its dependence on mostly American supplies of vital raw materials and to become self-sufficient, like the United States. The junta of generals who now totally dominated Japan overreached themselves, just as Hitler had done six months before: they opened a second front before closing their first, taking on the United States and its allies without ending their war in China.

What follows is a factual account, derived strictly from the enormous abundance of primary and secondary historical sources, of how that stupendous error came about, how Japan went to war and won all its objectives in short order, how it was checked and finally driven back across the Pacific – how the genie which Commodore Perry let out was (for a time) forced back into the bottle. It was only to emerge again in a different and much more subtle form a generation later, but that is another story.

For Japan the Second World War was a classical tragedy which it brought upon itself by its own defects. For the United States, which was not altogether blameless, the war became the triumph that confirmed it as the greatest power on earth. For both it was an avoidable conflict made inevitable by mutual underestimation and misunderstanding derived from arrogance and racial prejudice on either side. It is a story of error and heroism, terror and just occasionally humor. It also includes an astounding American intelligence coup which brought a decisive victory, yet had proved wanting when it was most needed – and too clever for its own good before that. But in telling the tale as it has not been told before – from both viewpoints, with the accounts of participants set in a framework of historical analysis and narrative – the rapacious and brutal Japan of the generals is not equated with the United States, which did not want war yet became the avenging 'arsenal of democracy.' There was no comparison; there was only naked aggression met by irresistible force. No less morally than militarily, no less ethically than economically, it was a most unequal fight which Japan should never have begun.

All the sources are listed at the back, together with the acknowledgments. I have tried to follow the convention, prevalent at the time, of setting down the family names of Japanese individuals before their given names. As this is a book intended for the general reader, like all my others, I have sacrificed footnotes for the sake of narrative flow. But there is no fiction or faction in these pages, and all quotations are from historical records unless otherwise stated.

Introduction

In the first week of July 1941, the government of the United States made a historic mistake. It took an intercepted message from the Japanese Foreign Ministry at face value and concluded that the Second World War was on the point of spreading to the Pacific Ocean. The American response, economic sanctions backed by military precautions, and Japan's reaction to them, turned this misinterpretation into a self-fulfilling prophecy. The United States armed forces were therefore denied the full margin of time they believed they needed to be ready for combat – by March 1942 at the earliest. Indeed, Japan's determination to pre-empt American interference in its plan for expansion in East Asia plunged the USA into war three months sooner, by knocking out the American battle fleet at Pearl Harbor on 7 December 1941. The Pacific Campaign which ensued – the largest naval conflict in history – was inevitable in any case, because the ambitions of the two rivals in China were irreconcilable. Japan had already been at war for a decade, in pursuit of its imperialist goals on the Asian mainland. This fact, and the diplomatic duplicity of the military which then ruled the Japanese Empire, simply overrode the powerful American desire to stay out of the fighting.

So there was very probably no way in which the United States could have avoided an armed confrontation with Japan. But it could have delayed the inevitable for long enough to give Army and Navy the extra months they had said they needed to achieve full readiness for combat. The irony in all this, one of several in the historical background to the war, is that the error of July 1941 – precipitating the Pearl Harbor disaster, from which the US has arguably never psychologically recovered – would not have been possible but for a brilliant and sustained coup by the very intelligence agencies that were later widely blamed for failing to save the United States Pacific Fleet!

President Franklin D. Roosevelt's third administration did not lead the American people into war in order to rescue Great Britain, still less to salvage Europe's Asian colonies, but in response to Japan's attacks on America's Pacific possessions, without warning or declaration of war and in the midst of diplomatic negotiations. Japan struck because it correctly identified the United States as the only power capable of frustrating its aims in the Pacific Ocean region – and because it wildly misread and underestimated the enemy it thus chose to conjure up against itself. It felt compelled to act in order to replace the strategic raw materials needed for its unending struggle in China but no longer being supplied by the United States.

These and related matters will be considered in detail in part one. Meanwhile, this is how the Rising Sun of Japan attempted half a century ago to immobilize the world's most powerful nation for long enough to seize all the main European possessions in Southeast Asia and the southwestern Pacific.

In the first week of December 1941, Admiral Yamamoto Isoroku, commander-in-chief of the Imperial Japanese Combined Fleet, was aboard his flagship, the battleship *Nagato*, at Kure, on Hiroshima Bay in Japan's Inland Sea. He was awaiting news from the Mobile Force of aircraft carriers he had sent to execute an operation that he had personally insisted upon adding to Japan's war plans for southward expansion.

The First Air Fleet – core of the Mobile Force – was based on six carriers in three divisions (*Akagi*, fleet flagship, and *Kaga*; *Shokaku* and *Zuikaku*; *Hiryu* and *Soryu*) with a total of more than four hundred planes. There was a screen of nine destroyers led by a light cruiser, a supporting force of two battleships and two heavy cruisers plus a train of eight tankers and supply ships. Three submarines provided reconnaissance, and two extra destroyers were in company to watch out for American war planes based on Midway Island, the US outpost northwest of Hawaii. In command was Vice-Admiral Nagumo Chuichi, four years younger than Yamamoto and rather more nervous. Barely eight months in the job, his speciality was torpedoes, not aircraft, and he was prone to bouts of anxiety which prevented him from sleeping; even the smallest decision caused him stress. He tried to cover up these weaknesses by bluster. It is an extraordinary irony that this man was about to lead his force on an unprecedented rampage of maritime destruction lasting six months – until he lost his biggest battle and his nerve. Lieutenant Commander Fuchida Mitsuo, thirty-nine-year-old chief pilot on the *Akagi* and designated operational leader of the attack, was obliged to visit Nagumo several times during the preceding night and found his admiral in a condition not far short of nervous collapse.

The force assembled by 22 November 1941, at Hitokappu Bay in the then Japanese Kurile Islands, immediately northeast of Japan proper. It set sail four days later, running slowly eastward through fog and heavy swell at first, then gales and massive seas. On 3 December the formation reached a point some nine hundred miles north of Midway and turned southeast. On the morning of the 7th it altered course to due south. After parting company with the tankers, all the warships put on speed until the carriers reached the designated point of launch, 275 miles due north of Pearl Harbor. It was just before 6:00 A.M. Hawaiian time (then ten and a half hours behind Greenwich) on Sunday, 7 December 1941; in Washington it was already 11:30 A.M.; in Tokyo, far on the other side of the international date line, it was 1:30 A.M. on Monday the 8th. Aboard the *Akagi* speeches were made and traditional toasts drunk in saké on the broad, wood-lined flight deck. Behind the rows of impassive pilots' faces mixed feelings seethed: pride, fear, devotion to duty, contempt for the soon-to-be enemy, exultation. They were about to face the unknown; but if meticulous preparation alone could guarantee success, they had little to fear. The plan they were about to execute had been drawn up by Commander Genda Minoru, air-staff officer of the First Air Fleet, known to the service as 'mad Genda' for his aggressiveness (which must have been exceptional indeed to be marked in this way). He was now on hand aboard the flagship *Akagi* and recalled afterwards:

In February 1941, at the order of Admiral Yamamoto . . . I made an investigation and concluded the attack was difficult but possible; I thought the best plan was to attack by planes but not to land . . .

In June and July I began to check on the details, first investigating only the employment of torpedo attacks. With that in view I had a carrier pilot carry out test flights at Kagoshima Bay to acquire data on shallow-water attacks and torpedo arming. The carrier pilots' crews were trained for the attack beginning in September and going intensively into training in October.

At 1:30 A.M. Tokyo time (used by all Japanese warships wherever they might be) the carefully preserved signal flag for the letter 'Z' which had been hoisted by Admiral Togo at Tsushima to signal the attack on the Russians was unfurled over the *Akagi*. Nagumo sent the order to take off down the chain of command. Within fifteen minutes 183 aircraft from the six carriers obeyed: torpedo bombers, high-level bombers, and dive-bombers with Zero fighters as escort. Fuchida, wearing a blood-red shirt to conceal any wound from his men, was already up, in an aircraft piloted by Lieutenant Matsuzaki Mitsuo; this first was commanded by Lieutenant Commander Nakaya Kenju. Fuchida shouted, '*Tora, tora, tora!*' into the radio microphone built into the mouthpiece of his leather flying helmet at 0753 Hawaiian time – the code word to report that total surprise had been achieved (the word means 'tiger'). He recalled:

Closing in on our objective, it was tricky adjusting the course for the bombing run. It was time to drop now . . . We were already into Pearl Harbor and over Ford Island . . . We were skimming over the streets of Honolulu, turning again into the bombing run . . . There was a huge explosion close to our objective. A deep red flame burst up into the sky, followed by soaring dark smoke, then white smoke to a height of what looked like three thousand feet. That must be the magazine! A shock like an earthquake went right through our formation and my aircraft shuddered with the force of it. It was the *Arizona* going up.

It was indeed. The battleship lies on the bottom of Pearl Harbor to this day, a permanent memorial of an attack without warning – the traditional opening gambit of the Japanese military.

As I was feasting my eyes on this scene [Nakaya recalled] a terrific column of water flew up into the air from the side of a capital ship. It was an excellent hit on the capital ship's side. The column of water went up high enough to compete with the altitude of the clouds. Columns of water continued to gush forth one after another . . . The surprise attack was a complete success.

Aboard the light cruiser USS *Helena*, moored off the southern shore of the harbor, Chief Petty Officer Leonard J. Fox was up betimes, writing to his wife and daughter in Phoenix, Arizona. Hearing aircraft overhead at four minutes to eight in the morning, he stepped out onto the main deck, looked north toward Ford Island in mid-harbor close by, and saw the bombs raining down.

I suddenly realized that these planes were not our own . . . Within the next few moments all hell broke loose. The Japs had at last done the impossible and were attacking our island fortress . . . Torpedo planes swooped in from almost over my head and started toward 'Battleship Row' dropping their lethal fish. First the *Oklahoma* . . . then it was the *West Virginia* taking blows in her innards. . . . Now, as I looked on unbelievingly, the *California* erupted . . . and now it is the *Arizona* . . . Men were swimming for their lives in the fire-covered waters of Pearl Harbor.
A Jap plane . . . spied the *Helena*. He made a fast turn and headed our way . . . He was only about 100 yards away when he released his torpedo . . . A tremendous shuddering took place and billowing smoke and flame immediately shut all the sunlight out of my life and I believed that my ship was no more but . . . the ship did not blow up . . .
Virtually every ship in the harbor that was able to, was bearing guns on the enemy . . . We could look up and see the bombs released . . . One of these bombs made a very fantastic hit on the *Arizona* and went down her stack and blew up her forward magazine and she was a blazing inferno within minutes.

The second wave of 170 attackers, the mixture of types as before, arrived over Pearl Harbor at 8:40 A.M. local time. Like the first, it concentrated on ships and airfields, including aircraft on the ground, undeflected by stiffer anti-aircraft fire rising from the smoking shambles of Pearl Harbor and its environs. In all, eighteen United States ships were either sunk or badly damaged, including four battleships on the bottom and four crippled; 164 Navy and Army aircraft were destroyed and 124 damaged. For this the Japanese paid with twenty-nine planes destroyed with their crews and seventy-four damaged. The Americans reported 2,403 killed and 1,178 wounded, the vast majority from the Navy.

Amid the triumphant messages exchanged between the flagships of Mobile Force and Combined Fleet, a bitter dispute developed among the staff officers aboard each: whether, in view of the overwhelming success of the first two waves, the planned third assault should go in. A nervously jubilant Nagumo vetoed it, over the vehement objections of Fuchida; Yamamoto backed the senior officer. Thus were the fuel tanks and other vital shoreside facilities of Pearl Harbor saved from all but incidental damage: they were to have been the prime target of wave number three; their destruction would have rendered the base useless and forced the US Navy back to the West Coast, over two thousand miles to the east. Further, although the bulk of the American Pacific Fleet, its battleships, had been knocked out, the Japanese had just demonstrated how aircraft carriers would dominate the conflict so brutally begun. And of the three carriers then assigned to the Pacific Fleet, none was in harbor at the time. Finally and worst of all, Japan had roused up an invincible enemy against itself.

One of the absent carriers, the *Enterprise* (Captain George D. Murray, USN), flagship of Vice-Admiral William F. Halsey's Task Force 8, had delivered a squadron of twelve Marine Corps fighters to reinforce the garrison on Wake Island, an isolated American outpost about two thousand miles west of Pearl Harbor. She was due back at Pearl on Saturday, 6 December, but, lucky ship that she was, had the good fortune to be delayed by bad weather. On Sunday morning, less than two hundred miles west of Pearl, *Enterprise* routinely flew off some of her planes to precede her to Pearl Harbor. Their pilots reported breathlessly by radio that the base was under Japanese attack and joined the fray; four or five were shot down, some by American fire. Even so, they and their Pearl Harbor comrades were not the first Americans to shoot at the enemy, any more than Pearl Harbor was the first place outside China attacked by the Japanese.

That dubious honor fell to Kota Bharu, at the northern end of British Malaya, nearly two hours earlier; and at 6:37 A.M. Hawaiian time on the 7th, off the entrance to Pearl Harbor, the destroyer USS *Ward* sank one of five midget submarines launched by conventional submarines in support of the air raid (all five were lost).

A dogfight over Hawaii during the Pearl Harbor attack led to the first – and shortest – Japanese occupation of US territory. The brief air battle pitted seaplanes from naval squadron VCS-5, divided among three USN cruisers, in an unequal contest with IJN Zeros. But the Americans forced one Japanese pilot to land on the island of Niihau, northwest of Oahu, the main island.

He took the machine guns out of his plane and with the aid of two Japanese gardeners actually captured the island and held it for a couple of days [Lieutenant Commander Earl A. Junghans, USN, squadron leader, recalled later]. They had no radio or cable or telephone there. They didn't know what was going on. Just a bunch of natives. But this Japanese was finally killed by a native Hawaiian who resented this machine gun being set up in the square in the village. So he just charged it – and although he was hit three times he finally got in on this Jap and bashed his head in with a rock.

Two minutes after the attack began, Rear Admiral Patrick Bellinger, commander of the naval air station on Ford Island in the middle of Pearl Harbor, was the first to broadcast the electrifying message: 'Air raid, Pearl Harbor. This is not drill.' It went round the world by radio, telephone and teleprinter in minutes – dread proof that the struggle in the West and Japan's war in the Far East had merged into one enormous global conflict, much bigger than 1914–18, which now truly lost its label as the 'Great War.' The sneak, hit-and-run attack on the American fleet was not Japan's first move, but that is a technicality: the news of it was the first intimation America and the world received of Japan's intentions, and it could hardly have been a more dramatic shock. If the Japanese wanted to galvanize the Americans into action, they could not have chosen a more effective way of doing it. Such is the true historical significance of Pearl Harbor.

Handed Bellinger's message in Washington, Frank Knox, secretary of the Navy, exclaimed: 'My God, this can't be true, this must mean the Philippines!' He was premature, but only by a matter of hours.

The signal that so badly shocked Knox was intercepted by the US Marine Corps radio operator on night duty at the Marsman Building in downtown Manila, capital of the Philippines, at 2:30 A.M. on Monday, 8 December, local time (seven minutes after the attack on Pearl Harbor began). The radioman recognized the sender's 'fist' – no two operators have quite the same touch when using a Morse-code key – as belonging to a friend and therefore knew the monitored message was genuine. He alerted the duty officer, who first telephoned Admiral Hart, the Asiatic Fleet commander, to say he was coming over with a message that had to be personally delivered, and then rushed the news to him at the Manila Hotel, some three hundred yards away. At 3:00 A.M. Hart was up and drafting an alert to all his ships; he was hard at work in his office at the Marsman an hour later. He sent his chief of staff, Rear Admiral W. R. Purnell, to liaise with Army headquarters in the Walled City, a short drive away. This was a fortunate inspiration, because Major General Richard K. Sutherland, chief of staff to General Douglas MacArthur, American commander-in-chief in the Philippines, had not heard the news. He picked up the telephone and called his commander at the penthouse suite in the Manila Hotel, which MacArthur used as a residence.

The conflicting accounts of what happened next among the general's entourage have never been reconciled, but much precious time was wasted. Major General Lewis H. Brereton, his air commander, independently told of the Pearl raid, was denied access to the C-in-C for hours by Sutherland. Brereton had thirty-four operational B-17s (the largest concentration of America's most modern heavy bombers at the time) and eighty-seven fighters and urgently wanted orders for their deployment. His instinct was sound enough: he wanted to bomb the air bases on Formosa (now Taiwan), the Japanese-occupied Chinese island some seven hundred miles north of Manila, from which a land-based air offensive against the Philippines was most likely to be mounted. Sutherland could not give permission for such a step; MacArthur would not. Brereton therefore limited himself to using his bombers for reconnaissance. At 8:30 he sent up virtually all his aircraft, in response to a general alarm at Clark Field; this was canceled at 10:00 A.M. The planes landed again, and the crews went for refreshment as ground staff carried out maintenance and refueling. At 11:00 A.M. Brereton got permission from Sutherland to attack Formosa, and the armorers at last began to load the B-17s.

At 12:40 they were still on the ground when two V-shaped formations of twin-engined bombers, fifty-four in all, appeared high over Clark. Five minutes later, or ten hours and nineteen minutes after the attack on Pearl, the Japanese Naval Air Force began to bomb and strafe. At a stroke, twelve B-17s were destroyed and five crippled; the other half of the 19th Bombardment Wing took lighter damage. Just four American fighters managed to take off, but they were destroyed by the formidable Zero fighters escorting the bombers. In all, one hundred US Army aircraft were knocked out at Clark, ninety-six of them in their serried ranks on the airfield – 'on the *ground*, on the *ground!*,' as a furious Roosevelt exclaimed when the news was broken to him.

The Japanese attackers were handed this strategic triumph on a plate, despite the serious disruption of their operational plan, intended to be synchronized with Pearl and the other opening moves, caused by bad weather in Formosa.

Because of the fog we were delayed [said Commander Shibata Bunzo, staff officer with the Navy's 21st Air Flotilla on Formosa]. We intended to attack at dawn but actually delivered the attack on Clark Field . . . about 12:30. The purpose of the first attack was to destroy enemy planes and installations on the ground. The fighters were to provide coverage for the bombers and intercept any American planes which arose in the air. After the bombers successfully

completed their mission they were to withdraw and the fighters were
to strafe.

[The fog had] really no effect, although we were very worried
because we were sure [that the Americans,] after learning of Pearl
Harbor . . . would disperse . . . or make an attack on our base at
Formosa. At Formosa we put on gas-masks and prepared for an attack
by American aircraft. [Japanese fliers from Formosa half expected to
pass their American counterparts from the Philippines heading in the
opposite direction, the commander said.]

Lieutenant Howard W. Brown of the US Army Signal Corps
was sent to MacArthur's office by Sutherland at about 1:00 P.M.
on the 8th to explain why an important intelligence message had
been delayed to the point of uselessness. He had to wait in
front of the imposing desk because MacArthur was shouting at
Clark Field in his powerful, sonorous voice even as it was under
attack. 'The general had the commanding officer on the phone
giving him hell,' said Brown (MacArthur also blamed Brereton,
who heartily reciprocated). Brown duly made his report to an
expressionless C-in-C, who answered, 'Thank you, son,' and
dismissed him.

Lieutenant 'Tex' Wymer of the US Army Air Corps, in charge
of the radar station at Iba, on the west coast of Luzon Island,
on the morning of the 8th, reported unidentified aircraft coming
'from China' at a range of 140 miles at about noon. He
repeated his alert when the planes were eighty, sixty, fifty,
forty and twenty miles out. It was only when they were in
sight that the twenty-four P-40 fighters stationed locally began
to scramble. About fourteen were destroyed on the ground,
killing ten pilots. Six got into the air, of which four were
promptly shot down. 'This was our first lesson in the Air
Corps . . . refusing to believe that the Japs were coming
until [we] could see, count, and identify the enemy planes in
the sky,' Wymer said later. He relieved his frustration, caused
by the widespread mistrust of the new-fangled radar among his
colleagues, by taking his personal, high-powered hunting rifle to
a sandbagged position on the beach and firing at the Japanese
intruders. He was convinced that he hit a Zero with his third

shot: at any rate it flew out to sea and did not return to the attack.

Thus was American airpower in the Philippines effectively eliminated; and with it went the strategy of defending the island of Luzon at all costs until the Pacific Fleet arrived (after traversing five thousand miles of dangerous ocean in defiance of a stronger enemy fleet!). Three hours earlier, smaller Japanese formations had bombed and strafed two military bases in the north of Luzon, the principal island of the Philippines; no report of these attacks had reached Manila, in the south of that island. But this has no bearing on the salient fact that MacArthur and his senior commanders had known of Pearl Harbor for eight and a half hours by the time the bulk of their airpower was caught helpless on the ground, at a cost to the enemy of just seven planes. In the circumstances, it can only be seen as fortunate that the Americans had not had the time to build up their B-17 force in the Philippines to the planned one hundred aircraft. Two days later, only thirty-five American fighters were available to challenge a double raid on the airfields north of Manila and the Asiatic Fleet's main base at Cavite. They were swept aside.

Soon landings by Japanese amphibious troops secured two airfields elsewhere on Luzon, to be used as forward bases for the occupation of all the Philippines; the forty thousand men, two reinforced divisions, of Lieutenant General Homma Masaharu's 14th Army (equivalent to a corps in the West) were able to land without difficulty at Lingayen Gulf on 22 December. Notwithstanding the brave and bitter last stand of the rump of MacArthur's 140,000 ill-prepared Filipino and American troops (described later), the inability of American air and naval forces to obstruct Homma had sealed the fate of the archipelago in advance. It was an infinitely greater strategic setback than the loss of the US battle fleet; and, unlike Hawaii, the Philippines command had received an invaluable warning which it chose to ignore.

Yet it was Pearl Harbor, back in business almost overnight, that became the subject of eight official investigations, crowned by a vast congressional probe after the war; it arouses heated controversy to this day. The catastrophic loss of the Philippines for three cruel years (and the gigantic effort which went into recapturing them) passed forever unexamined in Washington. Instead, the administration was to make MacArthur the most famous American never to occupy the White House.

The prewar idea – 'delusion' might be a better word – of a majestic progress by the United States fleet from the West Coast and Pearl Harbor across the Pacific to the Philippines, wisely abandoned until rashly revived in modified form on the urging of General MacArthur, envisaged the use of Wake Island and Guam as staging posts. The Japanese took this as read and therefore included these isolated islands on their list of objectives for the opening phase of the war, well before Pearl Harbor was added. Both were to be taken by Vice-Admiral Inouye Shigeyoshi's Fourth Fleet, based at Truk, the principal Japanese mid-Pacific naval base in the Caroline Islands. The Americans had prudently removed all families, dependents, and other nonessential personnel some weeks before hostilities began.

After two days of heavy air attacks on Guam, the lone US possession in the otherwise Japanese Mariana Islands, the Japanese, led by Rear Admiral Goto, flying his flag on the heavy cruiser *Aoba*, landed five thousand men on 9 and 10 December – including one troop of cavalry complete with horses. To resist this sledgehammer, the Americans could muster 365 Marines and about three hundred Chamorros – island indigenes – in the local militia, three naval patrol boats, and a few machine guns. Congress had consistently refused to provide funds for a stronger garrison; but on the orders of the president himself an advance party had arrived a few weeks earlier to prepare the ground for the construction of a military airfield. When the Japanese forces found the garrison, the island's governor, Captain G. J. McMillin, USN, let the fighting go on for twenty-five minutes for honor's sake before ordering the Stars and Stripes hauled down, forty-two years after Guam had become an American possession. US military and civilian personnel were held for a month on the island and then shipped to detention in Japan. Lieutenant (j.G. – junior grade) Leona Jackson of the Navy Medical Corps got over the shock of the invasion (and of hearing from a smiling Japanese naval captain, 'So sorry to tell you – your fleet all sunk') and studied her captors:

There are a number of things I observed about the Japanese [Mrs Jackson reported after repatriation as a noncombatant in 1942]. I think they have staked practically everything on this war, and I think they realize . . . that if they lose they are lost for a long time. It is going to be a nasty fight, and it is going to be a dirty fight, because they don't fight according to our rules, they make their own as they go along and they aren't pretty. They are not a nice people to be at war with. I think it is going to take an awful lot, I think we all know. It is just a dirty job that we have to get done. [Mrs Jackson added that she was very keen 'to get back and get a few licks in.']

After Pearl Harbor on the 7th, the opening attacks on the Philippines on the 8th and the capture of Guam on the 10th, there remained one American possession in the Pacific to tackle: Wake Island, roughly halfway between Guam and Hawaii. The Japanese sent Rear Admiral Kajioka Sadamichi with a force of three old light cruisers and six destroyers, plus transports carrying 450 Special Naval Landing troops (Marines), supplies, and equipment, to occupy it on 11 December. Their gross overestimation of what was needed to subdue Guam was now offset by serious underestimation of what was required to take Wake.

   Wake Island, actually an atoll of three islands round a lagoon, lies about six hundred miles north of the Marshall Islands, one of which, Kwajalein, was at that time the most important air base in all the Pacific islands mandated to Japan after the First World War. Wake had been designated, like Midway Island, for development as an air and submarine base for the American Navy in 1940. Work was well advanced on the airfield late in 1941: its commander arrived at the end of November, and Major Paul Putnam's Marine fighter squadron VMF-211 was delivered by the *Enterprise* on 4 December. The military garrison consisted of thirteen officers and 365 Marines from the First Defense Battalion, commanded by Major James Devereux, USMC. Among their weapons were six five-inch ex-naval guns in three batteries, and three more of four three-inch anti-aircraft guns each. In overall command was Commander Winfield S. Cunningham, USN, head of the new naval air station.

The atoll was subjected to a series of air raids from Kwajalein for the three days preceding the invasion. In the first raid, not long after the attack on Pearl, seven out of twelve Marine fighters, Grumman 'Wildcats,' were destroyed on the ground (yet again) and an eighth was wrecked (patched up later); twenty-three air and ground crewmen were killed. In the early hours of the 11th the invasion force approached in a heavy swell, firing a short barrage which did little damage. The American coastal-defense Battery A hit Kajioka's flagship, the light cruiser *Yubari*, and drove her off badly damaged. Battery L sank the destroyer *Hayate* and hit three other ships in quick succession. Battery B caused three other destroyers to retire. The four surviving Wildcat fighters damaged another cruiser and sank the destroyer *Kisaragi*, as well as driving off another air attack with several losses to the Japanese (and two to themselves). Only one American died; the Japanese lost about five hundred sailors and airmen. Admiral Kajioka ignominiously withdrew to Kwajalein. But this was the only setback for the Japanese in the opening phase of their enormously broad southward advance. The gallant garrison meanwhile was jubilant, and looked forward to relief from the east.

Admiral Husband E. Kimmel, commander-in-chief of the Pacific Fleet, was still suffering from shock after the attack on Pearl Harbor, but his brain was functioning. He had urged a very strong garrison for Wake in the spring of 1941, big enough to oblige the Japanese to send a substantial force against it in case of war. He had argued that this might well earn the Navy a chance to destroy a significant element of the enemy fleet. Immediately after the Pearl raid Kimmel had emotionally torn off two of his acting full admiral's four stars, correctly anticipating demotion to his substantive rank of rear admiral. His last major decision as CINCPAC was to order all three Pacific Fleet carrier task forces to the relief of Wake. But overcaution, breakdowns, fueling problems and general confusion over the enemy's whereabouts caused the group due to deliver the augmented Marine fighter squadron of eighteen planes and more troops – Rear Admiral Frank Jack Fletcher's Task Force 14 including carrier *Saratoga* with three heavy cruisers and eight destroyers – to turn back 425 miles northeast of Wake, on the evening of 23 December. Kimmel had already been dismissed six days earlier. His designated successor, Admiral Chester W. Nimitz, was stuck in Washington; and the acting CINCPAC, Vice-Admiral William S. Pye, was not about to rush strategic forces to the relief of Wake without a direct order. The only guidance Pye had from Washington was permission to evacuate the island at his discretion – and the considered opinion of his superiors that the place was a liability.

By then too, Admiral Najioka had returned in the patched-up *Yubari* with two more light cruisers, eight destroyers, four transports, and two troopships. This time he was covered by a support force of four heavy cruisers and attendant destroyers led by Admiral Goto, recently in Guam, whose role was to engage any relieving force coming from the east. To make assurance doubly sure, a division of two fleet carriers with supporting heavy cruisers and destroyers under Rear Admiral Abe Koki was detached from Admiral Nagumo's Mobile Force, still on its way home from the Pearl Harbor raid. This formidable lineup completely vindicated Kimmel's prediction of what the Japanese would do against a heavily defended Wake. Faster and more resolute deployment by the Americans might conceivably have enabled them to confront the enemy with forces strong enough to secure an early victory. But Admiral Pye was more concerned with keeping the residual Pacific Fleet intact and with protecting Hawaii. His order to pull back was received with incredulity and rage bordering on mutiny in the *Saratoga* group. But it was obeyed; and since US naval intelligence had no idea where the bulk of Nagumo's ships were but did believe that at least two carriers, two battleships and two heavy cruisers were within reach of Wake and the *Saratoga*, it still seems the only sensible course. Had the fight for Wake become an option, the Japanese ultimately had six carriers and two battleships at sea to the Americans' three and none. Fortunately for the US Navy and public morale, the story of the on-off expedition fiasco did not get out at the time.

Meanwhile, Abe's aircraft had joined in the raids to which Wake Island had been repeatedly subjected from Kwajalein ever since the repulse of the first assault. The Marines were now without any serviceable aircraft, and the ground defenses were being steadily worn down. At 2:35 A.M. on the 23rd the Japanese landings began in several places at once. With commendable understatement, Commander Cunningham at 5:00 A.M. sent a signal to Pearl Harbor: 'The enemy is on the island. The issue is in doubt.' This helped to make up the agonized Admiral Pye's mind in favor of a general withdrawal. Having got Cunningham's permission to surrender, Major Devereux did so and then toured as many Marine strongpoints as he could reach, persuading his men to lay down their arms. The Japanese, having lost more than eight hundred dead and three hundred wounded against 120 American dead in the second attack on Wake, none too gently took 470 military and 1,146 civilian prisoners. The Marine Corps publicity machine, one of the most aggressive and skilled operations of its kind in a war that, for the United States, was very much a PR man's battlefield, immediately made a hero of Major Devereux. He was presented in the handouts, and even in a later motion picture, as the leader of the unquestionably heroic defense, to the apparent detriment and real disappointment of Cunningham, the island commander. The major was reported to have sent a signal during the first attack saying, 'Send us more Japs'; even though some Japanese saluted this as an excellent morale-boosting ruse by their enemy, Samuel Eliot Morison, the quasi-official war historian of the Navy, says flatly that it 'was never sent by him or anyone else.' More distortions took place when the two officers returned from their ordeal in Japanese prison camps after the war, but at least Cunningham was finally awarded a Navy Cross for his role on Wake Island.

The fall of Wake, though not as important as it might have been because of the neglect of the defenses of Guam, its strategic counterpart to the west, effectively put the Philippines out of reach of the US Navy. The archipelago was now wide open to a full-scale Japanese invasion following the annihilation of the local American airpower. The Japanese had not only incapacitated the US battle fleet; even if it recovered overnight, they had also deprived it of all the bases it needed for a serious counterstroke.

But the Japanese Navy did not content itself with pre-emptive strikes against its most formidable enemy in Hawaii and the Philippines. It also seized the opportunity to preclude any serious interference from America's strongest potential ally in the Far East – the United Kingdom, with its great naval base at Singapore and its airfields in Malaya. The first move in a series that was quickly to knock Britain out of the war in the Pacific was an exemplary piece of aggressive opportunism by the Japanese Naval Air Force – a fortuitous embellishment of the Japanese Army's long-standing plan to take Singapore. This principal British bastion in Asia, with its southward-facing, heavy coastal batteries, was to be attacked from the rear, which is to say from the north, through Malaya.

Before going to war, the British, like the Americans, had a one-ocean fleet. The Royal Navy was already fighting in the Mediterranean as well as the Atlantic and was therefore badly overstretched. Incapable since the First World War of doing more at sea than matching the strongest foreign navy (with difficulty), the British Navy had in effect strategically withdrawn from the Far East, despite the home government's promise to defend Australia and New Zealand. The odd cruiser and smaller vessels apart, the Royal Navy concentrated on home waters and the Mediterranean, through which passed the sea route to India via the Suez Canal. This was still the main artery of an empire whose days were numbered, not least because the 1914–18 war left Britain too poor to sustain it militarily. But between the wars the British built up Singapore as their main naval base in the Far East – 'the Gibraltar of the Orient' – to which a battle squadron was meant to be sent in the event of tension or worse with Japan. The much-vaunted 'fortress' was completed in February 1938 at a cost of £60 million (then worth nearly $300 million). The

Japanese chose to regard it not as a defensive measure but as an active threat to them: a naval officer published a book called *Japan Must Fight Britain* in 1936 – in English as well as his own language.

As the threat of war with Japan mounted in the summer of 1941, the British cast about for a means of fulfilling their undertaking to deter the Japanese fleet, by now immensely strong and with the unanswerable advantage of being concentrated entirely in the Far East. The Royal Navy had already lost six capital ships (three fleet carriers, two battleships and one battle cruiser) in two years of war; others had been knocked out by damage or breakdown. The British fleet had to guard against strong enemy naval forces in Norway, northwestern France, and Germany itself. Yet, when the time came to face Japan, London scraped together three big ships for its eastward-bound 'Force G' – the old battle cruiser *Repulse*, the new battleship *Prince of Wales*, and the latest carrier, HMS *Indomitable* (twenty-three thousand tons) – to provide air cover. The last-named however ran aground on 3 November 1941, off Jamaica, on her way east. Her diversion to Norfolk, Virginia, for twelve days of repairs ensured that she would be unable to join the other two before war broke out in the Far East. Her designated companions had to make do without the protection of her forty-five aircraft – which doubtless saved the planes, the pilots and their carrier from destruction by superior Japanese forces.

The *Repulse* (Captain W. G. Tennant, RN), at thirty-six thousand tons all up, was as imposing as a battleship but rather less well armored. Launched and completed in 1916 and twice modernized in the lean interwar years, she had been in action just once, in the North Sea in 1917. Her main armament was six fifteen-inch guns in three turrets, plus a dozen four-inch. Her anti-aircraft armament was poor: just eight four-inch high-angle guns and only two eight-barrel 'pom-pom' batteries firing two-pound shells. She carried a crew of 1,240 men and sixty-nine officers. Her captain was known as 'Dunkirk Joe,' because he had been senior naval officer ashore there in the great evacuation of the summer of 1940 and was one of the last to leave.

The *Prince of Wales* (Captain J. C. Leach, RN) was a 'Treaty' battleship, limited to thirty-five thousand tons' displacement in accordance with the Treaty of Washington of 1922. She carried ten of the latest, rapid-firing fourteen-inch guns with a range of eighteen miles, eight pairs of 5.25-inch heavy anti-aircraft guns, six pom-pom batteries, and an array of heavy and medium machine guns. Hit by a German bomb before her launch, she ran aground soon after joining the fleet in January 1941 and was damaged in the Battle of the North Atlantic in May, when the British fleet flagship, battle cruiser *Hood*, and the German fleet flagship, battleship *Bismarck*, were sunk. She was therefore regarded by superstitious sailors as an unlucky ship (although she was surely lucky to escape from the *Bismarck*). She had been repaired in haste before going east with her crew of 1,502 men and 100 officers. Her captain had a large nose and was therefore called 'Trunky' by the British sailors, who were wont to find a nickname for everybody.

The flag officer responsible for the British Navy's forlorn hope in the Far East was known to the bluejackets as 'Tom Thumb,' from the coincidence of his first name and his small stature. Admiral Sir Thomas Phillips, vice-chief of the Naval Staff, had last seen action in 1917, like *Repulse*, and had most recently been in a seagoing command in 1939, just before the war began. He was naturally delighted with his orders to take command of the British Eastern Fleet. Force G was to be its nucleus, with more capital ships to follow when necessary and/or possible, together with cruisers, destroyers and other supporting vessels. The *Prince of Wales* and the *Repulse* joined up in Ceylon (now Sri Lanka), where they were restyled 'Force Z,' at the end of November. They reached Singapore with their escort of four destroyers on 2 December.

At lunchtime on Monday, 8 December, Phillips took the chair at a council of war on the *Prince of Wales*, his flagship. The preceding twelve hours had brought a bewildering series of shocks as the Japanese executed their plan of coordinated attacks across six thousand miles. At forty-five minutes past midnight on the 8th (Tokyo time, one hour ahead of Singapore) Japanese troops took possession of the international settlement at Shanghai, on China's east coast. At 1:40 A.M. they landed troops at Kota Bharu, northeastern Malaya, and at 3:05 they invaded southern Thailand; at 3:26 they were bombing Pearl Harbor, and two hours later they sank a British gunboat in Chinese waters. Not until 6:10 had Singapore been attacked by bombers – late and ineffectually because of poor visibility. At 8:05 the Japanese landed on Guam. At 9:00 A.M. the Japanese Army attacked the unfortified British crown colony of Hong Kong, in southeastern China. Only at 11:40 Tokyo time did Emperor Hirohito get round to issuing his rescript declaring war, by then of academic interest. Phillips decided to take his two big ships and four destroyers on a sweep up the eastern side of the Malayan Peninsula as far north as the Gulf of Siam that night, to see if he could disrupt Japanese landings. He asked the local RAF command to have land-based fighter cover standing by. A squadron of twelve ancient Royal Australian Air Force Brewster Buffaloes was allocated.

Phillips duly went north, found nothing, and put about on the evening of the 9th. On his way back to Singapore overnight, he received a radio report of another Japanese landing – at Kuantan, halfway down Malaya's eastern coast. He decided on a surprise attack at dawn. But there was to be no surprise: the Japanese submarine *I58* (Lieutenant Commander Kitamura Sohichi, IJN) had seen Force Z at 11:52 P.M. on the 9th, Singapore time. His torpedo attack failed so completely that the British did not notice it or his boat, but his sighting report was gratefully received by the 22nd Air Flotilla of the Japanese Naval Air Force, based north of Saigon in French Indochina since late October and commanded by Rear Admiral Matsunaga Sadaichi. The unit had been sent on the orders of Admiral Yamamoto, who realized that the Army, originally responsible for local air cover, had neither the long-range aircraft nor the maritime skills needed for operations in the Singapore area. 'This is an opportunity which comes but once in a thousand years,' Matsunaga told his pilots. 'Put forth everything that you have. All of you, come back dead.' Only the crews of three bombers were forced to obey this death-or-glory order. Seventy-six aircraft, fifty-two armed with torpedoes and the rest with bombs, took off from the Saigon area and flew south as other aircraft ran intensive reconnaissance in poor weather, looking for the British ships. If these managed to evade discovery, the plan was to go on to bomb Singapore, in which case many of the attackers would not have enough fuel to get back to base. This had no effect on the enthusiasm of the naval pilots for the chance of a fight with the Royal Navy. One squadron of nine planes sighted a merchantman, dropped its bombs, missed and returned to base. The rest flew on, past their 'point of no return' where they had only half their fuel left, still without a sighting. But at 10:45 A.M. Singapore time the British were located from a reconnaissance aircraft flown by Ensign Hoashi Masame; the bomber force reversed course and approached them from the south, sighting them at about noon. Lieutenant Commander Nakanishi Niichi, the wing commander, ordered the attack. The Japanese pilots were astonished that there were no enemy planes to be seen.

A series of attacks swamped the weak British anti-aircraft defenses. At 12:30 the *Repulse* turned over and sank. Forty-eight minutes later, the *Prince of Wales* followed her tragic example. Ensign Hoashi broadcast a blow-by-blow account of an unprecedented battle, ensuring that the Japanese knew for certain they no longer had to reckon with the British 'battle squadron,' which had comfortably outclassed the two old Japanese battleships in the area. The report of the 'landing at Kuantan,' fifty miles to the west of the scene, had, sadly, proved false. Only when the Battle off Kuantan was over did eleven of the RAAF Buffaloes from Singapore drone over the spot. One of the few messages from Phillips as his ships shuddered to the blasts of bomb and torpedo had been: 'Send destroyers.' Another plea from Force Z asked for tugs. Nobody requested fighters during the battle, in which 840 British sailors died. A total of 2,081 survivors were rescued by the British. The Japanese captured three. One of them, James Milne (rank unknown, a pom-pom gunner from the *Prince of Wales*), told Japanese reporters:

Admiral Sir Thomas Phillips stood on the bridge from the beginning to the end. A destroyer drew near just before his ship sank and signalled, 'Come aboard'. But he answered, 'No thank you'. Raising his arm in salute from the bridge, he went down with the ship. Captain Leach stayed with Admiral Phillips at salute [sic] and . . . they disappeared into the sea as if swallowed by it.

With them went what should have been America's strongest

support in the new theater of war, Britain's strategic naval presence in the Far East. Also sunk was the strategic role of the battleship, unequivocally exposed for the first time as fatally vulnerable to maritime airpower, even when maneuvering with great skill and high speed on the open sea. Admiral Yamamoto, the inveterate gambler who had decided to prove his point about airpower at sea by using planes rather than battleships against the British, happily gave his air staff officer, Commander Miwa Yoshitake, a case of beer to honor a ten-to-one bet. The aide had wagered that his fellow pilots would get both ships; his chief bet that only one would be sunk. Yet, just eleven days later, HIJMS *Yamato*, new and gleaming, joined the Combined Fleet as its flagship: seventy-two thousand tons all up with nine monstrous guns of 18.1-inch caliber, the biggest battleship ever built. Even she was not to restore her class of warship to pride of place at the heart of the fleet. Those days were gone forever, as the United States Navy was soon to confirm.

On the day after the Battle off Kuantan a Japanese aircraft piloted by Lieutenant Iki, leader of a torpedo-bomber squadron, flew over the scene and dropped a wreath. Whether it was for the two comrades he lost to British gunfire in the action or for a routed enemy was never made clear.

The first few days of Japan's war of expansion to the south and east had been, the strategically insignificant underestimation of Wake aside, a sensational success. Divided forces inflicted crushing defeats on the two leading navies of the world, opening the way to the spectacular territorial conquests described in chapter three. But we should now forsake the field of battle to consider the political, economic, diplomatic, and social developments which led Japan to go to war – and made America its enemy number one in the Pacific Campaign.

PART ONE
Collision Course

1

The View from the East

Japan's southward advance, even though it was in the opposite direction from all its previous expansion, derived directly from its military adventures, political scheming and economic ambitions on the Asian mainland. This is not to say that the move south was immutable fate, either for Japan or for its victims: the Japanese were and are as responsible for their own actions and choices as everyone else, regardless of foreign provocations and errors. Nevertheless, the short but brutish and nasty story of Japanese imperial expansion has features only too familiar to the students of past empires, whether the ancient Roman or the modern Russian. A power on the make begins to expand by 'absorbing' its immediate neighbor (in Japan's case, Korea in 1910); to protect its acquisition, it conquers its neighbor's neighbor (Manchuria), sets up a buffer state (Manchukuo), creates another buffer (northern China), and uses that as a base to move against its next victim (China), and possibly its most deadly rival (the Soviet Union). We see imperialism imitating scientific principles such as Newton's first law of motion whereby movement continues unless halted (imperial inertia); the abhorrence of nature for a vacuum is parodied by imperialist opportunism, which drew Japan first into China, then down upon the Asiatic empires of the European powers involved in the war with Hitler's Germany.

It is not customary to refer, in the context of the Second World War, to 'Tojo's Japan,' or even Hirohito's; nor do we equate the Imperial Rule Assistance Association, formed in 1940 to absorb all Japanese political parties, with the National Socialist party, the only legal one in Hitler's Germany, even though the former was in some respects a conscious imitation of the latter. The truth is that the Japan which took on the world at war and lost was run by a military junta of no fixed composition – a shifting, authoritarian oligarchy rather than a totalitarian dictatorship.

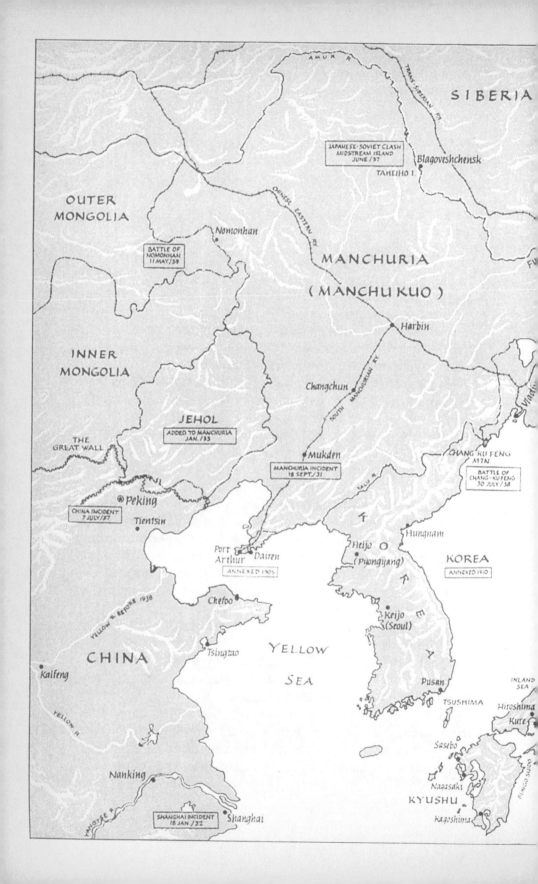

SIBERIA

AMUR R.

TRANS-SIBERIAN RY.

JAPANESE-SOVIET CLASH
MIDSTREAM ISLAND
JUNE /37

Blagoveshchensk

TAHEIHO I.

OUTER
MONGOLIA

CHINESE EASTERN RY.

Nomonhan

BATTLE OF
NOMONHAN
11 MAY /39

MANCHURIA

(MANCHU KUO)

INNER
MONGOLIA

Harbin

Changchun

SOUTH MANCHURIAN RY.

Vladi

THE
GREAT WALL

JEHOL
ADDED TO MANCHURIA
JAN. /33

Mukden

CHANG-KU FENG
MTN

Peking

MANCHURIA INCIDENT
18 SEPT /31

TALU R.

BATTLE OF
CHANG-KU FENG
30 JULY /38

CHINA INCIDENT
7 JULY /37

Tientsin

Port
Arthur
Dairen

Heijo
(Pyongyang)

Hungnam

KOREA

ANNEXED 1905

ANNEXED 1910

YELLOW R. BEFORE 1938

Chefoo

Keijo
(Seoul)

K O R E A

CHINA

Tsingtao

YELLOW

SEA

Pusan

INLAND
SEA

Kaifeng

TSUSHIMA

Hiroshima

Kure

Sasebo

YELLOW R.

Nanking

Nagasaki

KYUSHU

SHANGHAI INCIDENT
18 JAN /32

Shanghai

YANGTZE

Kagoshima

BUNGO SUIDO

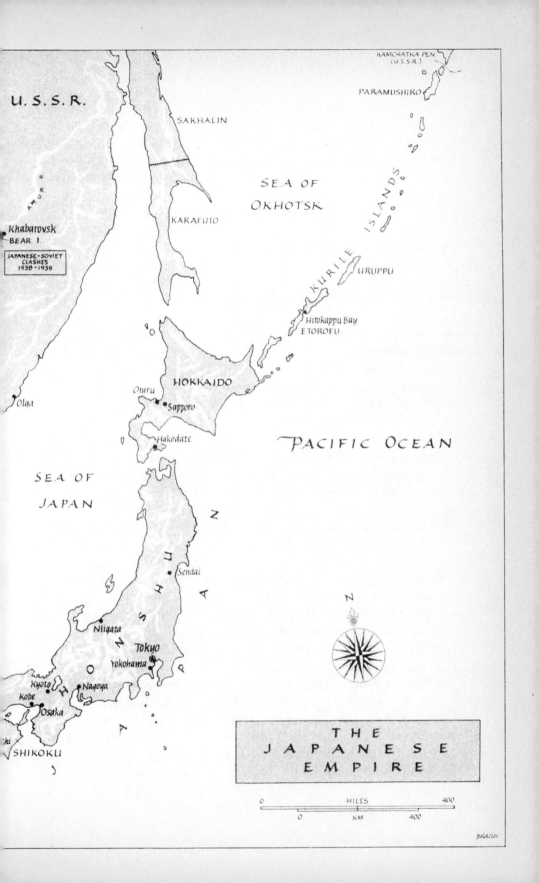

U.S.S.R.

SAKHALIN

KAMCHATKA PEN.
(U.S.S.R.)

PARAMUSHIRO

SEA OF
OKHOTSK

AMUR

KARAFUTO

KURILE ISLANDS

Khabarovsk
BEAR I.

JAPANESE-SOVIET
CLASHES
1938-1939

URUPPU

Hitokappu Bay
ETOROFU

Olga

HOKKAIDO

Otaru
Sapporo

Hakodate

PACIFIC OCEAN

SEA OF
JAPAN

HONSHU

Sendai

Niigata

Tokyo
Yokohama

Kyoto
Kobe
Nagoya

Osaka

SHIKOKU

N

THE
JAPANESE
EMPIRE

0   MILES   400
0   KM   400

palacios

It came to the fore in Manchuria in 1928, when the 'Kwantung Army,' as the Japanese garrison was called, killed an intractable local warlord by causing an explosion on the Japanese-controlled South Manchurian Railway (SMR). The junta won the support of most Japanese admirals in 1930, after the perceived 'humiliation' of Japan at the London Naval Conference, about which more later. Japan was easily humiliated: rejection of any of its demands was enough. Aggravated by Japan's severe suffering in the Slump, which helped to undermine moderate, civilian influence in government, the rising junta's Kwantung branch staged another explosion on the SMR at Mukden in September 1931 as an excuse for conquering the rest of Manchuria in a few months. This euphemistically named 'Manchuria Incident' led to the establishment of the puppet state of Manchukuo under the 'Emperor' Pu-yi, scion of the deposed Manchu dynasty, which had ruled China until 1911. Encouraged by this cheap success and undeterred by international condemnation, which merely provoked Japan to flounce out of the tottering League of Nations in 1933, the junta ran off the rails altogether in 1937. At the Marco Polo Bridge outside Peking, the Japanese 'China Garrison Force,' in place since the international suppression of the xenophobic Boxer Rebellion of 1900, engineered a clash with a Chinese Army patrol. This was then used as an excuse to attack northern China – all without consulting civilian or military superiors in Tokyo. The latter managed, however, to do what was expected of them: they sent reinforcements. The ensuing war, unwinnable for either side, spread across China; to the Japanese it always remained simply the 'China Incident.' It is not unreasonable to see in the manufactured clash of 7 July 1937, so similar to Hitler's ploy against Poland two years later, the true start of the Second World War, because these two participants fought each other continuously from then until 1945.

In its bid to become the USA of the western Pacific (a strictly economic ambition), Japan classed itself as a 'have-not' nation with a legitimate grievance. What it really 'had not,' like Germany and Italy among the larger powers, was territorial acquisitions to exploit – the only contemporary yardstick of greatness, even more important than a big navy. The rest of the world soon came to see Japan as an acquisitive aggressor, inordinately ambitious and completely ruthless. Japan came late – indeed, last – to old-style colonialism, but chose to learn nothing from its predecessors in this pursuit. Like them, it cared little for the feelings of the colonized; unlike them, it was never deterred by the views of the other powers, which it either ignored or used as grounds for more aggression while it built up its own empire. In this outlook it was very similar to Germany under Kaiser Wilhelm II, and even more under Hitler: unable or unwilling to distinguish between its needs and its wants, Japan helped itself to what it fancied and was quite often genuinely perplexed by the hostile reaction. Like Germany, where almost everyone who could walk and talk hated the Treaty of Versailles, Japan had an almighty bone to pick with the rest of the world. Most Japanese people regarded anyone who questioned their country's ambition as hostile and did not try to understand any other party's point of view. Where the rest of the world went wrong was in foolishly underestimating the unique capacity for self-sacrifice with which ordinary Japanese supported their country's aim to be a first-rate power.

There was much less disagreement among the Japanese (or in Germany) on the end than on the means of achieving the fulfillment of their country's 'just demands.' Hitler came to power on the back of the German national sense of grievance, and was as conscious as the Japanese military of the lessons of 1918. Like the Japanese, he thought his country was overcrowded and needed more territory, a rationalization of imperial ambition throughout the ages. The Nazis, like the Italian fascists, were a mass movement that rose to power from the grass roots under a populist leader, whereas the Japanese junta manipulated a complaisant emperor to impose its will from the top. But each Axis regime drew the same conclusion from Germany's defeat in 1918: the next war would be long, and therefore autarky, economic self-sufficiency, was the key to national security, military success and world domination. That was the only way to avoid a repetition of the blockade by sea and land which defeated Germany in 1918.

So, while Hitler schemed to acquire *Lebensraum* and Mussolini concentrated on empire-building in northeastern Africa, the Japanese were busy inventing the 'New Order in East Asia'

(1938) and the 'Greater East Asia Co-Prosperity Sphere' (1940), both designed to subordinate the region to the perpetual benefit and glory of a self-sustaining, greater Japan. Tokyo had some success at first in presenting this as a crusade against Euro-American domination of Asia. It won over many indigenous nationalists in British, French and Dutch colonies – at least until the Japanese Army arrived and lent new vigor to the old military customs of rape and pillage. The Germans made exactly the same error in the Soviet Union: each army behaved as the master race in arms; each used the stratagem of surprise attack without declaration of war, and then *Blitzkrieg* tactics, to get its way. But whereas Hitler dominated his generals and admirals the Japanese general staffs dominated Japan. The consequences for their victims were remarkably similar. There was, for example, not much to choose, except in such matters as climate and language, for the doubly unfortunate Dutch between life in the Netherlands under Nazi rule and in the East Indies under the Japanese.

Small wonder that Reich and Empire were to become allies regardless of reciprocal racial disdain. The first concrete sign of things to come was Japan's decision to sign the Anti-Comintern Pact with Germany in November 1936 (the Comintern – Communist International – was the Soviet mechanism for controlling foreign communist parties). A secret provision required each signatory not to help the Soviet Union if the other went to war against it; the published text was a vague commitment to oppose communism and all its works wherever they might be found. The future Axis partners had identified their overwhelming common interest: the Soviet Union, principal potential enemy of each.

For Japan this was just one of many fateful decisions that led to its disastrous war with the United States. The Slump became a time for taking tough measures at home – and taking sides abroad. The Pacific Campaign cannot be properly understood unless it is seen in the context of Japan's prewar domestic and foreign policies and the links between the two, as summarized below.

*   *   *

Foreigners had (and have) great difficulty in understanding how Japan worked as a state and who was really in charge. The Japanese had gone so far as to imitate the West in having a symbolic head of state and an executive, a legislature, a judiciary, an army, and a navy all formally answerable to him. The fact that the Army and the Navy were, as centers of power in the state, at least equal to the civilian organs of government rather than subject to their authority was not outside Western experience. In making this ultimately disastrous arrangement in the constitutional changes of 1889, the Japanese were only copying the Prussians who dominated Europe as the world's strongest military power for more than half a century, until 1918, on just such a basis (the Japanese chose to copy the British in establishing a House of Lords and a battle fleet and imitated the French in such areas as law and education). The independence of the military dated from the creation, in 1878, of general staffs for Army and Navy directly under the emperor and outside the control of the Diet (Parliament) or even the Cabinet. The paradox was that the emperor, unlike the Kaiser, did not feel free to intervene in government. He exercised his influence through his personal advisers or in private meetings with those, such as key ministers and chiefs of staff, who had the right of access to the throne. Thus his divine status was protected by noninvolvement in day-to-day policy with all its disputes, errors, and corruption; by the same token, those with real power could hide behind the façade of imperial rule whenever convenient, an excellent incentive for irresponsibility on all sides.

This gave very broad latitude indeed to leaders whose actions were rendered immune from challenge by the simple device of being declared as done 'in the name of the emperor.' A general could tell Hirohito, with the customary groveling and outward respect, what he was planning; the emperor had no power to stop him, so the general could then inform the Cabinet of what he was about to do, overriding any objections by laying claim to imperial sanction. From the turn of the century, the ministers responsible for the Army and the Navy had to be officers from the relevant service. After 1936 they had to be on the active list, to prevent the appointment of men from the retired list as a means of getting round the wishes of the serving generals and admirals. This gave the general staffs not only the decisive say (or veto) on individual appointments to these posts but also the power to prevent the formation of a new government, simply by refusing to supply serving officers to fill them. If they did not like a prime-ministerial nominee, they would decline to provide a general (as the Army did in 1940, for example) or an admiral as Army or Navy minister – even if the would-be premier had found favor with palace advisers and been recommended by them to the emperor. The three key men in each service – minister, chief of staff, and inspector-general of education and training – were thus free to pick their own successors without consulting any outsider, whether emperor, prime minister or the rival service.

The two armed forces were not required to inform the Cabinet of their strength and dispositions, in peace or even in wartime. Thus the claims by such as ex-Prime Minister Tojo and ex-Foreign Minister Togo at the Tokyo war-crimes trial that they were not told in advance of the Pearl Harbor plan (or of the great American victory at Midway for weeks after the event) are not as ludicrous as they seemed when they were first made. With this kind of contemptuous conduct as the norm in the highest ranks, it is hardly surprising that the Japanese forces were more Prussian than the Prussians, not to say medieval, in their approach to discipline. Brutality was institutionalized to a degree probably unparalleled anywhere in the modern world. Boy officer-candidates were put in harsh premilitary academies and cadet schools with narrow curricula, hard physical routines and very little intellectual training (something the Germans did not neglect). Free discussion and intelligent questioning were forbidden on pain of severe punishment, ensuring that the Japanese military elite was unimaginative, rigid, undemocratic, inflexible and totally lacking in initiative. This goes a long way toward explaining the sheer, all-embracing inadequacy of the Japanese leadership, overwhelmingly military in background as it was, before and during the war.

Training was aimed at producing in all ranks total, unquestioning obedience to orders, including standing prohibitions against retreat, surrender, and being taken alive. If captured wounded or unconscious, the Japanese officer or enlisted man was expected to kill himself for shame when he was able to, even if he had managed to return to his own side. Such was the 'Japanese spirit' inculcated at all levels, the mind-over-matter approach which persuaded hundreds of thousands to fight on beyond reason and throw their lives away. This, the unimportance of the individual in Japanese society at large and the psychic and physical explosion which took place when the constraint of total obedience was lifted from victorious troops licensed to rape and pillage after a victory, goes a long way toward explaining the peculiar horrors inflicted by the Japanese upon the troops, civilians and even the children of the enemy. Life was cheap in Japan; those in its services were as unlikely as anyone else to place greater value on an enemy than they did on themselves.

Japanese military leaders chose to believe that Germany's 1918 defeat was overwhelmingly caused by lack of raw materials. The idea that constitutional flaws such as overexaltation of the Kaiser or the assignment of undue weight to the opinions of the general staff might have contributed to driving Germany into an avoidable war did not occur to them. It was one of those errors Japanese officers were not intellectually trained to identify. Officers accepted no blame for their actions because they were obeying orders or executing the emperor's will (theoretically the same thing). There was nobody in a position to correct them, even if the emperor occasionally would not conceal his displeasure over military mistakes. But failure, if identified and made public, meant shame, and shame entailed ritual suicide in the samurai's Bushido code. Further, generals and admirals, exhausted in late middle age by a lifetime of repression and out of touch with the lower orders, fell under the influence of the younger, more vigorous middle-rankers. The captains, majors and colonels commanded the individual ships, battalions and squadrons or did all the work on the staff; they often came from rural backgrounds and were in touch with the peasants in uniform who constituted the majority of their men. These officers too were unable to act on their own, and drew courage from combining in various right-wing societies and clubs to impose their collective will on their flagging superiors. It was these middling commanders who increasingly saw the force at their immediate disposal as the instant solution to Japan's growing problems between the wars. Senior commanders, clinging to office to avoid sinking into obscurity on a poor pension, not only encouraged them but were also prepared to use them in furtherance of disputes with rivals among their own contemporaries.

Internal pressures had more to do with Japan's drift into the Second World War than external factors. Between the revolution of 1868, which formally restored rule by the emperor, and 1930, the population of the Home Islands rose from thirty to sixty-five million; by the end of 1941 it was about seventy million. Small wonder that Japan became an importer of food for the first time at about the turn of the century and felt insecure as a result. It had never been dependent on the outside world before, yet became even more so when its new industries demanded fuel and raw materials from abroad. Japanese interest in expansion on the Asian mainland was based as much on a desire to ease its population problem by emigration as on colonialist emulation of the West.

By the time of the 'incident' at Mukden in 1931, therefore, at least one million Japanese had migrated to Manchuria (ex-servicemen were preferred, on the ancient Roman colonial model). This substantial figure was, however, dwarfed by the huge migration from China proper into the region. According to Japanese sources, in the quarter-century from 1907 to 1931 the population very nearly doubled, from seventeen to thirty-three million. Even after allowing for incoming Japanese and natural increase, this represents a massive influx – one of the great migrations of the century – which contemporary Japanese officials naturally attributed to the orderly conditions and flourishing economy of the southern part of the region, under their control since 1905. In 1937, as Japan went to war with China, Tokyo planned to settle one million Japanese households – five million people – in Manchuria and northern China in the twenty years until 1957: some five hundred thousand actually emigrated in a couple of years; of these, half were farmers and one-fifth teenagers. Whether the Japanese (today 125 million) actually needed *Lebensraum* is debatable;

but they certainly did their best *ex post facto* to justify their claim to it.

In Japan itself, the decade that followed the end of the First World War was relatively stable, especially when compared with the thirties, despite increasing diplomatic and economic difficulties. Nominally at least, and a strongly authoritarian social structure notwithstanding, the civilians were in charge under a two-party system: they even defeated the Navy in forcing acceptance of the Washington Treaty of 1922 and managed to impose cuts in military and naval budgets. There was not much to choose philosophically between the Minseito party, financed mainly by the Mitsubishi corporation, and the Seiyukai, backed by the Mitsui concern; both were middle-of-the-road and by and large took turns governing in the broad interests of their backers and the new urban middle class. The absence of a parliamentary tradition only served to encourage generalized corruption on a huge scale. This bred a general contempt for politicians, their big-business backers in the Zaibatsu (the cartel of the leading conglomerates) and the political process in a country only recently emerged from feudalism and still strongly agricultural. Indeed, the level of tension between the traditional Japanese way of life and the swift spread of many aspects of Western civilization – economic problems, jazz, modern clothes, rapid urbanization, women office workers, mass media, political ferment on left and (especially) right, trade-unionism, an incipient youth culture, class conflict, cocktail parties and even potatoes – had no parallel in any other society. It was a powerful and dangerous social brew, and it soon went to people's heads – especially when Japan was forced to import the effects of the 'Wall Street Crash.'

As Crown Prince, Hirohito caused one sensation after another in 1921 with his unprecedented overseas tour by battleship to various parts of the British Empire, Britain itself, France and Belgium (including the First World War battlefields), the Netherlands, Italy and the Vatican – and the tastes he brought back with him. These included the great British breakfast of eggs and bacon, to which he remained loyal, except in wartime as a gesture to austerity, for the rest of his life. He also learned to like horse-racing, nightclubs and golf, all Japanese passions to the present day. When he came back, he was cheered to the echo by huge crowds which had been following his travels through the press, newsreels and radio: it was an all too brief suspension of the xenophobia and intolerance endemic in contemporary Japan. His taste for Western dress was made harder to satisfy by his divine status, which prevented tailors from measuring him except by the unreliable eyeball, applied fleetingly and at a suitably respectful distance. Hence the famous baggy clothes of so many early photographs. But his best attempts to dilute the stifling formality of palace life failed. He was wont to say in later years that his visit to England, especially his time in Oxford, gave him the happiest days of his life. He envied the informality (strictly relative) of the British House of Windsor, of which he was reminded when the Prince of Wales (later briefly King Edward VIII) returned his visit in 1922: they played a lot of golf together.

Hirohito was born on 29 April 1901, the eldest son of Crown Prince Yoshihito and Princess Sadako, and was, inevitably, brought up at the court of the Emperor Meiji. Considering the stuffiness associated with the imperial court, Meiji was a surprisingly convivial, uninhibitedly bibulous man: Hirohito is said to have been put off alcohol for life when he was made drunk by his father, and given an appalling hangover, before he was of school age. He was placed under the tutelage of General Nogi Maresuke, the intellectual war hero who defeated the Russians in 1905. Nogi saw to it that the always frail-looking Hirohito became a competent all-round sportsman as well as a conscientious student at the special school for the offspring of the Japanese peerage. The general was an ascetic and instilled the virtues of austerity into his pupil: displaying these qualities became the youth's way of rebelling against the licentious example set by his grandfather and even more so by his father. Meiji died of cancer in 1912, whereupon Yoshihito became the Emperor Taisho, Hirohito became Crown Prince – and Nogi committed ritual suicide. The general thus kept the promise he made when he lost both his sons in the bloody struggle with the Russians for Port Arthur in 1905, a commitment deferred on Meiji's order. It was a terrible, pointless example for such a gifted man to give to lesser mortals; it was also a trauma for the reflective Hirohito.

The new chief tutor was the other superhero of the Russo-Japanese War, Admiral Togo Heihachiro, victor in the Battle of Tsushima. But as a teacher the admiral was as disappointing as the general had been inspiring. Because his science tutors were better than his arts teachers, and probably also because he had an inquiring mind, Hirohito took most interest in the natural sciences. He eventually became an amateur marine biologist of repute – largely on the strength of discovering a hitherto unrecorded species of prawn at the age of seventeen. His strongest academic interest proved to be history, especially military; his heroes were Napoleon, Lincoln and Darwin, whose portraits were always to be seen on the walls of his study.

The year 1921 proved to be the most traumatic in the peacetime experience of the earnest Crown Prince. He became engaged to Princess Nagako – his own choice – after a convoluted dispute at court, ostensibly over the hereditary color-blindness in her family but actually between two powerful aristocratic clans for domination of the imperial household. A leading extreme-nationalist group (Japan had hundreds) was used to mobilize public opinion in favor of Nagako. As soon as the row was settled, Hirohito was almost bundled out of the country on the foreign tour already mentioned, disconcerting his British hosts by starting a week early and staying longer than originally planned. Nonetheless, he always remembered the warmth of his welcome abroad at all levels, especially in Britain. He was nothing if not sentimental.

While he was still away, a right-wing extremist assassinated the prime minister, Hara Takashi, the first untitled occupant of the post, in protest against Japan's participation in the Washington Conference, negotiating the relative strengths of the world's leading navies. This was just one instance, and not the most dramatic, of the extraordinary fanaticism evoked by the stresses and strains of Japan's struggle to excel in a world dominated by the West and its values. Society's readiness to understand if not condone such extreme reaction was symbolized by the assassin's sentence of twelve years on a capital charge, as if his had been a sexual rather than a political *crime passionnel*.

On the right, the most radical (a large and growing element) took the view that all 'eight corners of the world' should be united under Japanese domination. To the left, liberal, socialist and communist groups also existed. But they were having an increasingly hard time making themselves heard, let alone exercising the basic freedoms taken for granted in democratic societies but increasingly hard to come by in an instinctively authoritarian Japan. Hirohito had glimpsed such social and political freedoms being enjoyed during his grand tour. Though there was not much he could do about politics when he got home, he did what he could to ease the social atmosphere by getting rid of as much palace protocol as he dared.

His wishes soon carried rather more weight among the nebulous groups of courtiers whose self-appointed role was to 'protect' the throne (mainly from scandal, as they chose to define the term). On 25 November 1921, Hirohito, still not twenty-one years old, became regent when the eccentricities of his father merged into madness. The Crown Prince's taste for informality was allowed to run to one fairly rowdy palace party in December. After that it was back to the old ritual, with one major if super-ficial change: Western dress now effectively became compulsory except on the stuffiest ceremonial occasions. But it was Western attire of the most sober kind. Even so, for a few years, until Hirohito's marriage in 1924, the palace became the venue for an open-ended association of high-flying younger officers, bureaucrats, and other 'junior leaders,' who would gather at the Regent's behest to debate the issues of the day or listen to lectures from leaders in the academic, administrative, and military fields. This was a uniquely elitist club-cum-secret-society of the kind to which so many Japanese, with their strongly developed sense of 'family,' loved to belong.

But life in Japan was becoming no easier. At lunchtime on 1 September 1923, the worst earthquake in Japanese history, which is saying a great deal, laid waste the Tokyo-Yokohama region, causing huge fires and deaths in six figures. Millions were made homeless, and Hirohito's wedding was postponed. The superstitious saw this disaster as punishment for Japanese flirtation with Western decadence. They took out their feel-ings on left-wingers, with their foreign ideas, and on Korean immigrants, who did the most menial jobs; thousands were massacred. On his way to open the new Diet on 27 December, Hirohito narrowly escaped being shot by another of Japan's plentiful supply of extremists. This one, who was executed, was officially said to be a left-wing revolutionary, but he had more obvious links with the court faction that had lost the battle over Hirohito's fiancée. Unwilling to face another postponement, and undeterred by his brush with the violence never far from the surface of his simmering nation, Hirohito married his princess thirty days later, on 26 January 1924.

The omens notwithstanding, it proved to be a happy marriage. Its first three years were also a period of unusual calm, despite the death of Taisho in December 1926. Thereupon Hirohito, now the 124th emperor, followed ancient custom by choosing 'Showa' as the name by which he was to be officially remembered. The word means 'enlightened peace': hindsight enables us to savor the irony in full. The underlying, authoritarian social trend, however, continued unabated. It was fostered by a frustrated military which felt its marginal role in the First World War had caused it to fall behind, both in the international league and in the estimation of the nation. From 1926 onward, education was militarized, even at the elementary level, a reversion to past strictness after an unconvincing dabble with well-diluted liberal ideas. Emperor worship, aimed at the institution as the source of all legitimacy rather than the person, was nurtured; small boys put on uniforms and drilled with wooden 'rifles.' Discipline among adults was fostered by the foundation in 1928 – twenty years before George Orwell made the term famous – of the 'Thought Police,' whose role was to stamp out Western ideas such as communism, socialism, liberalism, materialism and individual rights.

But if the authorities automatically assumed that the really dangerous ideas came from the left, the most dangerous people, as so often in history, were to be found at the other end of the political spectrum. The wild men in the Kwantung Army killed Marshal Chang Tso-lin, the Manchurian warlord who obstructed Japanese domination of the region, by blowing up his train on the Southern Manchurian Railway in June 1928. The civilian government of the Seiyukai party, led by retired General Tanaka Giichi as prime minister, had wanted to use Chang as a counterweight to Chiang Kai-shek's Chinese nationalists in the Kuomintang, in a divide-and-rule strategy for northern China. The fire-eating colonels and majors were not interested in such subtleties. Their stupidity was made manifest when Chang's son, the 'Young Marshal,' took over, had two Manchurian officers shot for collaborating with the Japanese and then declared for the Kuomintang. Few assassinations can have proved quite so counterproductive quite so quickly.

The middle-rankers were unfazed by such setbacks; as will soon become clear, their invariable remedy for the failure of violence was more violence. They were also unable to appreciate the prudence which led Tokyo to give diplomatic recognition to the Soviet Union, their 'enemy number one,' five years before Washington did so. The Chang murder was ascribed to 'bandits' and two officers were suspended for failing to guard the railroad adequately. The scapegoat for the cover-up, which held until after the war, was Premier Tanaka, who lost his nerve in the behind-the-scenes political row about high-level indiscipline in the Army. He resigned early in 1929, once the elaborate and protracted ceremonies attending the formal enthronement of Hirohito at the turn of the year were out of the way.

It was the Minseito party's turn to form a government, and the task, a bed of nails as usual but also a poisoned chalice on this occasion, was awarded to Hamaguchi Osachi, a notably moderate civilian; his foreign minister, Baron Shidehara Kijuro, had a reputation for flexibility on China policy. Overseas observers of the increasing ferment in Japan were commensurately relieved, but the issue that was to determine the struggle between diluted democracy and mounting militarism for primacy in Japan was already looming: the strength of the Navy.

The Washington Conference of 1921–22 had, among other things, led to a naval treaty fixing the ratios of the American, British, and Japanese navies as 5:5:3 (and setting a tonnage limit on capital ships: one already commissioned first-rate Japanese battleship had to be scrapped as it was four thousand tons over.) All this stuck in the craw of the ultranationalists. So did such serious grievances as the principle of 'inferiority' imposed or accepted by treaty, the ban on Japanese migration to the Western hemisphere (South as well as North America) and Australasia (New Zealand as well as Australia), and the failure of Japan to extract a commitment to racial equality from the Western powers, whether at Versailles, the League of Nations or other international gatherings. The younger officers in the middle ranks and a surprisingly large proportion of the general public regarded all these setbacks as insults or outright humiliations. The Navy itself – or, more precisely, the officer corps – divided into 'fleet' and 'treaty' factions, a maritime reflection of the Army's extremist and purportedly moderate tendencies. The latter, confusingly, has also been labeled the 'total-war faction,' a most revealing clue to the true meaning of 'moderation' in prewar Japanese history: in this context, moderation meant keeping out of war – but only until Japan was totally ready to wage it! The senior bureaucrats in the civil service, the men who really ran the country, were similarly divided. It cannot be stated too often that agreement on the country's right to pre-eminence was not far short of universal in Japan. Factional differences on the issue were concerned only with the means of realizing this oriental version of manifest destiny.

In the Navy the fleet faction had fought to the limit in 1921 for a ratio of 10:10:7 and against 5:5:3, claiming that this tiny margin made all the difference between subservience and superiority vis-à-vis the Americans, whom they already saw as the obvious potential enemy at sea: a prime example of the Japanese capacity to extract the maximum heat from a minimal divergence of opinion. Behind the ridiculous and irrelevant demand for '70 percent or bust' lay a determination to hang onto the colossal one-third of the national budget the Navy was consuming at the time (the Army's share took military expenditure up to 60 percent of the total). The Navy's enthusiasm for southward rather than northward expansion was intimately related to the fact that this course required Japan to be ready for war with America and Britain, which entailed a very large, modern fleet. The Army's preference for the opposite course was no less closely related to the consequent need to muster huge new armies, over and above those already in Korea, Manchuria and China, for war against the Soviet Union, in which the more costly, oil-thirsty Navy would have a minor role. The naval argument revived in 1929, as contacts took place among the powers in preparation for the London Naval Conference of 1930, where the Japanese delegation was to include Rear Admiral Yamamoto. He already believed that 5:5:3 was perfectly adequate for Japan's purposes: it was not the tonnage but the types of ship, especially carriers, that mattered in his eyes – quality rather than quantity – even if he saw no harm in taking up a 10:10:7 position for negotiating purposes. But by this time the hawks had raised the ante: nothing less than parity would do. Indeed, any limitation on armament was (or could conveniently be represented as) an attack on the imperial prerogative to determine Japanese policy: divinity does not compromise or take orders from foreigners.

So, when the Japanese delegation got as near as made no difference to 10:10:7 for cruisers and other noncapital ships, Prime Minister Hamaguchi's reward for association with this solid diplomatic success was to be shot in the stomach on 14 November 1930. Perhaps his real offence had been to force a cut of 25 percent in the 1931 naval budget. At any rate, he took nine months to die from his excruciating wounds. Before he did so, he bravely whispered his argument for the reduced naval allocation on his last appearance before Parliament. Any samurai would have approved. But it was a pointless act of defiance: the Army, itself smarting from Hamaguchi's fiscal razor, still took the view that the Navy was getting far too much. His assassin, in the pay of radical officers who were never fully identified, was sentenced to death three years after his crime – and pardoned three months later, another extraordinary example of the dangerous Japanese practice of turning a blind eye to politically motivated crime.

Wakatsuki Reiijiro of the Minseito served as acting premier. His brief term was remarkable for two things: he survived it by many years, and he was in office over the period when the military hawks planned and provoked the 'Manchuria Incident.' When rumors of the Kwantung Army plot, aided and abetted by the Army of Korea, to extend Japanese control over the whole of Manchuria spread round Tokyo in the summer of 1931, Wakatsuki sent a major general to warn off the trio of colonels at the core of the conspiracy. They belonged to the 'Cherry Blossom' secret society, the biggest and most important right-wing association among Army officers ranking from captain to full colonel. It had many friends and supporters wherever Japanese Army units were stationed. The general, himself a sympathizer who was on public record as favouring the outright annexation of Manchuria, made haste as slowly as possible and delivered the written warning on 19 September – the day after the Japanese blew up a section of the South Manchurian Railway at Mukden. When a Chinese Army patrol went to investigate the blast, it was fired upon. So, at the same time, were other installations belonging to the Chinese nationalists and their local ally, the 'Young Marshal' Chang. The Chinese Army was blamed.

It may be noted for the record that the Japanese Army code of the time decreed, in article 35: 'A commander who opens hostilities with a foreign country without provocation shall be punished by death.' Article 37 provided the same penalty for one who moved troops out of his defined area without permission. Thus the commanders of both the Kwantung and Korean armies should have been court-martialed and executed. They were not put on trial. Hirohito cautiously limited himself to expressing displeasure; the military command in Tokyo took no action; the government was no more energetic. By November, all Manchuria was under Japanese control, and preparations were well in hand to establish the puppet state of Manchukuo under the 'Emperor' Henry Pu-yi, the last of the Manchus. The League of Nations Council urged a negotiated settlement by a margin of thirteen to one (Japan) and sent a commission of inquiry under the British Lord Lytton to investigate. When it uncompromisingly found against Tokyo in February 1933 the League as a whole accepted the Lytton Report by forty-two votes to one (Japan again), the Japanese delegation was led out of the chamber at Geneva by Matsuoka Yosuke, the future foreign minister, never to return. Japan formally withdrew from the League one month later.

Faced with this level of military intransigence, Wakatsuki saw no alternative but to resign over the Manchuria Incident in

December 1931. The next prime minister was due to come from the Seiyukai, which put forward the frail party-leader, Inukai Tsuyoshi, aged seventy-five. Appointed by Hirohito at the beginning of 1932, he lasted barely four months before being fatally shot in the face by a naval officer, one of a group of mainly military extremists who burst into his official residence on 15 May 1932. His financial minister and a leading banker, both moderates, had fallen victim to the new wave of terrorism before him. This '15 May Incident' was, sadly, far from unprecedented. When viewed alongside the Manchurian affair, it was also a turning point: it effectively marked the end of civilian government in prewar Japan. The military was now unquestionably in command of Japan's destiny.

Yet, rather than diminishing, the political violence in Japan increased as the impatience of the radicals knew no bounds. Nor did the territorial ambitions of the Kwantung Army and its many supporters in high places at home. They overreached themselves in the spring of 1932 by staging the first 'Shanghai Incident' as an excuse to enter China's largest city and main center for both industry and foreign, especially Western, settlement. Their purpose was to put down an effective local boycott of Japanese goods organized by the Chinese nationalists. The League of Nations managed to negotiate a Japanese withdrawal, a 'humiliation' that was the immediate provocation for the murder of Prime Minister Inukai. In January 1933 the Japanese helped themselves to the Chinese province of Jehol, adding it to Manchuria. Inner Mongolia and Hopei Province, inside the Great Wall of China, were next on the list; what the Kwantung Army really wanted was the mineral resources of Shansi, in northern China. A truce was arranged by international diplomatic intervention in May 1933 and a demilitarized zone temporarily established between Peking and the Great Wall. Even so, Japan's naked aggression and rampant ambition in China were not seriously challenged by the outside world, which was in the depths of the Depression, even when American and British nationals or other Westerners were insulted and assaulted by Japanese soldiers.

It was in 1933 that the first maps appeared in Japanese schoolbooks showing French Indochina, Siam (Thailand), Malaya and Singapore, the Philippines, and the Dutch East Indies – all the military objectives of 1941–42 – under the Rising Sun flag, as the new American ambassador to Tokyo, Joseph C. Grew, soon reported with alarm to Washington. In December the Empress Nagako gave Hirohito a son and heir, Akihito (the present emperor), after nine years of marriage. Any hopes that this long-awaited event would calm the ultra-imperialists, who had constantly bemoaned the lack of a crown prince, were soon dashed.

Internationalism, never much in fashion in Japan, now went out of style elsewhere as individual nations tried to find one-country solutions to the threats posed by creeping economic paralysis. In Japan the silk industry, which had been the mainstay of exports, collapsed as American and other foreign customers ran out of money for such luxury. The result was acute misery and deprivation in the Japanese countryside, from which so many Army recruits, including officers, were drawn. In the cities the country's comparatively small new industrial base, not yet strong enough to stand up to competition from the major powers after the First World War, was also weakened, but the paternalism of the big employers in the Zaibatsu cartel eventually cushioned some of the worst sufferings of the workers. A large gap opened between the urban and rural economies and standards of living. But policies such as the American New Deal and the British Imperial or Commonwealth Preference only underscored the arguments of the moderate expansionist majority in Japan as well as the extremists (and the Nazis in Germany): autarky was the only guarantee of national security.

With the London Naval Agreement due for renewal in 1935 and the Washington Naval Treaty by 1936, the exploratory talks convened in London late in 1934 on extending naval disarmament assumed great importance. Yamamoto, now a vice-admiral, was at the head of the Japanese delegation for this preliminary round. His job was made no easier by Tokyo's self-contradictory desire to see Washington, the principal pact, lapse and yet to continue with naval arms limitation. Admired both at home (sacks of fan mail were forwarded to him in London) and in the West for his bluff-sailor approach with no trace of dissimulation, Yamamoto set out to obtain Anglo-American consent to parity for the Imperial Navy, triumphantly transcending the pitfalls of a relentless social round (and relieving the British First Sea Lord of £20 at bridge). The Western powers were naturally hostile but wanted Japan to be the party that broke off the talks. They got their wish at the turn of the year. Japan gave due and proper notice of intent to abrogate both London and Washington and was soon in a position to build some of the finest warships ever sent to sea. By 1939 the Japanese had the strongest naval presence in the Pacific, and two years later they constituted the most formidable challenge ever faced by the US Navy.

But the wildest of the wild men were still to be found in the Japanese Army, and they finally ran amok in Tokyo itself in 1936, on 26 February (the '2/26 Incident'). The trigger was the trial of a lieutenant colonel who had murdered a general on the staff of the Army Ministry in revenge for the dismissal of another general, the colonel's hero and a rabid nationalist like himself. Elements of the Imperial Guards and first divisions – some twenty-four officers backed by twelve hundred men with heavy weapons – tried to stage a military coup in support of the accused. They issued a manifesto demanding immediate expansion abroad and death for imperial advisers classified (by the rebels) as 'disloyal.' But for the bloodshed, the incompetence of the ringleaders would have made a good plot for the Marx Brothers. The conspirators thought they had killed the prime minister, Okada Keisuke, at his home, but he hid in a linen closet after his brother-in-law had been mistaken for him and gunned down. A moderate general and the Lord Privy Seal, a key palace official, were also murdered. The Grand Chamberlain, Suzuki Kantaro, another important servant of the emperor and a future prime minister, was left for dead but survived; the office of Japan's leading newspaper, the moderate *Asahi Shimbun*, was briefly occupied. Hirohito, faced with the most serious domestic threat to his throne in a reign that endured for two-thirds of a century, acted with a degree of resolution and ruthlessness not seen from him before or after. Loyal army units sealed off an area in the center of Tokyo of about one square mile. A new prime minister was appointed (Hirohito learned of Okada's survival only later), and when all was ready, on 28 February, the emperor ordered the rebels to withdraw. On the 29th tanks and aircraft were called in to decide the issue, and the mutinous units surrendered. A purge of the officer corps followed; but, prevented at home from taking direct control of the state, the many extremist officers still in the Army looked abroad for the means to get control of imperial policy – specifically to China.

A series of generals now succeeded one another as premier, as if through a revolving door, until Hirohito found himself a more durable incumbent – Prince Konoye Fumimaro, a retired admiral, appointed in June 1937 and destined to hold office, with interruptions, for more than four years. Ten years older than the emperor, Konoye was a generally popular radical nationalist, on terms with Hirohito as close to informal as the monarch and prevailing custom ever permitted. Unfortunately he was also lazy, temperamental and chronically indecisive. The first week of the following month brought a long overdue, if also uneasy, alliance against Japan between warring nationalists and communists in China, a chance for those Japanese officials of the 'reds-under-the-bed' persuasion to say, with as much satisfaction as fear, 'I told you so.' General Douglas MacArthur came to Tokyo at this juncture for his first meeting with Hirohito, over lunch at the palace, in company with President Quezon of the Philippines.

But even in Japan, Konoye's appointment, like the execution by firing squad of nineteen rebel ringleaders on 17 July, was completely overshadowed by the latest and most fateful violent 'incident' of them all, on the 7th. The clash between patrols carefully engineered by the Kwantung colonels at the Marco Polo Bridge outside Peking plunged Japan and China into war, and large reinforcements were soon on their way from the Home Islands. For more than eight years, an apparently endless conflict in the apparently infinite spaces of China was to be the principal drain on Japan's resources of manpower, arms and money. China came to play the same role in the Far Eastern war as Russia did in the war in Europe. China became the giant, indestructible shock-absorber which tied down the bulk of the enemy armies, while the United States and its allies built up their aerial and naval firepower for amphibious counterattack on the other front, with much investment in technology – and much fewer casualties.

Japan's involvement, first in Manchuria and then in the rest of northern China, put it on a collision course with the Soviet Union, whose long, indeterminate Asiatic borders now touched territory controlled or claimed by the Japanese. The Russians, ever mindful of the Germans, were remarkably circumspect, virtually abandoning their internationally recognized sphere of influence in Manchuria. But they were not prepared to appease Japan to the extent of overlooking serious incursions. Not content with plunging their country into an unwinnable war with China, the hotheads on the Japanese Army staff were more than ready to provoke the Soviets at any time of day or night. Amid the chain of events that led Japan into conflict with the West in the Pacific, there occurred an all but forgotten war between the Japanese and Red armies in 1938–39. It is important because it made both sides resolve not to repeat the experience, a decision that had obvious implications for each in the coming war on its other front.

There were nearly three thousand exchanges of fire between Soviet and Japanese forces in the two decades after the Russian Revolution, and nearly two hundred armed skirmishes along the straggling Manchurian border. The worst of these thus far began in June 1937 over the ownership of some midstream islands in the Amur River (which have since been fought over by the Soviet Union and the Chinese People's Republic). Amid ill-concealed panic in both Tokyo and Moscow as rapid escalation threatened after the sinking of a Soviet gunboat, the Red Army pulled back into Siberia and the Kwantung Army reoccupied the islands. The affair was barely settled before the Japanese provoked their 'China Incident' on 7 July. The clashes between Russians and Japanese and their respective clients continued. Meanwhile, in August 1937, Moscow concluded a nonaggression pact with Chiang's nationalists.

As the Sino-Japanese War broadened and deepened, Moscow officially maintained a wary policy of nonintervention, because of the potential threat from Hitler on its western front. But the

Red Army quietly built up its infantry, armor and airpower in the Far East. There were unconfirmed rumors in the West that Soviet fliers were aiding the Chinese, just as the American Colonel (later Major General) Claire L. Chennault's volunteers – the 'Flying Tigers' – were already doing. But on 30 July 1938, there erupted a Russo-Japanese border incident of a much greater order of magnitude, at Chang-ku Feng. The name belongs to a mountain on the then disputed borders of Korea, Manchuria and the Soviet Union, close to the coast of the Sea of Japan. Remarkably inventive propaganda from both sides and a lack of impartial foreign observers forever obscured precisely what happened, but Soviet troops are known to have entrenched themselves on top of the mountain. It was left to the Japanese 'Army of Korea' – reduced to one thin but tough division by the China Incident – to deal with the crisis; wisely, it chose at first not to exacerbate the situation, regarding the matter as one for Moscow and Tokyo to settle by diplomacy. The Kwantung Army, aggressive as ever, fumed. The frustration spread to the usual coterie of malcontent middle-rankers among the Korean Army officers, who decided to 'reconnoiter in force' up the mountain and promptly got involved in a battle that ended in the displacement of the Russians. Several Soviet counterattacks failed, despite the Red Army's three-to-one numerical advantage and local superiority in air, armor, and artillery. Fighting went on for two weeks.

Both sides suffered disproportionately heavy casualties – more than one in five. Their military weaknesses, in the region and in general, were laid bare for those in a position to see. The Russians were seriously incompetent in several branches, the Japanese strategically exposed by long and vulnerable supply lines. Hirohito took the extraordinary step of rebuking his war minister, General Itagaki, who had also been deeply involved in provoking the Manchurian and Chinese 'incidents,' and the Army high command. Japanese anxiety, because of the war in China, to settle the matter without further ado was revealed to Moscow by Richard Sorge, the anti-Nazi German journalist who was the most effective Soviet spy of the period (he later warned – vainly – of the Nazi attack on the Soviet Union). The Russians therefore played hard to get diplomatically and were rewarded with a Japanese military withdrawal, transforming Moscow's tactical discomfiture into a political victory (which did not save the Soviet commander from execution in Stalin's still continuing, paranoid military purge).

The outcome of the Battle of Chang-ku Feng, grossly under-reported in the outside world, was read in various ways, according to taste, by the other powers. But it was eclipsed altogether only nine months later and seven hundred miles to the northeast. Lieutenant General Ueda Kenkichi, Kwantung Army commander, had worked himself into a lather over the perceived disgrace of the Korean Army at Chang-ku Feng. On 11 May 1939, after a series of border clashes which had been going on all year, he disobeyed standing orders from Tokyo to avoid major incidents: he let a cavalry regiment cross into the Soviet puppet state of Outer Mongolia at Nomonhan, in 'hot pursuit' of insurgent Mongolian horsemen who had fired on a Manchukuo Army detachment. Some one thousand Japanese infantry followed. Mongolian tanks and guns massacred the intruders in a carefully prepared trap.

Recognizing the serious potential consequences of an escalation, the Japanese, even the Kwantung Army, stayed their hand. To rub in the humiliation, the Russians, guided once again by Sorge's priceless inside information, warned that an attack on Mongolia would be treated as an invasion of the Soviet Union itself. Sorge was accurately reporting that the Japanese were determined not to be provoked; the Red Army therefore crossed the frontier from Mongolia into Manchukuo on 18 June. The Japanese colonels and majors who contributed so much to the ruination of their country got the upper hand in the Kwantung Army command in the ensuing, heated debate on what to do next, forcing through a plan for revenge. A reinforced infantry division with air, armored, and artillery support assembled against a numerically inferior Soviet-Mongolian force and reopened hostilities with a preemptive air strike against the Red Army's air strength – once again without waiting for sanction from Tokyo. A heavy clash on the ground at the beginning of July, however, led to an uncharacteristic Japanese withdrawal after a couple of days. The Kwantung Army's tanks proved to be seriously outclassed.

The Japanese also lost a furious exchange of artillery barrages a few days later. As Tokyo struggled with mixed success to hold back its wild men – especially from aerial bombardment, which was rightly regarded as the ultimate provocation of the Russians – the latter built up their forces in Mongolia to five infantry divisions and five tank brigades under General Georgi Zhukov, the future marshal and hero of the Soviet Union. Outnumbered nearly five to one, the Japanese were routed and fled back into Manchuria. The Russians showed good discipline by halting and digging in on what they claimed was the true border.

The Japanese, with willful stupidity, took this as proof that the Russians had lost their stomach for a fight – otherwise they would surely have followed the ancient military injunction to make the most of victory by pursuing the beaten enemy. They scraped together five divisions for a counteroffensive, which the Kwantung Army at least saw as round one of an all-out war against the Soviet Union, to start in mid-September 1939. But diplomatic events supervened: the Russians and the Germans signed a nonaggression pact on 23 August 1939. This supreme piece of political cynicism by two totalitarian regimes with diametrically opposed long-term interests shook the world in general and Tokyo in particular. Japan was totally isolated in the midst of a war with Russia which could hardly be regarded as small and might easily become vast now that the enemy had no potential second front to worry about. Therefore, despite the seventeen thousand Japanese and Manchurian casualties (over 30 percent) sustained in more than three months of fighting, Imperial General Headquarters (Army) decided on 2 September, the day after Hitler attacked Poland, to cut its losses and order a general withdrawal. In the meantime, the Japanese government of Baron Hiranuma Kiichiro fell victim to the Nazi-Soviet Pact and resigned on 29 August, taking the pathologically aggressive War Minister Itagaki with it. The new Cabinet under ex-General Abe Noboyuki set out to end the Soviet war at once by negotiation, leading to a cease-fire on 16 September. In the talks that ensued until formal settlement in July 1940, the anxious Japanese conceded in all directions and a border commission was set up to forestall future clashes.

Frontier incidents inevitably continued, but none was allowed to escalate into a major conflict again.

The consequences of the Nomonhan campaign were momentous in the extreme. The Japanese Army lost its bloodthirsty enthusiasm for the 'northern option' in imperial strategy. The Soviet Union was now an immediate threat for which serious provision (arbitrarily fixed at sixteen divisions on permanent standby in Manchuria) must be made, over and above the draining commitment to China proper and the extra needs of a southward strategy. The Nomonhan 'dress rehearsal' deterred both Japan and the Soviet Union from a war on two fronts and therefore made a major contribution to Hitler's fate, sealed when he attacked an undistracted Russia. When he did, Zhukov was free to rush west and save Moscow at the turn of 1941–42. As long as the China Incident continued, war with the Soviet Union was something Japan had to avoid at all costs. It was only when Stalin lost faith in his pact with Hitler, early in 1941, that Moscow and Tokyo suddenly signed the Russo-Japanese Neutrality Pact on 13 April 1941 – Tokyo's revenge for the tawdry deal between Hitler and Stalin in 1939. Ironically, Japan signed because it regarded a war between Germany and Russia as highly unlikely. Had it thought otherwise and been in a position to attack the Soviet Union at the height of the German inroad from the west, the world war and subsequent history might have gone very differently indeed. Even so, the pact held until the last week of the war.

The Nomonhan disaster, actually a blessing in disguise for Japan, broke the Kwantung Army both as a fighting force and (rather too late) as a disruptive factor in Japanese military planning and international relations. Finally, and most significantly for our narrative, the Japanese Army henceforward took a rather more positive and sympathetic view of the Navy's preference for the southern (and therefore largely seaborne) strategy: only if the Russians were thoroughly beaten by the Germans would the northern option regain its appeal for the Japanese Army General Staff.

*   *   *

The general yearning among military leaders for a Japan as self-sufficient as it had been during the centuries before the West forced it to take cognizance of the outside world was really a rationalization of their desire to have their cake and eat it, an option humanity tends to go for whenever it seems to be within reach. They wanted autarky, a closed economy. Having won access to Korea – seen as a Chinese dagger pointed at Japan from only a hundred miles away – by defeating China in 1895, and having annexed the peninsula outright in 1910 (subsequently industrializing it at breakneck speed), the Japanese established themselves in neighboring southern Manchuria. Once drawn into that power vacuum, left by a weak and divided post-Manchu China, they were duly caught up in the classic cycle of imperial inertia already described. Their holdings in what had been or what legally still was Chinese territory expanded until they collided physically with Russia's to the north and west, and diplomatically with the interests of a weakened West in China as a whole. Continued expansion, therefore, now entailed an armed confrontation to the south and east.

The underlying aim was an economic zone within which Japan could support itself without recourse to outside, meeting all its needs in oil and other minerals, rice and other foods, steel and other industries. But in order to stabilize this new yen-zone – which meant not only fighting the inefficient Chinese, who nevertheless stubbornly refused to concede, but also guarding against Russian intervention – Japan had to import more and more from the United States and other territories in or controlled by the West. The fight for autarky was making Japan even less self-sufficient than it had been when it started. The deeper it sank into the Chinese mire, the more raw materials, oil and food it needed from America and Western possessions in Asia. Thinking officers who recognized this strategic Catch-22 often found it frustrating to the point of exploding with rage. Even Hirohito was moved to complain more than once, not about the war in China as such but rather about the generals' repeated failure to deliver the victory so confidently promised. Yet, the harder they tried to win it, the further it seemed to recede into the future.

One possible Western response in such circumstances – to

find a suitable form of words, cut one's losses and withdraw from an unwinnable struggle – was not available to the Japanese, who would have lost face throughout Asia and the world. 'Losing face' is of itself no more than an oriental synonym for injured pride, and fear of it is not confined to the Far East. The difference is that it could prompt suicide in Japan, where public shame far outweighs private guilt as a social inhibitor. A 'withdrawal to prepared positions,' meaning a forced retreat to somewhere that looks defensible, was an option not available to the Japanese soldier, any more than surrender or captivity. The result for the Japanese in China was to follow the example of the despairing gambler who doubles his stake every time he loses. They were still trying to 'bring an early end to the China Incident' well into 1945 – and even then, formidable fighters that they were, they nearly succeeded.

The autarkists found natural allies in the senior levels of the bureaucracy, who were subject to a familiar inertia of their own: administrative difficulties, of which Japan had more than its fair share, could best be dealt with by increasing the power of the administrators in the civil service. If force did not work abroad, try more force; if regimentation did not work at home, try more regimentation. If Japan was to be self-sufficient, such resources as it had, especially the people, should be properly mobilized to the best effect – as if, indeed, they were soldiers. This lay behind the bureaucrats' call for 'reform,' which was their harmonious counterpoint to the military demand for autarky. The Japanese fully understood that in a total war everything and everybody had to be thrown into the struggle and there was no such place as 'behind the lines.'

Japan's military leadership, however, had nothing more to offer than an unthinking, unchallengeable discipline of the barrack square, whence it had come, with which to motivate the emperor's subjects: civilians too were only there to obey orders. The result was that the phenomenal capacity for individual self-sacrifice, readily made, among ordinary Japanese citizens was never forged into an instrument representing the will of the nation, a fact for which, in retrospect, Japan's enemies had every reason to be thankful. One of the greatest ironies of the Second World War is that the democracies proved much better than the totalitarian countries, including their own ally Russia, at total mobilization; the secret of their success was that they had elective legislatures to debate and vote on such measures, and that in general they treated the public as adults. Britain, the most exposed of the undefeated Allies, went furthest, introducing limitations on the liberty it was fighting for that would have been unthinkable in peacetime.

Both purported solutions to Japan's self-imposed problem – autarky and reform – gained much ground in the Depression. In a country with little experience and less faith in Western-style democratic institutions – and here the parallel with the Weimar Republic, which gave way to Nazism in Germany, is almost total – the general population found it easy to blame Parliament, politicians and the civilian establishment in general for what was really a worldwide malaise. Meanwhile 'one country' solutions were increasingly favored round the world, from America with its New Deal and Britain with its abandonment of the gold standard (slashing the prices of its exports), to Russia with its new five-year plans and Germany with its regimented labor-force and its dream of *Lebensraum*. Even in the democracies there was widespread support for authoritarian solutions. With its twin aims of autarky and reform, the ruling elite in Japan was thus marching to the same tune as the governments of the other leading powers; nor was it lagging behind in the broad drift toward totalitarianism – on the contrary.

Increasing regimentation of the population through modern media and communications was common to the democracies and the dictatorships alike. In Japan, where the economy was reeling in the Slump, strict exchange controls and government expenditure cuts were imposed in the early thirties. In 1935 the Cabinet Deliberative Council was set up to bring political and business leaders together; its bureaucratic instrument was the Cabinet Investigative Bureau, which coordinated the work of senior and middling ministry officials. Out of this apparatus in 1937, after the China Incident, grew the Cabinet Planning Board or CPB (and ultimately out of that the Ministry for International Trade and Industry – MITI – which has done so much to make the postwar Japanese economy the world's most powerful). A Welfare Ministry was established at the same time to improve the health of the people; the Army, worried by the poor quality of recruits as more men were needed in China, was deeply involved from the first. On the consciously adopted model of the Tennessee Valley Authority, electric power was nationalized in 1938; on the Nazi model, an agricultural relief program was introduced. The CPB took an interest in all major aspects of the nation's life, including not only business, finance and industry but also communications and culture. Industry was to make Japan self-sufficient in liquid fuels by manufacturing synthetic petroleum products, as a result of which a huge investment in unperfected technology produced barely one-sixth of the planned output. Even the Germans, who led the world in this field and helped the Japanese, got nowhere near being able to dispense with the real thing.

Nonetheless, all these measures looked like a convincing first set of steps toward the long-term goal of autarky – if only Japan could have five years of peace to complete its program. But, thanks to the impatience of the extremists in the Army (the autarkists, as ever, representing the moderate tendency), the China Incident put that, and with it self-sufficiency in the yen-zone of Japan-Korea-Manchukuo, beyond reach. Autarky was the prerequisite for future expansion; now expansion was the prerequisite for autarky. The oil of the Indies, the rice in Indochina, the rubber and tin of Malaya had to be added immediately to Japan's resources if it was to solve the China problem without continued dependence on an increasingly irritated America. Those officers who feared the Soviet Union most of all advocated coming to terms with Chiang Kai-shek's nationalists and persuaded their colleagues to drop the idea of a second puppet state alongside Manchukuo in northern China. But a negotiated settlement of the China Incident came to nothing, because of Japanese arrogance and Chiang's determination to use anti-Japanese sentiment to unite his country.

In Chicago in October 1937, soon after the Kwantung Army had badly injured the British ambassador to China and damaged Nanking, Chiang's capital at the time, by bombing, Roosevelt made his renowned 'quarantine' speech, advocating the isolation of 'lawlessness' to prevent an epidemic. It was a message meant for Tokyo above all, but it was not accompanied by threats or any practical suggestion as to how to force patients suffering from congenital aggressiveness into a straitjacket. The US administration was deeply divided, White House versus State Department, over whether to begin tough measures against Japan. From his increasingly difficult outpost in Tokyo, Ambassador Grew suggested mediation, but Washington feared being drawn into an appeasement policy on the Anglo-French model. The nations that had signed the Nine-Power Treaty guaranteeing the rights of China met in Brussels at the end of 1937 and adjourned without agreement – forever, as it turned out. On 13 December 1937, the American gunboat USS *Panay* was sunk by Japanese Navy war planes as she escorted three Standard Oil tankers on the Yangtse River. On the insistence of Admiral Yamamoto, now C-in-C of the Combined Fleet, who was not involved in the incident, Japan apologized and paid compensation: but that was only one more 'humiliation' abroad for the fire-eaters to add to all the rest. Meanwhile, on the same day, the Japanese Army tried a new method of putting a quick end to the China Incident by unleashing its victorious troops on the surrendered soldiers and civilian population of Nanking. Nobody knows how many people died in the ensuing orgy of mass murder, rape and looting, but the total ran well into six figures, in one of the worst single atrocities of a bloodstained century. General Matsui Iwane, the Japanese C-in-C in China, was executed for it by the Allies after the war, even though the massacre had flouted his express orders. The only step taken by Tokyo was quietly to reassign eighty staff officers to duties elsewhere. It was in China too that Army Unit 731, ostensibly an engineer outfit concerned with water purification, conducted unspeakable experiments in chemical and biological warfare on civilians and prisoners of war, equal in horror if not in scope to anything undertaken by the Nazis at Auschwitz and elsewhere.

But the growing list of Japanese war crimes prompted nothing more from Roosevelt at this stage – early 1938 – than a call for a 'moral embargo' of Japan by US exporters, especially of arms and the means to manufacture and use them. The moral embargo was clarified as a policy in July 1938, following the Japanese bombardment of Canton. When Cordell Hull, the secretary of state, sent a long letter in protest against Japan's conduct in China in October, Prime Minister Konoye replied in November with his declaration of a 'New Order in East Asia,' under which Japan would entrench its domination of Manchuria and China regardless of anyone else's 'rights' in China. The president knew that the American public, despite this loud and contemptuous slamming of the 'Open Door,' would not abide sanctions, for opinion polls were already in common use. US neutrality legislation prevented him from acting against Japan on its own: any formal embargo would also have to be applied to China, which desperately needed US aid, as well as Japan, which did not (though it needed imports from America more than ever).

Even so, the moral embargo was a turning point in the unfolding tragedy of mutual incomprehension that was already pointing to war. It was the first concrete expression of American displeasure with Japan, and it soon proved palpably if patchily effective. Some major US companies refused to sell to the Japanese, despite the 1911 Treaty of Commerce and Navigation between the US and Japan, which was another legal obstacle to mandatory sanctions. The Americans gave the requisite six months' notification of termination of this accord at the end of July 1939. Japan had already made a mockery of it by closing the Open Door, but Washington quite rightly took the view that it should adhere to the 'no-reprisals' clause. Frustrated American officials were reduced to hoping that a Japan indifferent to international law would fall victim to the laws of economics: surely Tokyo had bitten off much more than it could possibly chew in Manchuria, let alone China? It had, but chose to ignore the fact, an attitude to which there was no effective diplomatic or even logical response. Japan was simply immune to argument, persuasion, moral appeals, sanctions, all measures short of war, and ultimately war itself: only superior force would bring it to change its ways; only destruction, over and above complete military defeat, could put an end to its territorial ambition.

Yet, just as Britain could unilaterally have stopped Mussolini's expansion in Africa by closing the Suez Canal to him, and just as France could have stopped Hitler by overturning his reoccupation of the Rhineland, so the United States could have forced the Japanese war machine to grind to a halt with a handful of selected sanctions in the late thirties. Japan at this time depended on America for half its copper, two-thirds of its machine tools, three-quarters of the scrap metal from which it made new steel for arms, four-fifths of its fuel oil, and over nine-tenths of its high-grade petroleum products (including machine oil, gasoline and aviation fuel); and the Japanese were using up their stocks at an alarming rate in China. Japanese fears, premature though they turned out to be, that Washington would make full use of its economic leverage to save its diplomatic position in China, caused Tokyo to cast its eye on the conveniently located resources of Dutch and British territory in the Far East long before the European owners became embroiled with Germany. A clear warning of what could lie in store was given in July 1940, when the President of the United States, under the new National Defense Act, banned exports of aviation fuel and some other special petroleum products just after huge Japanese orders had been placed for these items to meet needs in China.

With typical opportunism, the Japanese Army swung fully round to the southern strategy in June 1940, when Hitler's European *Blitzkrieg* had triumphed, Holland and France had fallen and the defeat of Britain seemed imminent. The junta innocuously titled its new program an 'Outline for Dealing with the Changes in the World Situation.' The aim was autarky; meanwhile there was to be an alliance with Germany, maximum austerity at home, and the establishment of a Japanese lien, by diplomacy or by force, on French and Dutch possessions in Southeast Asia and the Pacific. Force having failed to settle the China Incident, the Army now seriously proposed dividing Japan's already stretched resources by risking the opening of a huge and complex new front. But at least the generals had the wit to limit their designs to French, Dutch and (only if unavoidable) British possessions, leaving out all American territories such as the Philippines. They were clearly gambling that a German defeat of Britain would force America to concentrate on its Atlantic front. The Navy, understanding more about the inherent strategic advantages of large islands, did not think a British collapse likely. The admirals therefore disagreed with the generals. The residual Dutch and minimal British maritime presence in the Far East was not enough to justify the Navy's voracious budgetary plans, past, present or future. Conveniently, their staff war games late in 1940 were supervised by Yamamoto himself, and showed that an attack on French and Dutch possessions in the Far East would bring first the British and then the Americans into the ensuing war. The Navy argued that the Netherlands, Britain and the United States were inseparable, that leaving the Philippines untouched would expose the flank of a Japanese southward advance – and therefore that, instead of the risk of ignoring the US, the much bigger risk of attacking its interests should be taken from the first! The student of Japanese strategic thinking at this period cannot fail to be struck time and again by the numbing, mountainous stupidity of the generals and admirals and those who supported them. It was on such a grand scale that the human mind is hard put to encompass the magnitude of the error. Seen in this light, the last-minute addition of Pearl Harbor to the finalized plan at the insistence of Yamamoto, the serious gambler, looks like a minor embellishment of an elaborate program for national suicide. The phenomenal success of postwar Japan – as dependent as ever, in an unstable world, on imports for oil and nearly all its raw materials – makes the logic of the time look even more bizarre.

Yet the Navy came to terms with the Army on a program that required, first, a move into northern Indochina (rapidly completed in September 1940) to cut off Chiang and to be well placed for a move farther south, thus keeping all options open; second, diplomatic pressure on the Dutch to supply more oil, rubber and tin (which failed: Japan had to accept the half or thereabouts which the Dutch and US oil companies in the Indies offered, and to pay for it in cash); and third, the Tripartite Pact with Germany and Italy, signed on 27 September 1940.

In the Army the extremists were now completely in charge: the more calculating 'total-war' faction was not extinct, but it was deeply depressed to the point of impotence. Autarky had been seriously deferred by the war in China; to get the resources to win that enervating conflict, the Army now proposed to move southward, which would lead to immediate suspension, before the targeted territories could be acquired, of deliveries of the vital materials needed for their conquest. This would therefore have to be completed, and secured against counterattack, with such stockpiles as Japan had managed to accumulate. Autarky would thereupon vanish over the horizon; the original Catch-22 of the China adventure was now to be squared. The only answer the extremists could produce was supplied by their friends in the bureaucracy – even more austerity at home, in the form of a maximum diversion from civilian to military use of such items as fuel, steel and shipping, and of course money. Such was the background to the speech by Admiral Konoye, just back as prime minister after a short absence, in August 1940, announcing a 'New Economic Order' and national self-reliance for Japan. On the day before Japan threw in its lot with the Axis, the United States announced an embargo on all exports of scrap iron and steel except to the Western hemisphere and Britain, on the grounds of its own national security. There was no need to explain that Japan was the true target, thus to be punished for its precipitate advance into northern Indochina.

The Hawaiian Islands in the mid-Pacific, roughly halfway between the West Coast of North America and East Asia, had been taken over as a territory by the United States in 1898. They offered the ideal advance base for US naval operations in the Far East – specifically against Spain in the Philippines shortly afterward. Pearl Harbor, on the southern coast of the main island of Oahu, rapidly became an important facility for the expanding US Navy. It was therefore an obvious target for the Japanese Navy as the Empire nurtured its expansionist ambitions, fully aware that the only power capable of challenging it was the American fleet in the Pacific. All this was so self-evident that the British journalist Hector Bywater published his ideas about a naval attack on the base as early as 1925. Two years later a mere lieutenant commander in the Imperial Japanese Navy – Kusaka Ryunosuke, a staff college instructor in aviation – drew up a plan for an air attack. The possibility was no less obvious to the United States Navy by 1932 at the latest: Captain Ernest J. King, USN, then commander of the aircraft-carrier *Lexington*, launched a 'successful' mock air attack on Pearl during maneuvers. Its importance as a base (and by the same token as a potential target) increased exponentially when the administration decided to make Pearl rather than the West Coast the principal base of the US Fleet in May 1940. The motive was political, to deter the Japanese from opportunistic ventures in the Pacific. The Netherlands had just been overrun by the Germans, presenting the oil-obsessed Japanese with a very strong temptation to make a move against the Dutch East Indies, whose home government had fled to England. The ensuing fall of France also drew acquisitive Japanese eyes to French Indochina, as we have seen.

The attack on Pearl Harbor on 7 December 1941, which achieved total surprise despite a series of alerts from Washington, including a war warning on 27 November, will forever be associated with the name of Yamamoto. It was he who conceived the operational plan in the latter part of 1940 and sent a nine-page memorandum on it to the Navy Ministry on 7 January 1941. He then used his unique personal prestige to force its adoption by Imperial General Headquarters as the final embellishment, Operation Z, of the overall strategy for southward conquest. The rationale was simple: given that Japan required the resources of the Indies and was resolved to seize them, it needed a defensive perimeter large enough to protect them and the Home Islands of the Empire from attack by sea or air, including naval airpower. This perimeter would be much more secure, and Japan would gain a priceless margin of time, if the only force strong enough to disrupt the Empire's grand design – the United States Navy – could be crippled at the outset. After that the American will to fight a difficult campaign for the recovery of European colonies in Asia, taken over by a strong and entrenched opponent who held the initiative, would be so weak that Japan would be able to exact a profitable peace. The hawks were only encouraged in this belief when the House of Representatives voted on 13 August 1941, by a margin of one – 203 to 202 – to renew the military draft.

The Pearl Harbor operation, like the Imperial Navy itself in its formative years, drew heavily on British ideas, and not just those of Mr Bywater, which were well known to Yamamoto. There was the classic example of Nelson, who destroyed the Danish Fleet in port in 1801 and thus gave the English language the verb 'to Copenhagen' for a naval pre-emptive strike. Rather more recently, Admiral Sir Andrew Cunningham, British commander-in-chief in the Mediterranean, had launched a devastating torpedo-bomber attack – the first in history – from the carrier HMS *Illustrious* against the main base of the Italian Navy at Taranto on 11 November 1940. Royal Navy pilots eliminated three out of four enemy battleships and tilted the balance of naval power in the ocean in Britain's favor. There was thus, as so often, nothing new under the Rising Sun. But there was one fundamental difference: the British were already at war on each occasion; Japanese diplomats in Washington were still negotiating with the United States at the time of the attack on Hawaii.

Yamamoto Isoroku, the man who issued the order to go ahead with it, was the sixth son of a former samurai. Born in 1884 at Nagaoka, Yamamoto entered the Naval Academy and graduated in 1904, just in time for the war with Russia. He was badly wounded at the decisive Battle of Tsushima, won by the Japanese Navy's big guns in 1905. He went to the United States as a junior naval attaché just after the First World War and then visited Europe. He was already forty years old when he presciently switched his professional specialism from gunnery to aviation in 1924; nobody was to play a greater role in educating the Japanese Navy in the importance of the new seaborne airpower. He became senior naval attaché in Washington in 1926–28 with the rank of captain, then commanded a cruiser and next the carrier HIJMS *Akagi* before distinguishing himself as a delegate to the London naval disarmament conference of 1930, an event that made him a public figure in Japan. As a rear admiral, Yamamoto served as head of the technical division of the Navy's Aeronautics Department and then commander of the First Carrier Division (on the *Akagi* again). Promoted to vice-admiral, he became chief of the entire Aeronautics Department, and then deputy Navy minister in 1936. Paradoxically, his strong opposition to war, which exposed him to death threats from extremists, led to his appointment as commander-in-chief of the Combined Fleet, the main body of the Japanese Navy, on 30 August 1939. His moderate friends in high places thought he would be safer if protected by the forty thousand sailors then manning the fleet. He was promoted to full admiral in 1940 at the age of fifty-six.

He was short and slight, even by the diminutive Japanese standards of the time: five feet three inches tall with a noticeable stoop and weighing barely 130 pounds. The index and middle fingers of one hand were missing, shot off at Tsushima; as his biographer, Hiroyuki Agawa, reveals, he was therefore known to the geisha girls whose company he liked so much as '80 sen' – the price of a geisha manicure then being 100 sen (1 yen). His spotless uniform concealed many other scars from the battle, but he prided himself on his fitness: even when past fifty he demonstrated it by doing head- or handstands on minimum provocation or none at all. His neatness extended to the calligraphy for which he was also well known (even the toughest Japanese officers were wont to cultivate gentler talents such as brushwork and poetry, in keeping with the samurai tradition; Yamamoto's poems are, however, described as plodding). But his taste for order did not extend to his private life. Married in 1918 and the father of two sons and two daughters, Yamamoto was never close to his wife, Reizo, preferring the geisha Kawai Chiyoko, whose professional name was Umeryu. This relationship endured from 1934 until his death; when it was revealed by her in 1954, it caused a sensation in Japan. There were occasionally other women too, giving the lie to the forbidding public face.

He was quite capable of enjoying himself, but he was essentially a loner inclined to melancholy. Presented by the slavish Japanese press on his elevation to supreme command as the strong, silent type, Yamamoto was in fact outspoken to the point of rudeness and no respecter of persons. He liked to drink Scotch whisky and smoke expensive cigars in late-night sessions with colleagues or journalists, or at parties in the geisha houses. Those who knew him in relaxed mood, when the stony mask of the Japanese officer could be set aside, found his character most attractive. His greatest off-duty passion was gambling, on everything from *shogi* (Japanese chess) to Chinese mah-jongg and Western billiards, roulette, poker and bridge.

In his years in Tokyo from 1935 until his final appointment, Yamamoto was frozen out by the militant tendency in Navy, Army and government. The hard-liners did not have the moral courage to sack him as deputy Navy minister but did all they could to prevent him from converting his personal prestige into political power as a force for moderation. He found it acutely depressing to have little or nothing to do. Both the enmity and the prestige derived from his high-profile performances as a delegate to disarmament conferences, assiduously reported by his journalist friends. He was thought by the militants to be too accommodating to foreign opinion because he was that rare contemporary manifestation, a Japanese leader who understood the art of compromise and appreciated the need for it. But, like most influential Japanese, he also thought his country should have a place at the top table of world powers – even if he opposed force as a means to that end and regarded war with the United States in particular as suicidal. Not until he had reluctantly concluded that such a war was inevitable did he argue that Japan should make the strongest possible pre-emptive strike against the United States: the only strong enemy should be knocked out at the beginning. If, as Japanese strategists insisted, Holland, Britain and America were strategically inseparable because an attack on the Far Eastern interests of one inevitably entailed war against all three, Yamamoto's gamble was not even a calculated risk but a move that the rules of the game required him to make.

In September 1940 he told Prime Minister Konoye: 'If we are ordered to [go to war with America] then I can guarantee to put up a tough fight for the first six months but I have absolutely no confidence about what would happen if it went on for two or three years . . . I hope you will make every effort to avoid war with America.' And as late as October 1941 he told a friend: 'I find my present position extremely strange – forced to make up my mind and unswervingly follow a course [war with America] which is exactly the opposite of my personal views.' These observations, made in private and never intended for external consumption, clearly represent the

true feelings and opinions of Admiral Yamamoto on war with
the United States. The intercepted signal of early July 1941
mentioned at the beginning of this book, which convinced the
US administration that Japan would go to war in the Pacific in
a matter of weeks if not days, was meant to be just as private,
a fact which did nothing to deter the US eavesdroppers from
overreacting – on the contrary! The text on which the American
deduction was based seemed clear enough – clearer by far than
the elaborately oblique and bland documents and statements
supplied on purpose by the emperor's representatives to foreign
governments. Leaving aside the content and its implications, the
message, by definition not intended for foreign eyes, might well
have seemed positively refreshing to the Americans precisely
because of its candid duplicity and cynical opportunism. It said
in part:

In order to guarantee the security and preservation of the nation,
the Imperial Government will carry on with all necessary diplomatic
negotiations concerning the southern regions [relative to Japan]. . . .
In case diplomatic negotiations break down, preparations for a war
with England and America will also be carried forward. First of all,
the plans which have been made with regard to French Indochina
. . . will be followed through in order to consolidate our position in
the southern lands . . . We shall not be deterred by the possibility of
becoming embroiled in a war with England and America . . . All plans
will be executed so as to place no serious obstacles in the way of our
basic military preparations for a war with England and America . . .
We shall turn our attention at once to putting the nation on a war
footing and shall take special steps to strengthen national defense.

The text summarized a resolution passed on 2 July at an
'imperial conference,' a meeting of military and government
leaders in the presence of Emperor Hirohito. It began by
setting out the need to force an end to the inconclusive war
of expansion in China so foolishly provoked by Japan four
years earlier. This was one of two reasons for moving south
from the bases in northern Indochina acquired from a helpless
Vichy France in September 1940: such a stroke would serve to
isolate the stubbornly struggling Chinese nationalist regime of
Chiang Kai-shek. The other reason, also unmentioned in the
message but so familiar as to be taken for granted by all its

recipients, was to gain forward bases from which to be able to attack British and Dutch possessions in Malaya, Borneo and the East Indies, with their abundant oil and other strategic raw materials. The whole world knew that this was now the central aim of Japanese diplomatic, economic, and military policy. The Japanese, ever opportunistic, would stay out of the war with the Soviet Union begun by their Axis partner Germany just ten days earlier, the dispatch went on to say – unless a chance presented itself to eliminate the Russian threat from the north to Japan's interests on the Asian mainland.

Japan, it seemed to Washington, had at last resolved its long-standing strategic dilemma between a move north against Russia and a move south against Western positions in the Far East. The Americans could put two and two together like anyone else. Unfortunately on this occasion they came up with an answer considerably greater than four. They had been seriously misled by their own cleverness in cracking Japanese codes and ciphers. How they achieved that will be described in chapter two; it is enough to observe here that the intercepted dispatch concealed far more than it revealed about the state of mind of Japan's deeply divided leadership. The text seen by the Americans was precisely translated and completely correct – as far as it went. But it did not go anywhere near as far as Washington thought. For the few American policymakers in the know, it was effectively the point of no return, beyond which war with Japan was inevitable. But for Japan a war with the United States became inevitable only as the result of the Americans' acceptance at its face value of a message written for a different audience altogether. Washington thus ignored the long-running, strategic north-south dispute in Japan, of which it was fully aware, and which it might have known, on reflection, could not have been resolved overnight or by a single meeting. And Tokyo never got the slightest inkling that the Americans had read what in any case was only a summary.

The Foreign Ministry's smooth *résumé* of the conference naturally gave no hint of the abiding divisions between advocates of the northern and southern strategies, within and between the Army and the Navy, and between the armed services and the government, over peace and war. It was merely reporting the

latest round in an intractable debate which both sides assumed
would go on for a long time yet. There was no need to record
all the well-worn arguments, even more familiar to Japanese
diplomats than to the Americans who themselves had heard
them times out of number from orthodox as well as clandestine
sources. Indeed, Washington itself might well have come round
to a more relaxed assessment of the text in due course – but for
the Japanese ultimatum in the middle of the month to Vichy
France. This demanded the right to station troops and aircraft
in southern Indochina, to which the Pétain regime acceded on
21 July. The consent for bases in the north had been won by
diplomacy in 1940 (although Japan knew it could not be denied,
because it held all the cards while France, newly defeated by
Tokyo's Nazi allies, had none).

For Japan, therefore, the ultimatum in July 1941 was a
watershed, akin to Hitler's occupation of what remained of
Czechoslovakia in March 1939, just a few months after winning
the Sudetenland by negotiation at Munich. It too represents
the transition from acquisition of foreign soil by diplomacy,
however aggressive, to seizure by force or the threat of force.
For the United States it was confirmation that the intercepted message had indeed revealed a Japanese decision on
an imminent advance southward. The Americans, still deeply
isolationist, did not want war any more than Prime Minister
Neville Chamberlain's appeasement-minded Britain had wanted
it in the spring of 1939; but, like the British after the German
takeover of the rump of Czechoslovakia, the Americans now
resolved that aggression had gone far enough. Washington's
equivalent of the fateful British guarantee to Poland was its
decision, announced on 25 July, to ban exports to Japan of
high-octane gasoline (suitable for aircraft), to limit all other
petroleum exports to normal levels – and to freeze Japanese
assets in the United States.

The latter measure, originally intended to deny funds to the
Japanese spies assumed to be everywhere and to retaliate for
tight Japanese exchange controls, had devastating implications.
The freeze effectively prevented Tokyo from buying any oil at
all, or other raw materials needed to sustain the war in China,
for which it was still overwhelmingly dependent on the United

States. This was a far more serious sanction than the limited steps – notably a ban on exports of new steel and reusable scrap – which had followed upon Japan's initial advance into Indochina ten months earlier. For Tokyo it was the realization of its most persistent nightmare: America was about to bleed Japan to death by forcing it to rely on its precious oil reserve. The oil embargo, supported by the British and the Dutch with their huge oilfields in Borneo and the Dutch East Indies (now Indonesia), was the *casus belli* both feared and desired by the all-powerful Japanese military. We shall look more closely at the thinking of the generals and admirals below. Japanese troops began to occupy southern Indochina bases on 23 July, and forty thousand reinforcements began to land there only five days later. The events of the next four months drove Japan and the United States into a war that both had intermittently seen coming for more than twenty years.

At this time General George C. Marshall, chief of staff of the United States Army, publicly announced that Washington intended to reinforce its ground and air garrisons in the Philippines, hitherto privately regarded by Washington as indefensible in a war against Japan. General Douglas MacArthur was recalled from the United States Army's retired list to command in the great archipelago, which had been under the American flag since the war with Spain of 1898. Boeing B-17 'Flying Fortresses' – the latest high-level strategic American bombers – started to arrive at Clark Field in the Philippines at the end of July, immediately after delivery to the US Army Air Force. Some seventy-five thousand troops were earmarked for transfer from the United States, the Philippine Army was incorporated into the US Army, and eventually MacArthur was meant to have two hundred thousand men under his command. The underlying strategy, to which MacArthur had won over a skeptical Marshall, was to defend Luzon island for long enough to enable the American main fleet to fight its way across the Pacific and use the Philippines as its forward base for operations against Japan. Admiral Thomas H. Hart, USN, commander-in-chief of the Asiatic Fleet, redeployed his meager forces from China to the Philippines area, with Manila Bay as his principal base. On 7 August, the Senate extended the period of service for draftees in the armed forces by six months, to eighteen.

At this point an alarmed Japanese government proposed a summit conference between Prime Minister Konoye and Roosevelt. Secretary of State Cordell Hull finally got around to rejecting the proposal on 2 October. Although diplomatic contacts were to continue even beyond the last moment of peace, the failure of the summit proposal seemed to both sides at the time to indicate that all efforts to achieve an accommodation would be in vain. At this stage the US government was playing for time, at the urgent request of Marshall and his opposite number, Admiral Harold R. Stark, chief of naval operations, so that Army and Navy could complete their preparations for war by the spring of 1942 if all went well. But their own government's miscalculation of Japan's mood in July ensured that they did not get their wish. In Japan, by contrast, the generals and admirals contemplated the embargo and their steadily dwindling oil stocks and pressed for war as soon as possible. This recourse had become a case of 'when' rather than 'whether' on 6 September 1941. On that day an imperial conference of military and government leaders with the emperor agreed on a statement that bore the innocuous title 'The Essentials for Carrying out the Empire's Policies.' It had been adopted by a 'liaison conference' (the same cast of characters without the emperor or palace officials, the mechanism for coordinating military and government policies) on the previous day.

Protocol at imperial conferences was stiff to the point of immobility. The emperor wore a long black morning coat with striped trousers, and all present were required to do likewise unless entitled to wear uniform. Hirohito sat on a plain throne on a dais, looking down the length of a pair of long tables covered in silk cloth that reached to the floor. The inner sides of the tables were unoccupied, while the outer sides were each lined with a row of officials sitting to attention. The emperor left the chore of questioning ministers to the president of his Privy Council, Hara Yosimichi. There was no debate; participants took it in turn to deliver prepared statements and answer Hara's questions in staccato fashion, as if on parade. The document from the liaison conference was, as usual, adopted without amendment.

Its main effect was to harden into policy, complete with deadlines for a decision on peace or war and for actually going to war, the tentative program formulated on 2 July, described above. The 'humiliation' of a surrender to pressure was not an option for the Japanese junta, as United States officials should have known. Washington had therefore made the gross psychological error of leaving its antagonist no means of preserving his pride except by going to war. American sanctions, intended to prevent the war that the 2 July message foreshadowed, only served to make it inevitable. The key sections of the 6 September resolution were as follows:

In view of the current critical situation, especially the offensive attitudes that such countries as the United States, Great Britain, and the Netherlands are taking toward Japan . . . we will carry out our policy toward the south . . . as follows.

1) Our Empire, for the purposes of self-defense and self-preservation, will complete preparations for war, with the last ten days of October as a tentative deadline, resolved to go to war with the United States, Great Britain and the Netherlands if necessary.

2) Our Empire will concurrently take all possible diplomatic measures toward the United States and Great Britain and thereby seek to gain our objectives . . .

3) In the event that there is no prospect of our demands being met by the first ten days of October through the diplomatic negotiations . . . we will immediately decide to commence hostilities against the United States, Britain and the Netherlands . . .

And so, with a few unimportant delays and adjustments, it was to turn out. There was only one surprise at this portentous meeting, where Japan's chiefs of staff and most senior ministers formally decided, in the divine presence of their emperor, to go to war with the West – unless diplomatic means succeeded after all, against the perceived odds. The sensation was provided by the emperor himself, and only after the adoption of the policy statement: Hirohito spoke. His unprecedented intervention, not to be repeated until all was lost in August 1945, took the form of quoting an ode composed by his grandfather the Emperor Meiji, which translates approximately as follows:

All men are brothers, like the seas throughout the world;
So why do winds and waves clash so fiercely everywhere?

All present were deeply moved by the piping voice, which they otherwise heard only during the private audiences with the emperor to which their exalted offices entitled them. A few manly tears were shed – but the policy as dictated by the military was not modified by so much as a comma. Yet the two chiefs of staff, General Sugiyama of the Army and Admiral Nagano of the Navy, rightly took this recitation *ex cathedra* as a rebuke and disingenuously promised, contrary to the general drift of the policy statement, that the emphasis would be put on diplomacy rather than preparation for war.

When the deadline came, the Cabinet split over the war issue, Konoye resigned, and General Tojo Hideki, the Army minister, succeeded him on 17 October. Thereafter air forces (military and naval) and eleven Army divisions were assembled for a remarkably complicated multiple thrust southward across a six-thousand-mile front; so was almost every ship in the formidable Japanese Combined Fleet, commanded by Admiral Yamamoto. The war plan for the southern option had existed, subject to various adjustments, for years; the one major change to it, added on Yamamoto's insistence, was officially less than a year old. To execute it, Vice-Admiral Nagumo Chuichi's mighty Mobile Force, six aircraft carriers with powerful escort, sailed into the bleak North Pacific in the utmost secrecy on 26 November 1941, under strict orders to maintain wireless silence. It was Japan's turn to make a historic mistake, and this was to be a much greater one than the American error, which did not cause but did precipitate the execution of Japan's long-standing program of conquest.

2

The View from the West

The belated and forced entry into the war of the United States, the last and most powerful of the major combatants, was neither as sudden nor as clear-cut as it may appear.

What we know as the Second World War is generally held to have begun two years and fourteen weeks earlier, on 1 September 1939, when Hitler sent the Wehrmacht into Poland after faking a raid by 'Polish troops' across the border. Two days later, Britain and France, in keeping with their guarantee of Poland's borders, and having had no answer to their demand that the Germans withdraw, declared war on the Third Reich. (A contemporary inhabitant of Manchuria would, however, have regarded these dates as eight years after the true start of the war.) Hitler's real motive was to acquire Polish territory (some of it had been German until 1918) as a jumping-off point for a later invasion of the Soviet Union. Stalin, meanwhile, cynically consented to accept a share of the spoils – eastern Poland – from the quick German victory over the Poles and held to his alliance with the Nazis, in keeping with the nonaggression pact concluded by the two totalitarian rivals on 23 August 1939. Having for the moment settled Eastern Europe to his satisfaction, Hitler was free to turn west and put an end to the deceptively quiet 'Phony War,' as the British called it, in which the French Army, reinforced by a relatively small British Expeditionary Force, waited without enthusiasm for something to happen. Having won the race against Britain and France to occupy Norway and Denmark in April 1940, Hitler launched his *Blitzkrieg* (lightning war) against the Netherlands, Belgium, Luxembourg and France on 10 May. Superbly trained tank forces, closely supported by the dive-bombers and fighters of the Luftwaffe and the world's best infantry, took a mere six weeks to force the French into an armistice on 21 June 1940. Italy entered the war on Hitler's side on 10 June. The

British evacuated most of their Army, without its equipment, in 'the miracle of Dunkirk,' but an invasion seemed imminent until mid-September 1940, when the Royal Air Force won the Battle of Britain. Too weak to assail what was rapidly becoming 'Fortress Europe,' Britain fell back on her traditional stratagem of the flank attack and took on the Italians in the Mediterranean and North Africa. Troops from Britain and the Commonwealth won easy victories on land and sea until Hitler came to the rescue of his Axis partner, sending Rommel with a corps to Africa, and submarines, E-boats, paratroops and strong air forces to the Mediterranean.

Meanwhile, Japan, Italy and Germany signed the Tripartite Pact formalizing the Berlin-Rome-Tokyo Axis on 27 September 1940. The spreading conflict underwent its mightiest shift on 22 June 1941, when Hitler, his western defenses secure and his Mediterranean commitment manageable, launched Operation Barbarossa against the Soviet Union with 143 divisions. Everything he had done in power since 1933 and in the war up to now was for Hitler secondary to this, his main thrust for *Lebensraum* in the east, as foreshadowed in his book, *Mein Kampf*. For the four years until the German surrender, this was to be the bloodiest killing ground of the war. The Germans got to the outskirts of Moscow in the opening months; what saved the Soviet capital was Zhukov's arrival from Soviet East Asia with seasoned troops. Stalin, though surprised on the day despite countless Allied warnings (shades of Pearl Harbor!), knew that a war with Hitler's Germany was well-nigh inevitable, whereas war with Japan was avoidable. Tokyo was inclined southward and wanted to shield its northern front in China from the Russians. Japan thus sealed the ultimate fate of its Nazi ally by its diplomatic accord with Moscow.

The Americans, under the wily master politician Roosevelt, sat out these momentous events even as he initiated his unique correspondence with Winston Churchill on 11 September 1939, which was kept up until the president's last illness. His clear understanding that a German victory was against his country's interests conflicted with the palpable desire of the American people not to get involved in the murderous conflicts of Europe for the second time in less than a quarter of a century. The

US administration was also worried about the possibility of a war on two fronts. So many Americans or their forebears had emigrated to the United States to get away from Europe's rival nationalisms and the often oppressive regimes that represented them. There was a widespread and deeply held feeling among ordinary citizens that the United States had been tricked into the First World War to bail out the British and the French (and their empires). Whether this was true or not, it was certainly undeniable that the Europeans had hardly made the best of the short interlude of peace between the previous conflict and the present one – even if the collapse of the peace was also partly due to the errors of President Woodrow Wilson, his successors and the US Congress in 1918–20 and after.

Nonetheless, under Roosevelt a neutral United States moved step by step to a position from which it was doing everything except actually fighting Germany – 'all aid short of war,' as the president defined it. America soon became, in another of his well-remembered phrases, 'the great arsenal of democracy' which supplied the British with what their leader, Churchill, called 'the tools to finish the job' – armaments of all kinds, aircraft, equipment, ships and food. American aid enabled the British island fortress to hold out and to defy Hitler by bombing German cities. The Nazi leader's greatest error was now exposed: he had opened a second front without closing the first. Britain mounted a colossal naval effort to fend off the U-boats attacking the transatlantic supply line, which was already the main artery of the war against Nazism. On 3 September 1940, Roosevelt agreed to exchange fifty old destroyers for bases in British colonies in the western Atlantic; on 11 March 1941, he signed the Lend-Lease Act, whereby America 'loaned' Britain vast amounts of war materiel. The arsenal of democracy was also the arsenal of Chinese nationalism and of Soviet communism. The United States Navy defined and patrolled a huge protected zone in the western Atlantic, closed except to powers with territory in the Americas (thus open to Britain and its ally Canada). In 1941, well before they officially went to war, the Americans sent Marines to relieve the Anglo-Canadian garrison in Iceland guarding the North Atlantic route, began to escort convoys, set up bases in the

United Kingdom and Newfoundland, and even got involved in skirmishes with U-boats.

At the beginning of the year, from 29 January to 27 March, elaborate and detailed staff talks took place between British and American military leaders in Washington – 'American-British Conversation Number One,' or ABC-1. These took the broadest possible international and strategic view, anticipating American belligerence against not only Germany and Italy but also Japan, with which war was regarded as very likely. These talks agreed on the 'Germany-first' policy, which was to last until 1945. It was officially confirmed at the Anglo-American 'Arcadia' summit conference in Washington at the turn of 1941–42, which set the seal on the Anglo-American alliance.

'Germany first' may have been the policy of the United States from before its baptism of fire, but the war with Japan took and held first place in American public interest. The Pacific was overwhelmingly the main front for the protagonist in our narrative, the United States Navy and its Marine Corps, and the chief drain on American industrial resources. We have seen how Japan came to be at war with America; we now need to recall how America slid into war with Japan.

No sooner was the First World War formally ended at Versailles in June 1919 than the signatories, including America and Japan, began to fall out among themselves. Congress kept the United States out of the new League of Nations, even though it had been invented by President Wilson. The powers disagreed not only over European issues but also over their future relations, above all about armaments, especially naval, the battleship being the ultimate strategic weapon of the period. Exhaustive efforts were made to settle these differences by negotiation, at a series of international conferences. The first and most important of those that concern us was the Washington Conference of 1921–22, the earliest serious postwar attempt to defuse international tension and rivalry by negotiation.

This complex gathering, called by the newly inaugurated President Warren Harding, toured the entire horizon of relations among the major powers, though defeated Germany and the new Soviet Union were not present. Among the dozen resolutions and nine treaties from a conference that came remarkably close to achieving long-term harmony among the powers were agreements on foreign interests in China, the relative strengths of the leading navies and, by implication, Japan's position in the world, the underlying key issue. Britain, forced to choose between good relations with the United States, a potentially insuperable naval rival in the Atlantic, and her now embarrassing alliance with Japan of 1902, picked the former. Japan's 'compensation' was a four-power treaty with America, Britain and France of mutual respect for one another's rights in the Pacific region.

A treaty signed by these four and Italy fixed the relative capital strengths of the world's five largest navies at 10:10:6 for Britain, America, and Japan and 3.3 each for France and Italy. A limit of thirty-five thousand tons was set for individual vessels, and a ten-year 'naval holiday' from new construction was declared. As a concession to Japan, it was agreed that no power would increase the strength or number of its outlying bases in the Pacific; Hawaii and Singapore were excepted, but the provision favoured Japan as the mandated occupant of most of the ex-German islands in the western Pacific.

A nine-power treaty sought to regulate and reconcile the Chinese interest of the signatories, including America, Britain and Japan. The pact piously looked forward to Chinese equality and the return of all enclaves and extraterritorial rights held by foreigners. Respect was promised for Chinese independence and territorial integrity as the foreigners clung to their concessions in Shanghai, in Manchuria and on China's coast. This commitment to Augustinian 'virtue but not yet' was a perfect example of imperialist rationalization. As one of the nine signatories, China, with saintly forbearance but also mindful of its recent experience of Japanese ambition, went so far as to accept the Open Door, the American policy of equal rights in China (for all except the Chinese). The main effect of all this was to deny Japan a free run while increasing American interest in China, thus ensuring the aggravation of a rivalry that turned to hostility and finally to war. The clash between America and Japan over China was a root cause of the war in the Far East.

Another was racial prejudice. It is a pity that the distinction once made between 'racism' (the idea that ability is determined by race) and 'racialism' (the idea that one race is superior to the rest) has faded, for the former suits prewar white imperialist attitudes and the latter Japanese. At Washington, as at every other prewar international forum, including the League of Nations, Japan demanded (but never got) true equality while firmly believing that the Japanese people, descended from the gods, was superior to all others. Like Europeans, Americans tended to regard the Japanese as stunted, bow-legged crea-tures with bad teeth and pebble glasses – self-evidently inferior. Illogically, however, they also feared these 'stunted' aliens, who had established themselves by immigration in large numbers in Hawaii and the West Coast states over the preceding half-century. The US Supreme Court cravenly decreed in 1922 that ethnic Japanese could not become citizens, and in 1923 that they could not own or lease land. In 1924 Congress passed the Exclusion Act, banning further Japanese immigration. This was also the heyday of the 'White Australia' policy. Such white prejudice was excellent ammunition for the many xenophobes in ambitious Japan, a society which so suppressed aggression that it was all the more terrible when unleashed. White racism underestimated the Japanese, just as their own racialism led them to underestimate their future American enemy.

American disapproval of Japan's designs on China marched hand in hand with US domestic policy against Japanese immi-grants, but there was as little desire among Americans for war in the Pacific as in the Atlantic. The pendulum had swung toward foreign adventure at the end of the nineteenth century, aggressively extending manifest destiny, which had brought expansion to the West Coast, across the Pacific to Japan, the Philippines and China. In 1918 it swung the other way, to isolationism, and stayed with it for the twenty-three years between the Armistice and Pearl Harbor. This was one of many reasons why the Japanese regarded America as lacking in 'sincerity,' a term they defined differently from Americans. In Japan it meant total commitment to a goal, regardless of cost or means (including deception); in America it meant genuine straightforwardness. Thus the Japanese regarded American reluctance to fight, or otherwise pay the full price for what they wanted, as insincere, whereas thc Americans saw the deviousness of Japanese diplomacy as no less so.

Had Japanese diplomats been doing their job, or had they exercised more influence over diehard elements in Tokyo, Pearl Harbor and the Philippines might have been left out of the grand-slam strategy with which Tokyo went south. The diplomats would have interpreted Roosevelt's support for Britain as the preferred alternative, rather than the prelude, to belligerence. Avoiding US possessions would have left the administration with a choice between antagonizing the isolationist majority by a decision for war over Europe's Asian colonies or else procrastinating as before. Roosevelt was notorious for deferring unpleasant decisions, a tactic that had become a policy. Further, since secrets had such a brief lifespan in Washington, it must have been well known to the Japanese Embassy that the American forces did not feel ready for war. The fact that the Nazis were stoking up isolationism with slush funds, lobbying, and propaganda through such organizations as the German-American Bund and the Make Europe Pay Its War Debts Committee does not disprove the sincerity (in every sense) of grassroots and highly placed isolationists alike.

These included the Republican Senator Burton K. Wheeler from Montana, who said Roosevelt was bent on dragging America into a war that would 'plow under every fourth American boy.' It was an idea he got at dictation speed from a German-American lobbyist for Hitler, and it infuriated the president like no other accusation, not even the senator's cry of 'warmonger' when Roosevelt appealed, at Christmas 1940, for all possible aid to Britain. Wheeler was the leader of the Committee to Defend America First, better known as the America-First Committee. Its most prominent supporter was Charles A. Lindbergh, still a national hero after making the first solo, nonstop flight across the Atlantic in 1927, but naïve and a knee-jerk conservative in politics (he took one look at what the Luftwaffe chose to show him on a prewar visit to Germany and concluded it was invincible). Roosevelt received him privately at the White House in April 1939 and completely failed to win him over, concluding that he was a dangerous man. On the outbreak of war in Europe, the president offered to create for Lindbergh the post of secretary for air if he gave up preaching isolationism. The eccentric aviator, a sad figure since his child disappeared in a sensational kidnapping, was not so easily bought.

Also beyond the president's reach in the Wheeler camp was Colonel Robert R. McCormick, proprietor of the Chicago *Tribune* and congenital hater of Roosevelt, the consummate populist, whose many enemies were singularly passionate in their loathing. McCormick's cousin Eleanor 'Cissy' Patterson owned one of the main Washington newspapers. As the Federal capital's leading hostess, she dominated a coteric dismissed by Roosevelt as 'the American Cliveden set' (British socialites and appeasers led by Lord and Lady Astor of Cliveden House). Cissy's brother, Captain Joseph Patterson, owned the New York *Daily News*. These three papers and the Hearst Press, founded by William Randolph Hearst (model for Orson Welles's *Citizen Kane*), formed a powerful and strident anti-FDR chorus which did not shrink from fighting dirty, as we shall see. They helped to turn Roosevelt against the press as a whole and toward radio as a direct line to the voters. The press tycoons' hostility, based on opposition to the New Deal with its welfare provisions and state intervention in the economy, turned to hate when the Lend-Lease Act, a monstrosity to all Anglophobes, became law.

The strength of anti-British sentiment in a nation full of people of Irish and German origins was illustrated by the popularity before Pearl Harbor of an isolationist diatribe, *England Expects Every American to Do His Duty*, by one Quincy Howe. Admiral Nelson, RN, was clearly as quotable in the United States as in Japan. Pacifists, communists, fascists and the mineworkers' leader John L. Lewis, prominent opponent of conscription, also belonged to the diverse cross-section of Americans who translated isolationist sentiment into a political tendency transcending party. No politician, least of all Roosevelt, could afford to take it lightly.

*   *   *

But opposing Roosevelt over something he held dear was also not to be undertaken lightly. For all the privileges that went with the nearest thing America has produced to an aristocracy, Franklin Delano Roosevelt was a graduate of the University of Hard Knocks. Born in 1882 at Hyde Park in upper New York state of an indulgent father and ferociously dominant mother, both of impeccable early-settler stock, Franklin was spoiled as a child. Privately tutored until fourteen, he was sent to Groton, modeled on the English 'public' school (meaning open to all with inordinately rich and influential parents) and founded only a few years earlier, complete with instant arcane 'traditions.' At eighteen he went on to Harvard to read history, literature, and politics – in breadth if not in depth. He played football, rowed, joined a glee club, and worked for the student press. In the long summer vacations he accompanied his parents to Germany, France, and Britain.

In 1901 his distant cousin, Theodore, already his partner in a mutual-admiration society, became president on the assassination of McKinley. This gave Franklin an early glimpse of life in the White House, which he loved. By the time he left college, if not sooner, the young Roosevelt was a consummate manipulator of people, not least young women, an expert at compromise with an instinct for seizing the middle ground. He was quite clearly inspired to develop these skills by the need to get round his mother, Sara. In 1902 he met the other principal woman in his life – Eleanor, niece of Theodore. They married in 1905 and had six children, one of whom died in infancy. Franklin completed his formal education at Columbia Law School in New York City in 1907 and became an attorney. But he was already committed to following in his illustrious Republican cousin's footsteps – as closely as possible, even if he was a Democrat. In the same year, he predicted that he too would become, successively, a New York state legislator, assistant secretary of the Navy, governor of New York, and finally president of the United States! And so he did, starting in 1910. A series of illnesses gave the lie to his general air of athletic fitness. He suppressed his distaste for pork-barrel politics when he realized he would get nowhere without the backing of the Democratic state political machine at Tammany Hall.

His stint in the Navy Department lasted the seven years until 1920; it was marked by feverish wartime activity, including a visit to England and the western front in France, and by an enormous, politically almost fatal scandal at its end, just as Roosevelt was preparing to run for vice-president. The row centered upon Roosevelt's investigation of a homosexual corruption network at the naval training station in Newport, Rhode Island. Liberal use was made of entrapment. In 1921 a Senate subcommittee concluded that Roosevelt must have known of the dubious methods and had therefore perjured himself before the court of inquiry. This low point in his career was swiftly followed by the low point in his life (no coincidence, we may surmise): he came down with infantile paralysis (poliomyelitis) in the summer of 1921 and never walked unaided again. In the seven hard years that ensued, he learned compassion and self-reliance and became a much more serious individual: life had ceased to be a game. His political comeback as state governor led to his nomination for the presidency in 1932; on 4 March 1933, just seventeen days after a near-miss assassination attempt at Miami, Florida, a second Roosevelt moved into the White House. He was to be its most enduring tenant, and began his twelve years in power with a record-breaking first hundred days of frenetic economic legislation. It was nothing less than a bid to outlaw the Slump, and it succeeded – not least because of the stupendous economic activity generated by Hitler's war, begun two hectic years before this glad-handing cripple led America into battle as its commander-in-chief.

Whatever the shortcomings of the United States and its C-in-C as Japan so brusquely catapulted them into combat, there was no shortage of information about the new enemy. Indeed, hindsight provides many reasons for believing that there was too much of this normally scarce commodity, a veritable cornucopia of intelligence, in the military sense of the word, not matched by intelligence of the other, higher kind: the wit to interpret the secretly acquired information correctly and draw the right conclusions. But the dearth of assessment and application skills should not be allowed to detract from the sheer brilliance of the American intelligence-gathering effort between the wars, all the more remarkable for being so haphazard – sometimes downright chaotic – and involving such extraordinary, not to say bizarre, individuals.

The story goes back to the First World War, when American forces did no more than lay the foundations of their own signals-intelligence ('sigint') efforts. First in the field was Colonel George Fabyan, who was not even a real colonel (the honorary title was awarded by the state of Illinois for his part in the American mediation that ended the Russo-Japanese War in 1905). Fabyan, a wealthy eccentric, ran a biological laboratory at Geneva, Illinois, where his amateur interest in cryptography led him to set up a cipher department before the US entered the war in April 1917. Fabyan had been decoyed up this intriguing linguistic byway while trying to prove by verbal analysis that the works of Shakespeare had been written by Francis Bacon. The first head of the department was one Elizabeth Smith, and it soon began to undertake assignments for the US departments of State, War, Navy, and Justice, none of which had any such facility.

Smith was soon succeeded by a young geneticist on the laboratory staff who also took an initially amateur interest in codes and ciphers: William F. Friedman, born in Russia in 1891 and brought to Illinois by his immigrant parents in the following year. He was a graduate in biology from Cornell University. Until diverted to cryptanalysis, Friedman had been experimentally mating fruit flies as part of a study of how to control them. A rare photograph of Friedman in later life shows a dapper, clever-looking man with a long nose and a short mustache over a shiny bow tie with diagonal stripes, all reminiscent of the film actor Adolphe Menjou. It thus comes as no surprise to learn that he was a snappy dresser fond of ballroom dancing; he also kept a carbon copy of every written communication he ever made, which may help to explain his reputation for phenomenal recall. He did much original intellectual work on the application of statistics and coincidence theory to cryptanalysis.

Meanwhile, late in 1917, the United States Army appointed Herbert O. Yardley, a young coding clerk in the Department of State, to head the brand-new cryptanalysis section of Army

Intelligence, MI-8. Yardley had come to official notice and proved himself up to the job by breaking into presidential cables in his spare time, even though he was not privy to the ciphers used for them. He was a 'natural' at this kind of thing, and his name soon came up when the Army was looking for a suitable candidate.

One of his very first colleagues was Friedman, who was taken into the US Army Signal Corps as a second lieutenant, to serve the last six months of the war in staff-section G-2 (intelligence) at General Pershing's American Expeditionary Force headquarters in France, in the Code and Cipher Solving Section. Friedman, relying strongly on intuition because his training in higher mathematics was weak, established himself by penetrating the 'unbreakable' new British Army encipherment machines intended for Anglo-American communications; eleven thousand of them had to be scrapped as a result. On his return to the laboratories at Riverbank after the war, he surpassed this feat by breaking into an AT&T cipher-teleprinter that was also, rashly, billed in advance as 'impenetrable.' Small wonder that in January 1921 the Army decided it could not do without Friedman's services. The Signal Corps gave him a six-month contract as a civilian employee, the start of a 'temporary' attachment which lasted for thirty-four years. Even after his retirement in 1955, Friedman continued as a consultant, most notably in 1957, when he wrote a secret critique of the Pearl Harbor conspiracy theory (file SRH 125 at the US National Archive) which can reasonably be described as crushing. His main claim to fame, however, is the penetration of the Japanese diplomatic cipher-machine code-named 'Purple' by the US, the basis of the entire 'Magic' campaign of cryptanalysis – a word coined by Friedman, the greatest American practitioner of this secret art (see below).

But in the immediate aftermath of 1918 the running was made by Yardley, still hard at work eavesdropping on German wireless traffic at the Versailles peace conference. No sooner was this concluded in June 1919 than Yardley proposed to his masters in the State and War departments the establishment of a permanent cryptanalytical service. The result was the clandestine creation, with secret funds from both departments, of what soon became known to those privy to the operation as the Black Chamber, at an office in lower Manhattan. It passed itself off as the Code Compilation Company. By the time of the Washington Conference, which opened in November 1921, its prime target was already Japanese diplomatic traffic. After a year's work and in time for the all-important conference, an outstanding diplomatic event in Japanese history between the wars, Yardley broke through. He was therefore able to tell Charles Evans Hughes, secretary of state, leading the American delegation, first that the Japanese representatives had been ordered to avoid a row with the US and then that Tokyo was prepared to go along with the 10:10:6 naval ratio for Britain, America and Japan. This gave the Americans the confidence at the talks to persist with their underlying aim of torpedoing the prewar Anglo-Japanese alliance while limiting the size of the world's leading navies (and thus the US budget).

The British, not in on the secret and unable at that time to penetrate Japanese ciphers, were alarmed by American persistence in attacking an apparently unyielding Japanese position. Unfortunately, as we have seen, this diplomatic victory proved to be something of a military boomerang. For the diehards in Japan 'compromise' was a synonym for 'humiliation' – and 60 percent of American or British worldwide maritime strength left Japan with *de facto* naval supremacy in the Pacific, even before she cast aside all limitations on her own armaments. The collateral ban on expanding ocean bases also left Japan with far more naval facilities in the Pacific than any other power. Nonetheless, Yardley was awarded the Distinguished Service Medal in 1922, ostensibly for his war work but in fact for his undoubted intellectual and technical triumph against Japanese ciphers after the war. Despite his disgraceful subsequent conduct, he cannot be blamed for the fact that the US administration too often misinterpreted or ignored the flood of information from Japan which he inaugurated.

The Black Chamber was mainly an Army enterprise. The Navy was slower off the mark, having relied almost totally on the superb British effort, run from Room 40 at the Admiralty, against the Germans in 1917–18. Washington's decision to allocate $100,000 to naval sigint in 1918 came too late to be significant for the First World War: $65,000 still remained in this secret war-chest when a singularly upright intelligence chief returned it to the US Treasury in 1931 in order to keep his hands clean!

It was only when Captain Andrew Long, USN, became director of the Office of Naval Intelligence (ONI) late in 1920 that the Navy began to acquire its own cryptanalysis section. Young naval-officer volunteers were sent to the American Embassy in Tokyo to learn the language. In 1924 the Navy's Code and Signal Section (CSS), part of the Office of Naval Communications (ONC), set up a Research Desk under Lieutenant Laurence L. Safford – the root of an interservice rivalry between ONI and ONC which was to have its grotesque culmination during the Pacific Campaign. Meanwhile Safford, originally supported by one cryptanalyst, Mrs Agnes Driscoll, née Meyer, and two typists, went to work on the 'Red Book,' a two-volume copy, with translations, of the Japanese consular code, bound in red cloth by the Americans (hence the name). This seminal aid to eavesdropping was put together in a series of illegal and secret raids by naval, FBI and police agents, abetted by a convicted safecracker, on Japanese premises in New York, including the offices of an assistant naval attaché and the Imperial Railways, during which the code was photographed page by page. The agents had followed several Japanese spies – usually Army and Navy officers pretending to be visitors – to these buildings, which were clandestinely collecting large quantities of raw intelligence of all kinds on the USA. In the process, access was also obtained to the main naval codebook. It was not the burglaries – in retrospect, despite the provocation and the far-reaching results, still a recklessly cavalier undertaking – but the hard work put in afterward that was the real achievement of 'Red Magic.' Armed with the code (a system where words, letters, or digits stand for names, phrases or references), the cryptanalysts had only to solve the essentially mathematical problem of the cipher (where one letter or digit stands for another). Even so it took nearly two years.

In October 1925 Lieutenant (j.G.) Joseph J. Rochefort, USNR, a graduate and former First World War enlisted man, joined Safford at the Research Desk. It is difficult to determine which of the two was the more eccentric. Safford, who graduated from the Naval Academy at Annapolis in 1916, was one of those people who are the despair of uniform tailors as well as orderly organizations. He wore his hair in the 'mad-professor' style and talked disjointedly because his mouth could not keep up with his mind; his forte was pure mathematics. Rochefort was mild-mannered, dedicated, and serious but also persistent, energetic and impatient of hierarchies and bureaucracy, his mind unfettered by orthodox officer-training. He eventually went to Tokyo for three years to learn Japanese. Forced to work in a dank redoubt at Pearl Harbor during the war, he habitually wore carpet slippers and a red velvet smoking jacket on duty to keep himself warm. This provoked senior officers without affecting the quality of his intuitive cryptanalytical work, which was the foundation of America's most important naval victory, as will be seen in chapter five.

Neither man would have achieved what he did without the inspiration and practical training provided by the no less extraordinary Agnes Meyer. Like Yardley, she was originally a humble coding clerk, but in the Navy Department; like Friedman, she could not resist the challenge of cracking an 'unbreakable' cipher invented and submitted to the Navy in 1921 by Edward Hebern, a manufacturer of encipherment-machines (Friedman was able to deduce the random internal wiring of one such from only ten messages passed through it). Hebern had the wit to offer Meyer a job in California, but within a year she married Michael B. Driscoll, a lawyer, and returned with him to Washington, where she went back to work for the Navy in 1924 at the age of thirty-five. Elegant, demure and aristocratic in bearing, Mrs Driscoll was unsurpassed in the United States, both as cryptanalyst and instructor. Not even a serious car-crash in 1937, which maimed her, could prevent her from playing a key role in sigint throughout the war.

Such was the original team of three which went to work on Japanese codes and ciphers for the Navy in Washington. Friedman's later conquest of the 'Purple' diplomatic cipher seems more glamorous and was at least as difficult but could give no clue to Pearl Harbor. Had more urgent attention been given to the humbler Red, Washington might have got its warning after all, as will be shown shortly.

Soon ONC set up listening posts ('Y-stations') at Guam, Shanghai, Peking, Cavite in the Philippines and Pearl to intercept Japanese radio traffic, but then rather negated the value of the exercise by not forwarding what it got to ONI on spurious 'security' grounds. It was all part of the internal struggle for control of the Code and Signal Section, which ultimately (see chapter five) became Op-20-G, directly under Admiral Ernest J. King as C-in-C, US Fleet (COMINCH), and chief of naval operations (CNO). In view of this, it will hardly come as a surprise to the reader that no thought had yet been given to cooperation with the Army in the new sigint.

The latter service all but retired from the field in 1929, when President Herbert Hoover's conspicuously honorable new secretary of war, Henry L. Stimson, reacted angrily to the Japanese intercepts appearing daily on his desk. 'Gentlemen do not read one another's mail,' he decreed, and ordered it stopped, together with funding for the Black Chamber, which was officially closed down in May. Stimson's naïve diktat was a blessing in disguise, because it prompted a long-overdue reorganization of Army sigint. By then the Chamber was already moribund, with little to do but still retaining a staff of six and a payroll of which Herbert Yardley was helping himself to three-eighths. The Stimson order coincided with the conclusion of a Signal Corps investigation which had uncovered Yardley's penchants for poker and philandering.

The work of this free-lance subcontractor was taken back under direct Army control, independent also of the State Department. The Signal Corps, apparently concluding that what Mr Stimson did not know would not hurt him, set up the Signal Intelligence Service to handle attacks on foreign codes and ciphers as well as provide the Army with homegrown ones for its own use. Friedman, still a civilian though an officer in the Army Reserve, was the obvious candidate to take charge of it. Starting with a staff of six, he built it up step by step to more than three hundred by the end of 1941 (and ten thousand by the end of the war). The SIS was the direct ancestor of the Armed Forces Security Agency (1949) and the National Security Agency (1952). Friedman was midwife to both, and retired as special assistant to the director of the latter, which is now the largest and most important American intelligence organization. The NSA listens to the world on a colossal scale – not least to forestall being caught out again, as at Pearl Harbor on 7 December 1941.

Unfortunately, Yardley was not prepared to allow his dismissal to pass unavenged. He coolly published a book entitled *The American Black Chamber* in 1931, as well as a series of articles in the *Saturday Evening Post* – but not before approaching the Japanese with an irresistible offer to sell them everything he knew for $7,000. Had they waited a few weeks, the information would have cost them no more than the couple of dollars needed to buy a copy of his exposé in a bookstore. Yardley planned a follow-up, to be called *Japanese Diplomatic Secrets*, but the US government managed to find a means of preventing its appearance. The Japanese Foreign Ministry complained bitterly of American deceit (perhaps twice over, once for the penetration of its communications and then for Yardley's confidence trick). One of the most surprising aspects of this extraordinary double betrayal was that the Japanese did not become permanently alert to American interest in their secret communications and the lengths to which US agents were prepared to go. Perhaps they preferred to regard Mr Stimson rather than Mr Yardley as truly representative of American attitudes.

Ten years after its foundation, the SIS under Friedman's personal leadership was hard at work on breaking into the cipher used with the new Japanese diplomatic encipherment-machine, Purple. It took the team eighteen months to build an analogue of the machine, during which the highly strung, depressive Friedman, who knew no Japanese but only English, came close to nervous collapse more than once. The foundation of his team's triumph, apart from very hard work, was a huge blunder by the Japanese. The Purple machine was technologically straightforward but intricate in structure and slow in manufacture. It was therefore introduced piecemeal, first on the Tokyo-Washington circuit in February 1939 and then gradually to Japanese embassies round the world. The Americans guessed correctly that general messages being sent by Purple to some recipients had to be sent by Red to the rest and were able to use the latter, already penetrated, to help break the former.

The Purple machine, like other electromechanical cipher-machines of the period, had a typewriter keyboard. When a key was pressed, a current passed through a series of mechanical parts with asymmetrical internal wiring (not wheels, as in other such machines, but selectors, as used in old-fashioned telephone systems) and electrical circuits to encipher the chosen letter. Each time a key was pressed, at least one of the moving parts inside the machine took up a different position relative to the rest, so that each letter of the message was separately and differently enciphered. To decipher the signal, the recipient had only to set his machine in the same starting position as the sender and run it through. The main flaw in this complex device was that it used different circuits for vowels and consonants, which helped the American cryptanalysts. Their version was actually superior technically to the Japanese original, because it used brass rather than copper contacts and therefore took much longer to wear out (when this happened, messages were garbled). The Friedman team got its first break into the system in August 1940 and decrypted its first complete message on 25 September, just in time to filch the text of the Tripartite Pact binding Japan, Germany and Italy, which was signed two days later.

This American achievement was the perfect counterpart to the British successes against the German 'Enigma' cipher-machine, to be crowned in May 1941 with the breach of naval Enigma, the best defended system. The British had not tackled Purple but had, like the Dutch, made considerable progress since the early 1920s, more than the United States, against Japanese naval and military ciphers. Thus, when the British Technical Mission arrived in Washington in August 1940 to arrange exchanges of technological expertise so as to offset dwindling exports and reserves, the foundation of a crucial swap in sigint equipment and ideas was laid more than a year before the United States actually went to war. Friedman was to have gone to Britain in the spring of 1941 as a result of the agreement to pool this vital information, but he collapsed from exhaustion. The Purple triumph was by no means the end of his enormous contribution to wartime US signals intelligence. He wrote at least sixteen training manuals, ran a training program, did much valuable work in such diverse areas as codes and ciphers for the US Army (Congress belatedly paid him $100,000 by way of lost patent-rights for these in 1956), location by radio direction-finding, and even invisible ink.

Perhaps most important of all was the personal link he established with Safford at Navy, a lateral move that was the basis of a highly unusual interservice cooperation before and through the war. The most obvious result was the avoidance of duplication by two already overstretched sigint operations. The Army carried on with its work on Purple while the Navy tackled other diplomatic traffic, as well as the main Japanese naval system and its variants code-named JN25 (JN25b was introduced six days before the Pearl Harbor attack, to the temporary discomfiture of the US eavesdroppers; they would, however, have been none the wiser had they been able to penetrate it at once). The two services pooled their work on Red, leaving the Americans weakest against Japanese Army signals, which proved exceptionally stubborn. This did not resolve the struggle for control of sigint within the Navy, which was beyond Friedman's powers, but opened the way to an exceptionally fruitful, not to say exemplary and much too rare, exercise in interservice cooperation in the United States. That the British sigint operation at Bletchley Park, northwest of London, was also remarkable for the degree of cooperation among Royal Navy, Army and RAF helps to explain why Anglo-American intelligence collaboration in the Second World War was in itself one of the most successful alliances in military history.

But the palmy days of total Allied superiority in sigint still lay in the future on the morning after Pearl Harbor. At that moment, the burning question in Washington and wherever the American flag flew was how such a devastating stroke could have been delivered with impunity. A huge row broke out in Hawaii and erupted in the United States proper as soon as the news of the ruination of the American battle fleet came through. The row has gone on ever since, and has become deeper and more complex with the passage of time. As the Americans went to war with the battle cry 'Remember Pearl Harbor!' they were already demanding to know how it could have happened and whether it could have been prevented. This natural reaction rapidly developed into a hunt for scapegoats (the easy part) and for the proverbial 'smoking gun,' which has still not been found, despite more and more ingenious theories and the outpouring of millions of words. Most of these have shed more heat than light on that most vexed of questions: was the Pearl Harbor debacle the result of confusion or conspiracy? That it was most obviously the result of a brilliantly planned, ruthlessly executed surprise attack by a bold enemy has not dampened half a century of fierce debate which no contemporary historian can afford to ignore.

It has been proved that the United States forces did possess intelligence information which could have alerted them to the specific threat to Pearl Harbor. It has been alleged on the basis of this established fact that President Roosevelt knew the Japanese were coming and did nothing to save the fleet, so as to ensure that America was forced into the war. It has also been claimed, more subtly perhaps, that Roosevelt did not know but Churchill did, and that 'perfidious Albion' withheld the information to make sure that the United States came into the war on Britain's side.

British furtiveness has given the latter charge a longer life than it might otherwise have merited. The morbid British obsession with official secrecy has allowed half a century to go by without the release of a single syllable of the intelligence material British agencies acquired about and from Japan. Indeed, London has never officially admitted to possessing any, even though American and other sources have shown otherwise. Professor Hinsley's official history of British intelligence in the war is no less coy. The field of historical speculation is thus abandoned to the diehard skeptics, the conspiracy theorists, and the paranoid. They have had a wonderful time arguing that, if the British are hiding so much, there must be something monumental to hide; that there can be no smoke without fire; that there are a few veterans whose belated revelations in old age, even if sometimes extracted by leading questions, 'prove' the British knew and never told. None of these people have logically proved the point to my mind, even though some have unearthed a great deal of circumstantial evidence of a kind flat denials or shifty silences from Whitehall will never demolish unaided. But as long as the British government is so childish about superannuated secrets, it positively deserves to be accused of perfidy in 1941 – and there will always be something of a case to answer. Indeed, the recent shift of target from Roosevelt to Churchill in the tireless hunt for the smoking gun may well derive from the fact that nothing has been found in the open US archives, whereas so many British records remain closed. As long as that is true, the speculators can say what they like without fear of contradiction.

Like Washington's accurate error of July 1941, the irrepressible suspicion that the United States could have forestalled Pearl Harbor derives from America's own brilliance in signals intelligence between the wars. We now know that the United States was able to read Japan's diplomatic traffic at consular and ambassadorial levels alike, with little delay and almost as if it were an open book. The US code word for the resulting intelligence was 'Magic.' The Americans had also made great progress in penetrating Japan's military codes and ciphers by 1941 (code word 'Ultra,' also used by the British for military signals intelligence, which they exchanged with the Americans). Sometimes information from one source filled out, clarified, or confirmed interceptions from another. It is hardly surprising that for some people the question has become, not 'Did we know?' but 'How could we not have known?' But is that fair?

First the facts. The Americans in Washington, Hawaii, and the Philippines, the British in Singapore and Hong Kong, and the Dutch in Bandung, Java, collectively intercepted a number of Japanese messages which indicated war was imminent in the Far East. The armed forces of the three powers exchanged information via liaison officers stationed in one another's Asiatic intelligence bureaux throughout 1941. Anglo-American cooperation in this field was agreed upon in July 1940 and formalized by a written agreement between the chiefs of staff of the two countries in December. The Dutch 'Black Chamber' or 'Room 14' decryption center had been making inroads into Japanese consular, diplomatic, and naval ciphers since 1932 but destroyed the records ten years later, when the Japanese invaded: the obvious reason why the Dutch contribution has usually been underestimated. The American Army completed its assault on the two ciphers used by Japanese consulates – Red, in 1936, and the much more sophisticated Purple used by embassies – in September 1940. The US Navy started to breach the main Japanese naval cipher, JN25, about the same time, and could read a significant and growing proportion of the traffic by a combination of hard graft and good guesswork.

The Americans gave the British three of their then extant stock of eight painstakingly constructed imitation Purple machines between January and September 1941. The third would otherwise have gone to the US Army at Pearl Harbor, which thus was unable to obtain its own Magic. Nor could the US Navy in Hawaii extract its own Ultra, as the copy of JN25 originally intended for its cryptanalysts at Pearl had also been given to the British in January 1941. US forces in Hawaii were therefore totally dependent on Washington for relevant Magic and Ultra material; from the crucial month of July 1941 onward the supply even of these predigested extracts was cut off, partly for fear of leaks and partly because the circuits were said to be overloaded. But since Purple Magic provided only diplomatic material (including priceless strategic, economic and political information about Germany, as transmitted from the Japanese Embassy in Berlin to Tokyo), it would not have given the slightest hint of planned or impending Japanese military operations against Pearl Harbor or anywhere else. The absence of a machine at Pearl, of which so much has been made elsewhere, is thus a purple herring: the presence of one in the Philippines, complete with an excellent team of cryptanalysts, demonstrably did MacArthur no good at all.

Since the Mobile Force of the Combined Fleet had been briefed on its Hawaiian mission while ashore and observed strict radio silence while at sea – no transmissions, only a listening watch – Ultra from the new JN25B would have revealed little of value to Pacific Fleet headquarters in Hawaii, even if they had been able to unravel the material. Its unavailability is thus another distraction from the main issue. Nagumo's carriers followed a route across the North and Central Pacific that was out of range of Allied land-based aircraft and hundreds of miles off the main sea-lanes. They were under orders to sink any Allied or neutral merchant ship sighted, to prevent her from sounding the alarm by radio. The carriers met just one cargo vessel, which turned out to be Japanese: she sailed blithely homeward as if alone on the broad ocean.

The Americans made an understandable error in handling Red Magic. Because it was derived from the humble consular material, it was decrypted last and often delayed in distribution or ignored on receipt as it was mostly routine. Although it too contained no operational information, since the civilian Foreign Ministry, and thus its diplomatic and consular officers, were systematically denied such matter by the military even more strictly in Japan than in other countries, it held clues that would have alarmed intelligence officers. For example, there was a long message on 3 December 1941, from Consul Kita in Honolulu listing US Navy ships in Pearl Harbor. Because of the pressure on the cryptanalysts, this was not deciphered until 11 December. The hyperactive Honolulu consulate, a round-the-clock espionage center, had earlier been ordered to impose a grid pattern on a map of Pearl Harbor and to report daily what ships were in each square. The resident chief spy was Ensign Yoshikawa Takeo of the IJN Reserve, alias Foreign Ministry clerk Morimura Tadachi, who garnered priceless information by such means as taking a job washing dishes in officers' clubs.

There were also the notorious 'winds' messages. The Foreign Ministry told its consulates on 19 November 1941, that certain phrases would be inserted in Tokyo's shortwave international weather forecasts to indicate the imminent outbreak of war. 'North wind cloudy' meant war with Russia; 'west wind clear,' war with Britain; 'east wind rain,' war with America. This was only translated a week or more later by the overworked Americans. But an alert was sent to all American listening posts to monitor the Japanese radio forecasts, and to listen for the follow-up message – the signal to execute orders previously sent, to destroy sensitive papers together with cipher-machines

and take other precautions. At least one senior American cryptanalyst came to believe such a 'winds-execute' message had been
picked up a few days before Pearl Harbor, but no second source
has confirmed it beyond reasonable doubt. Japanese diplomatic
traffic increased hugely from about 3 December, adding to the
workload of the American eavesdroppers.

One such – Lieutenant Howard Brown, the radio-interception
officer whom we met in MacArthur's office in Manila – recalled
receiving a message from Washington on or about 5 December,
asking for an immediate report, to be marked 'urgent,' on any
short Japanese language signal transmitted from Tokyo and ending with the English word 'stop.' He said such a signal was picked
up at 5:00 A.M. Washington time on 7 December (11:30 P.M. on
the 6th in Hawaii). If such messages were indeed sent, they
would in all probability have been intercepted by the Dutch, who
had to destroy their records, and/or the British, who have been
too devious, stubborn, or perhaps ashamed to say, although
one or two former Allied naval officers emerged in the late
1980s claiming to remember such interceptions. Even though
the memory of the very old for events in the distant past can be
exceptionally clear, corroboration is still required. And nobody
claims that such a message, if any, contained the slightest hint
of an attack on Pearl. Not even the exhaustive, fourteen-point
message from Tokyo to its negotiators in Washington on 6–7
December, read in full by the Americans and concluding with
precise instructions on when to hand over a quasi-ultimatum
(half an hour before the Pearl raid), gave such a clue.

Intriguing though all the above may be, there is no indication
that any of these Japanese signals (except the delayed Red
Magic decryptions from Honolulu) gave the slightest inkling of
a surprise attack on Pearl Harbor or any other specific place.
That being so, hard though it may be on the 'revisionists' – the
historians who believe Roosevelt and/or Churchill knew and did
nothing – all this is entertaining but completely irrelevant.

The only significant messages known for sure, apart from
the daily shipping report from the Honolulu consulate, to
have been transmitted to Nagumo after he sailed were from
Yamamoto and were not acknowledged. The first was sent
in the evening (Tokyo time) of 2 December, the date of the

last prewar imperial conference, which confirmed once again the decision to go to war. The message was from Yamamoto himself and said simply: 'Climb Mount Niitaka 1208.' There was no need for prefix, address or signature: only Nagumo and his staff could have known what this prearranged message signified. Further, it is hard to understand why Yamamoto or anyone else should have run the unnecessary risk of repeating a long and complex operational order, unaltered since given to its addressee in writing long in advance, as well as sending this cryptic confirmation. The reference to Japan's highest mountain meant that hostilities against the United States were to begin on 8 December (Tokyo time). The implication was that taking on America was the hardest task in the forthcoming general offensive; but no Allied cipher-expert would have been able to deduce a raid on Pearl Harbor from this apparently innocuous signal either, even though it could well have been taken as meaning that something big was going to happen on 7 December, Western time.

On that day, Yamamoto sent the traditional Nelsonian admonition to a fleet about to go into battle: 'The fate of the Empire rests on this enterprise. Every man must devote himself totally to the task in hand.' (Nelson would have recognized this as a distorted echo of his 'England expects every man to do 'his duty' at Trafalgar in 1805. Nagumo's own flag-signal to his fleet repeated the pithier version of the same sentiment by the no less imitative Admiral Togo to his battleships at Tsushima a hundred years after Nelson: 'The future of the Empire depends on this battle. All men must give their utmost.')

There remains Mr James Rusbridger, a former British intelligence officer and veteran of the Government Code and Cipher School at Bletchley Park, base of the highly successful British wartime decryption effort. His attempts to substantiate the revisionist theory (which was first aired in 1945 and is best known from the books of the historian John Toland) are rare in recognizing the need for logical argument as well as circumstantial evidence – of which he appears to have found at least as much as anyone else. According to him, on 27 November

1941, the Dutch broke into a JN25 signal to Nagumo's task force, sending it to sea (it left on the 26th) and giving 8 December, Japanese time, as 'X-day.' This was passed, it is claimed, to the Dutch naval attaché in Washington; he allegedly forwarded it to US naval intelligence, where it was disregarded. The Dutch decryption, as recalled long after the war (there being, of course, no written record of it), apparently echoed a paragraph in Admiral Yamamoto's operational order to Nagumo, a copy of which was found by divers aboard the Japanese heavy cruiser *Nachi*, sunk by US aircraft in Manila Bay in January 1944.

In 1989 Mr Rusbridger presented for publication in Britain the story of a ninety-year-old former Royal Australian Navy officer who was also the co-author. Captain Eric Nave, RAN, veteran of the Far East Combined Bureau (the British decryption station in Singapore and later Colombo), had no doubt that the Japanese intention to attack Pearl Harbor was known in advance to the British, who had got into the Japanese ciphers by or during 1925. The book claims that the departure of Nagumo's task force and its refueling at sea eight days later were detected by sigint, and speculates that Churchill withheld the information from Roosevelt to ensure American participation in the war. The temptation, if any, to laugh this out of court must be tempered by the scarcely credible reaction of the Britain government: it blocked the book by effectively challenging the publisher to an unequal courtroom battle under the recently revised Official Secrets Act. The only factor that prevents the present author from automatically assuming that Rusbridger and Nave must therefore be right is personal experience of the congenital paranoia of British officials when reluctantly storing away records for posterity. Papers infuriatingly locked up for as long as seventy-five years and then released have turned out to be inexplicably, even boringly, harmless. But – in this case especially – nobody can be certain. *Betrayed at Pearl Harbor* was eventually published simultaneously in the US and Britain in the summer of 1991.

In the absence of the last few vital pieces of the conspiracy theorists' elaborate jigsaw puzzle, we are reduced to logic as the only instrument for testing their belief that 'People in High Places' knew of Pearl Harbor in advance.

If we start with Churchill, it has to be asked whether he would connive at the destruction of the American battle fleet, on which Britain, as it fought for its life in the West, relied to preserve its own and Holland's large interests in the Far East. There is evidence (but not in their correspondence) that Churchill sent Roosevelt a vital message on 26 November; its contents are buried at the Public Record Office in London until the year 2016 – 'in the interest of national security.' This telegram might have been the alleged Pearl warning; or it might have been a tip that Japanese convoys were moving southward from French Indochina, which officially convinced the administration on 27 November that war was absolutely inevitable and imminent, leading to a *de facto* ultimatum from Washington (by which time the Japanese carriers were already on their way). Or it could have been something entirely different. The risk involved for Churchill in finding out, not telling, and later being discovered to have known – a very real risk, considering how closely Anglo-American intelligence staffs worked together and the short shelf-life of American official secrets – would surely have been too high even for such a scheming gambler.

Moving on to Roosevelt, we are asked to believe that this erstwhile assistant Navy secretary connived at the destruction of the bulk of the Pacific Fleet – its pride and joy, the battleships – to save Britain's bacon in North Africa, the Mediterranean, and the Far East (and to win the election in November, even though polling inconveniently preceded the attack). All this from a man who loved the service, of which he was now commander-in-chief, so much that he usually referred to the Navy as 'us' and the Army as 'them' during the war, to the lasting chagrin of Army chiefs. Hence, no doubt, the provocative transfer of the fleet from the West Coast to its exposed position in Hawaii. Given the strength of the isolationists, it was essential that Japan strike the first blow (a thought unwisely uttered in public by Henry L. Stimson, Roosevelt's stickler of a war secretary). To get away with it, the president would surely have needed the cooperation of Admiral Harold R. Stark, chief of naval operations, and General Marshall, perhaps the finest individual ever to grace an

American uniform, and at the very least their intelligence staffs, including the junior officers and enlisted men who worked on raw sigint.

Forewarned, Roosevelt could have set a trap for the Japanese, dispersed his own fleet, or proved Japanese aggressive intent to the world by having Nagumo caught red-handed on his approach. There was no need to throw away the battleships in order to use them as a lure – to say nothing of thousands of American lives. The very absence of the aircraft carriers can of course be used to support the conspiracy theory: Roosevelt's foresight told him this type of ship was to win the war, so he saved them by sending them to sea (unfortunately, as we have seen, the *Enterprise* should have been in Pearl on the 7th). Once again the risk of subsequent discovery looks prohibitive, especially in a country as unsecretive as the United States. And the most likely result of an American sally by a maximum of three carriers against Nagumo's six would have been a reverse Battle of Midway. The core of the United States Navy's air strength would have been outclassed, outgunned and outflown by veteran pilots from the war in China, attacking a fleet that had never fought in the air. Indeed, on this basis the raid of Pearl begins to look like a positive favor to the American Navy. A lost carrier-battle on day one would have made the wait to turn the tables much longer than the six months between Pearl and Midway.

So what is the truth about Pearl Harbor? The answer is not tidy but is certainly, if not wholly, psychological. Washington could not bring itself to believe that the Japanese would be so bold – indeed, so stupid – as to do what they did against the United States. No matter that all Japan's previous wars in half a century had opened with a surprise attack; no matter that Ambassador Grew sent a warning from Tokyo about a surprise attack on Pearl in the event of war as early as 27 January 1941, less than three weeks after the Japanese finally adopted the idea. Since he got it from the Peruvian ambassador, it must have been common gossip in the Japanese capital. No matter either that Japanese and American naval officers and amateur naval strategists round the world regarded a strike against Pearl Harbor as obvious. Such a move was still included in the assumptions made in American naval war plans as late as May 1941, although the possibility seems to have slipped out of sight by December. But not on one battleship, where officers had discussed at length what they would do in an air raid: the USS *West Virginia*, though badly hit twice, fought back with exceptional vigor when it came. That Washington was not entirely silly to regard such an attack as unlikely, which is no excuse for not preparing for the contingency, is shown by the initial resistance of the two Japanese high commands to Yamamoto's scheme. The Army's original plan to avoid American territory would have presented Roosevelt with the uncomfortable choice between continuing to appease the isolationists and going to war to recover European colonies in Asia.

The other important factor behind the humiliation, if such it was, of Pearl Harbor has to do with the processing of intelligence about a potential enemy before he has declared himself by an act of aggression. We have seen that the Americans were culling masses of information out of the airwaves and sending it up the line to Washington. Although interservice and international cooperation in this area was extraordinarily good, there was no coordinating American brain sifting everything that came in to get the full picture – not in the White House or the War or Navy departments, not at State, or on the staffs of the two services, or in any intelligence organization. Success comes not from collection, which is easy, but from collation and evaluation. From July 1941 distribution of Magic and Ultra was severely restricted; MacArthur was an exception, but he made no profit from this advantage, any more than he did from the many hours of grace he was given by the Pearl attack and the accident of fog over Formosa. Neither the British nor the Dutch were able to turn their sigint from Japan to good account. A British double agent, Dusko Popov, revealed in the summer of 1941 that the Japanese had asked the Germans to get them information on Pearl Harbor from their sources in the United States, but

J. Edgar Hoover, head of the FBI, would give no credence to this unusual, not altogether respectable source.

Farther afield, Stalin knew the Germans were coming but was still caught napping; the British knew the German battle squadron was going to run for home from France in 1942 and drew up a plan to deal with it but failed on the day. Good intelligence and the slogan 'Never Again!,' the corollary of 'Remember Pearl Harbor!,' did not prevent the American Army from being badly surprised at the Battle of the Bulge, or the British at Arnhem, in 1944. Admiral Kimmel and Lieutenant General Walter C. Short at Pearl Harbor were given war warnings on 27 November – the strongest of several alerts over the year – but took them as no more than a reminder to guard against sabotage. Thus the US aircraft came to be neatly lined up on the tarmac for the convenience of Commander Fuchida and his comrades. But as the American herald of seapower, Admiral Alfred T. Mahan, used to say, hindsight is always better than foresight.

There is no evidence that the Japanese diplomats involved in the last-ditch attempt to settle their country's irreconcilable differences with the United States at talks in Washington had any foreknowledge of the attack on Pearl Harbor. There is plenty of evidence that even the highest Japanese officials, even in Tokyo and even if they could still lay claim to an admiral's or a general's title, were not in on the plan: there has probably never been a stricter interpreter of the 'need-to-know' rule than Imperial General Headquarters. Indeed, IGHQ (Navy) saw no need to inform IGHQ (Army) of the final, fateful embellishment to the fleet's operational plan for day one of the southward-expansion strategy before the first week in November 1941. To see the last contacts between America and Japan in their proper context, we need briefly to review the relevant aspects of Japan's foreign policy between the start of the war in Europe and its extension to the Pacific more than two years later.

Matsuoka Yosuke (1880–1946), a radical nationalist barely in control of his unpredictable emotions and widely regarded even by other Japanese leaders as not right in the head, nevertheless tried to do his best for his country, with astounding results. It fell to him, as noted in chapter one, to lead the Japanese delegation in its histrionic walkout at the League of Nations when the atrocities in China came home to roost in 1933. He was appointed foreign minister in Prince Konoye's second Cabinet on 26 July 1940 (alongside General Tojo as war minister). Konoye's ambitious program included plans to curb the military by establishing the Imperial Rule Assistance Association as the single permitted political organization and thus an alternative center of power; to settle the 'China Incident' – and to improve Japan's strained foreign relations. Autarky was the central economic aim; there was to be a New Order in Greater East Asia, including Manchukuo and China and starting with 'reform' – the total mobilization and arming of the people in the Hone Islands. Japan was to have a 'National Defense Economy' (which in essence it still has, according to some observers).

Abroad, Matsuoka negotiated Japan's fateful accession to the Tripartite Pact with Germany and Italy, signed in September 1940. Originally seen by the Cabinet and the Army as a means of neutralizing the threat of the Soviet Union, the pact came to be regarded as insurance against Britain and America as the war in China dragged on and Japan developed its designs on French, British, and Dutch colonies in 'Greater East Asia.' A ministerial delegation went to Batavia (now Jakarta), capital of the Netherlands East Indies, in September and October 1940, with a view to obtaining Dutch 'cooperation' in the New Order – entailing unrestricted Japanese access to raw materials and the oil owned and exploited by Dutch, British, and American companies. The Dutch colonial administration, even though its home base was under Nazi occupation, summoned up all its notorious national stubbornness and held out. The Japanese, for the moment, decided to grin and bear what many of them inevitably saw as an unforgivable insult. For such people, the proposition that the other party to a negotiation had a right to different views or interests was not only inadmissible but utterly alien. This terrifying intransigence was more than a match for the blind arrogance it so often encountered on the Western side, where the emphasis was different: but Europeans and Americans too seldom tried to see themselves as others saw them, especially when the others came from another race. The collision between these two failings was a powerful psychological recipe for a catastrophe based on total mutual incomprehension.

Matsuoka now turned his attention to Japan's strained relations with her main strategic supplier, the United States. He decided a new ambassador was needed if the climate was to be improved, and made a considered choice in November: Admiral Nomura Kichisaburo (1877–1964), formerly of the naval staff, delegate at the Versailles and Washington conferences, and briefly foreign minister in 1939. He had served as a naval attaché in Washington during Roosevelt's term as assistant Navy secretary and was thus acquainted with the president, to whom he presented his credentials on 11 February 1941. Six feet tall, two hundred pounds in weight, and no less heavy in manner, he spoke poor English but genuinely liked America. Another ambassador, Kurusu Saburo, was to come and support him in time for the eleventh hour on 15 November 1941; meanwhile, there was Colonel Iwakuro Hideo, previously chief of the military-affairs section of the Ministry for War, to represent the Army's interests. These men conducted the last rounds of talks with the administration in Washington before the Pacific war.

In the spring of 1941 Matsuoka was off on his travels again. They culminated in another breathtaking diplomatic coup: the neutrality pact with the Soviet Union which he signed on 13 April. The Russians signed because they feared a war with Germany, and the Japanese because they feared there would not be one: thus do diametrically opposed interests and completely irreconcilable views of the world sometimes perfectly converge. The energetic foreign minister came back to a popular hero's welcome after his undoubted triumph in Moscow.

But there had been portentous and peculiar developments in Tokyo's relations with Washington while he was away. Exhausted by his efforts and keen to catch up on sleep lost courtesy of the Trans-Siberian Railway, Matsuoka was not best pleased to be presented on his return, as a *fait accompli*, with what purported to be nothing less than a breakthrough in relations with Japan's most important trading partner and main source of strategic materials. What was more, his prime minister, Konoye, was fully informed of these events and had not told him – even though it was one of Matsuoka's rare but intense bursts of charm amid his customary emotional rambling that had helped to bring it about.

Visiting Japan just before Matsuoka's momentous spring tour, Bishop James E. Walsh and Father James M. Drought of the Roman Catholic Maryknoll Fathers, a missionary organization, had the unusual honor of being received by the foreign minister. He managed to convince them of Japan's pacific, as distinct from Pacific, intentions. As soon as they got home, they approached a prominent Catholic in the administration, Postmaster General Frank C. Walker. He in turn went to Roosevelt to propose an unofficial channel of communication with Tokyo via the missionaries and, on the Japanese side, Ikawa Tadao, a banker with an entrée to Prime Minister Konoye, and Colonel Iwakuro, an intimate of War Minister Tojo (and a man of much experience in espionage). Ready to try even this unlikely 'paradiplomacy' in the hope of getting through to the junta, Roosevelt agreed. Secretary of State Cordell Hull was no more in the picture than Matsuoka, his Japanese opposite number. Walker provided the two clerics with government funds to hire a room in a New York hotel as headquarters for their spatchcocked but indubitably well-meant peace initiative.

On 19 April they surprised Hull with a proposal for a 'draft understanding' between Japan and America, personally delivered by Colonel Iwakuro on his arrival in the US. It was mainly a reiteration of well-known Japanese economic and political demands and positions, but it did seem to indicate that Japan would not use force in pursuit of its Pacific aims and would not go to the aid of Hitler unless the US attacked Germany – if Washington merely lifted sanctions, abandoned Chiang Kai-shek in China, and distanced itself from Britain in the Far East. Hull sat down and rewrote the 'draft,' summoning Ambassador Nomura to tell him that, if it were to be resubmitted in its thus amended form, it could lead to serious discussion. Nomura dangerously misinterpreted this as meaning the amended text was America's blueprint for a settlement, requiring only the redrawing of details.

In fact, Hull appended four 'points of principle' which made negotiation useless, because if Japan accepted them talks would be unnecessary, and if it did not they would be a waste of time. In brief, Japan was required to respect (1) the sovereignty and territorial integrity of all nations, (2) noninterference in the affairs of other nations, (3) equal commercial opportunity for all nations, and (4) the *status quo* in all Pacific territories unless changed by peaceful means. In other words, Japan should abandon all thought of expansion. Compounding the rampant misunderstanding, Nomura decided not to send this rather serious set of qualifications to the amended text until Tokyo had had time to digest the 'draft understanding' itself. On 8 May, however, Matsuoka rejected the unexpected fruits of his politeness to the priests out of hand, even before the 'but's arrived; one week later, he demanded a neutrality pact from Washington which would leave Japan a free hand in China and the Pacific. Five weeks after that, Hull counterproposed that Japan withdraw from the Tripartite Pact, China and Indochina.

The next day, Hitler invaded the Soviet Union. The volatile, pro-Nazi Matsuoka immediately called for war with Germany's new enemy, even though the ink was barely dry on his neutrality pact with Moscow. Konoye, however, viewed Hitler's scuppering of the Nazi-Soviet Pact as treachery to Japan and saw no profit in gratuitously taking on the Russians when the southward-expansion policy was gaining ground with the junta. America, Britain and Holland would be enough new enemies for the time being, when there was no sign of victory in China. On 2 July an imperial conference took the 'twin-track' decision to prepare for a military move southward while simultaneously pursuing the diplomatic option. Konoye determined on a spectacular initiative in the latter field, but first he had to dispose of his embarrassing foreign minister. This was smoothly done on 26 July, when, after a quiet word with the emperor and his advisers, he resigned. The rest of the Cabinet was as usual obliged to follow suit, including Matsuoka, the object of the exercise. Two days later, Konoye was back in office with the same ministers bar one: Matsuoka was replaced by Admiral Toyoda Teijiro. The prime minister then publicly proposed a summit meeting with Roosevelt.

By Magic, the Americans, as we have observed, were secretly able to follow some, but only some, of the feverish debates over external policy going on behind the scenes in Tokyo. They inevitably came to the conclusion that the junta, which they mistakenly regarded as monolithic, was up to its eyes in deceit, with a foreign minister privately telling his ambassador the opposite of what that envoy was saying to the administration (Roosevelt described Matsuoka's diatribes to Nomura as 'the product of a mind which is deeply disturbed and unable to think quietly or logically'). There was very nearly a disaster in April 1941, when the British Embassy relayed some Magic material about the impending German invasion of Russia, which it had been given by the State Department, to London. German intelligence broke into the message and told the Japanese that their diplomatic traffic was compromised; Matsuoka warned Nomura early in May, but, inexplicably, the Japanese took no further action beyond tightening handling procedures. It can only be deduced that, utterly confident in the security of their codes and ciphers, they must have assumed that the leak had occurred in Washington after decipherment. The Magic flowed on, and once Washington got over the shock of almost losing it, its distribution was reduced to an absolute minimum. Even the White House ceased to get more than a digest or extracts until November 1941, when Roosevelt specifically ordered the restoration of a complete service. Hindsight surely makes a moot point of whether the loss of Magic would have been more tragic for the Americans than its retention.

But the 2 July message was passed on in full, and more sanctions were therefore imposed at the very moment that Japan pounced on southern Indochina and its prime minister offered a summit conference. Roosevelt's main motive now in continuing negotiations was to buy time for the US Army and Navy to get ready for a war with Japan, henceforward regarded as inescapable. But the evidence suggests that the president would have gone to a meeting – Alaska being the favored venue – had Konoye been able to offer a single genuine and substantive concession on Japan's entrenched and hardening position over

China and the Far East. The danger of another 'Munich' unless a real success was guaranteed in advance was clearly appreciated. Cordell Hull was totally opposed to a meeting on any basis; he could not get over the contrast between Nomura's blandishments and the malevolence in his Magic extracts. Hull's formal rejection of the invitation was followed by Konoye's third and last resignation from the premiership on 16 October. This was the moment at which the junta stopped vacillating and determined that diplomacy had failed: to keep to the schedule for war by the end of November as laid down by the imperial conference of 6 September, the Army had to start moving troops that very day – or accept the unthinkable loss of face involved in a climbdown. The weasel suggestion from a Navy secretly unsure of victory in a long war (not so secretly in Yamamoto's case) to continue on the twin tracks of diplomacy and mobilization was rejected by Tojo: the scale of the military preparations could no longer be disguised as aimed at China and would render the diplomacy nugatory. And the oil reserve was beginning to run down. The 'civilians' (among whom we must count Konoye and Toyoda in this instance) and the Navy were still unable to choose between the two options at a ministerial meeting on the 12th. On the 18th Tojo was made prime minister and the die was cast.

Nomura, unable to straddle the gap between developments at home and the assurances he was still giving to Washington, offered his resignation four days later. He was overruled and therefore stayed abroad to lie for his country just a little longer. Even so, the support for the diplomatic option was not yet extinguished in Japan, as a liaison conference lasting more than twelve hours showed on 1 November – but not to the unaware Americans. The Army wanted to go to war on the 13th, and the Navy on the 20th; the civilians wanted a last chance for peace and a halt to mobilization while the chance was being explored. At midnight it was decided to give the negotiators until 1 December; war would begin a few days later if they failed. Mobilization would continue. An imperial conference confirmed the decision the next day. Two days after that, the first meeting was held of the Supreme War Guidance Council of the most senior military leaders in the presence of the emperor. Magic revealed no hint of all the tension or how hard-fought the decisions were – it never did – but it left no room for doubt that the political temperature was rising in Tokyo; Nomura, however, did not know how far things had gone at home.

The Cabinet tried to bolster Roosevelt's confidence on 6 November by concluding unanimously that the American public would support a naval campaign against Japan, even if he precipitated it rather than await an incident. On 18 November Tojo's new envoy, Kurusu, and Nomura had a three-hour session with Hull which broke down over Japan's adherence to the Axis alliance. The next day, Bishop Walsh of Maryknoll reinserted his toe in the heated diplomatic water (or put his foot in it) by calling upon Nomura and asserting that US oil would flow again if Japan withdrew from Indochina. This right-reverend démarche was no more productive than its predecessor.

On 20 November the Japanese Ambassador communicated to us proposals for a *modus vivendi* [wrote Roosevelt to Churchill on the 24th: the phrase could just as well have been replaced by the single Latin word 'ultimatum']. This Government proposes to inform the Japanese Government that . . . [the] proposals contain features not in harmony with the fundamental principles which underlie the proposed general settlement and to which each government has declared that it is committed. It is also proposed to offer to the Japanese Government an alternative proposal for a *modus vivendi* . . . [to] remain in force for three months. . . . This seems to me a fair proposition for the Japanese but its acceptance or rejection is really a matter of internal Japanese politics. I am not very hopeful and we must all be prepared for real trouble, possibly soon.

Japan's five-point last word was a summary of its final diplomatic position (its irreducible, nonnegotiable minimum) as thrashed out in the Tokyo meetings in the first week of November: no troop movements by either side in Southeast Asia; Japan to withdraw from southern to northern Indochina and to leave altogether on peace with China; Japanese-American 'cooperation' on access to raw materials from the Dutch East Indies; unfreezing of assets; an end to American aid for China. There was no 'or-else' clause, but it was obvious that Japan would neither yield nor offer anything more: if Washington could not swallow this list whole, there would be war. For *modus vivendi*, or accommodation, read *sine qua non*. For the sake of form, the Americans countered on 26 November with their five-point alternative: joint commitment to peaceful means; no further advance by force or threat; Japan to withdraw from southern to northern Indochina and to reduce its forces there; resumption of trade on a month-by-month basis for civilian needs; Japan to negotiate fairly with the Chinese nationalist government in Chungking. Nobody in Washington thought for a moment that this would be acceptable in Tokyo. Once again the Americans were utterly convinced that war was imminent. This time they were right.

Of course it is for you to handle this business [Churchill replied on the 26th] and we certainly do not want an additional war. There is only one point that disquiets us. What about Chiang Kai-shek? Is he not having a very thin diet? Our anxiety is about China. If they collapse our joint dangers would enormously increase. We are sure that the regard of the United States for the Chinese cause will govern your action. We feel that the Japanese are most unsure of themselves.

This response appears to show a deeper understanding in London of the profound tensions and divisions within the Japanese leadership which lay behind the outwardly hardening intransigence of their 'negotiating' position; but it also overlooks the 'loss-of-face' factor, which decrees that a position, once publicly taken up, cannot be abandoned without 'humiliation.' Given that the Americans, like the French in 1914, were absolutely determined not to strike the first blow, the issue of peace or war now lay entirely with the junta. For Tojo and the generals the choice looked significantly different: war, or abject surrender with no room for compromise, a position that they were psychologically incapable of seeing was entirely their own fault. But the president was absolutely right in observing that war or peace depended not on external factors but on the internal politics of Japan.

On 26 November (the day, let us recall, on which the Mobile Force set sail for Hawaii) Roosevelt was given hard military intelligence that a large Japanese convoy with some fifty thousand troops was at sea south of Formosa (the force that was to make the landings in Thailand and Malaya), even as the Nomura-Kurusu double act was discussing the possibility of a military withdrawal from the Indochina area. Hull therefore met the envoys again that evening and made them an offer he knew they could not accept – an accommodation dependent on a Japanese withdrawal from Indochina and China. The two envoys suspended their customary bowing and smiling and gaped: there was no point in forwarding such an idea to Tokyo, they said with uncharacteristic bluntness.

The next day, the US Army alerted its forces, and Admiral Stark, chief of naval operations, sent this unequivocal but verbose message to Admirals Kimmel and Hart of the Pacific and Asiatic fleets in Pearl and Manila:

This dispatch is to be considered a war warning. Negotiations with Japan looking toward stabilization of conditions in the Pacific have ceased. An aggressive move by Japan is expected within the next few days. The number and equipment of Japanese troops and the organization of naval task forces indicates an amphibious expedition against either the Philippines, Thai or Kra [Malaya] Peninsula or possibly Borneo. Execute appropriate defensive deployment preparatory to carrying out the tasks assigned in Warplan 46.

Convinced beyond argument that the Americans could not be persuaded to make any concession to Japan's 'just demands' and plans for East Asia, the imperial conference of 1 December formally resolved to go to war one week later.

On the evening of Saturday, 6 December, Roosevelt decided on one last effort – a direct appeal as head of state to Emperor Hirohito himself. The State Department sent it by radio to Ambassador Grew that evening. War broke out before it could be delivered. At the same time a very long message, in thirteen parts, was coming in the opposite direction, from Tokyo to the Japanese Embassy in Washington. The transcript, all of fifteen pages, restated Japan's position in completely intransigent terms. 'This means war,' said Roosevelt when he had scanned the Magic intercept. On the Sunday morning, a fourteenth and last part of the message was intercepted, telling Nomura to present the entire document at precisely 1:00 P.M. Washington time. At noon Nomura duly telephoned to ask for an appointment at that time with Hull. Shortly afterward he asked for a deferment. Some serious Saturday-night drinking among the embassy's cipher clerks had delayed the transcription and translation of a text the Americans had already seen!

Roosevelt and his closest adviser, Harry Hopkins, were at lunch at the White House at 1:40 P.M., when Navy Secretary Frank Knox telephoned the news that Pearl Harbor was under air attack. Twenty-five minutes later, Nomura and Kurusu arrived in Cordell Hull's outer office and were made to wait while the secretary of state completed his telephone conversation with the president, who had just told him of the raid. Hull pretended to read the text he already knew when the envoys presented it to him – and then orally blasted them out of the building. The attack had started seven minutes early, and Japan's message heralding it had been delivered sixty-five minutes late. Tokyo's plan was to give half an hour's notice of the outbreak of hostilities (without saying where), so as to be able to deny attacking without warning. The point is a quibble: an ultimatum with an intended duration of half an hour is indistinguishable from an undeclared war. The raid on Pearl Harbor and all the other opening moves in the Japanese plan were sneak attacks. But that was the Japanese way: the Bushido code says all is fair in war.

PART TWO
Japan Rampant

3
Japan Attacks

Winston Leonard Spencer Churchill, British prime minister since 10 May 1940, and Roosevelt's partner in the personal correspondence which was already the linchpin of the Anglo-American alliance, endured in the four days after Pearl Harbor a range of emotions that would have unhinged most people. His first reaction to the raid was huge relief bordering on jubilation: there could be no doubt now that the United States would fight, and therefore its already burgeoning industrial capacity would expand at top speed and in all directions to sustain the joint war effort. Churchill placed a telephone call to Roosevelt as soon as he heard. It was not recorded verbatim, but as soon as the president had confirmed the news, the prime minister expressed the view that the Japanese attack had in a way cleared the air. They exchanged promises to ask their respective legislatures for a declaration of war the next day. In Washington, before a joint session of Congress, Roosevelt expressed the nation's outrage by describing 7 December as 'a day that will live in infamy.'

The Senate passed the all-out declaration of war eighty-two to nothing [he wrote to Churchill that evening] and the House has passed it three hundred eighty-eight to one. Today all of us are in the same boat with you and the people of the Empire and it is a ship which will not and cannot be sunk.

Churchill's joy turned to anxiety when the declaration of war proved to be limited to Japan. What price 'Germany first' now – would Washington go back on the agreed Anglo-American policy of treating Hitler as enemy number one, to be defeated first? And would Hitler honor his promise to Tokyo to declare war on the United States (even though Japan had got its own back on the Nazis by attacking in the Pacific without alerting them, just as they had done when they invaded the Soviet Union)? And if Hitler remained silent, as he did on the

7th, the 8th, the 9th, and the 10th of December, could Roosevelt divert Congress and people from their rage against Japan ,for long enough to win their support for hostilities on the other front at the same time? Hitler's decision to declare against the Americans on the 11th, dutifully imitated by Mussolini in Italy, is one of the outstanding minor mysteries of the war. The only immediate advantage it brought him (not an inconsiderable one, however) was the opportunity for Admiral Dönitz's submarines to move into the American neutrality zone in the western Atlantic, doubling their chances of finding targets, and to fire on naval and merchant ships everywhere without having to worry whether they were American.

But his diplomatic and intelligence sources must have told Hitler that the American people, if not their president, were reluctant to fight Germany, and that they were no more likely to favor it after the Japanese surprise attack. The Nazi dictator also had a ready-made excuse for holding out on his fierce Asian ally, because the Japanese had put their nonaggression treaty with Stalin above their membership of the Tripartite Pact (on the grounds that they were obliged to help Germany if it was attacked, not if it did the attacking). Why should he rush to call down the might of the United States on his head when he was so deeply committed in the Soviet Union? And how would Roosevelt have played the situation had the European Axis partners hung back? All the evidence of his handling of the war in Europe up to this moment suggests that the president would have stayed his hand. He had complained of having only a one-ocean navy for a two-ocean war, and now, although the shipyards and aircraft factories were already working flat out, the US battle fleet was in ruins. Germany posed no direct threat to the United States and its possessions; Japan manifestly did. Meanwhile Britain, generously supplied by North America, was less and less likely to go under with every day that passed.

Before Churchill was relieved of his dreadful doubt, he was plunged into despair by the bad news for the British from the Pacific region on 10 December, when the Royal Navy's bluff was so tragically called off Kuantan. But a rout decidedly worse than

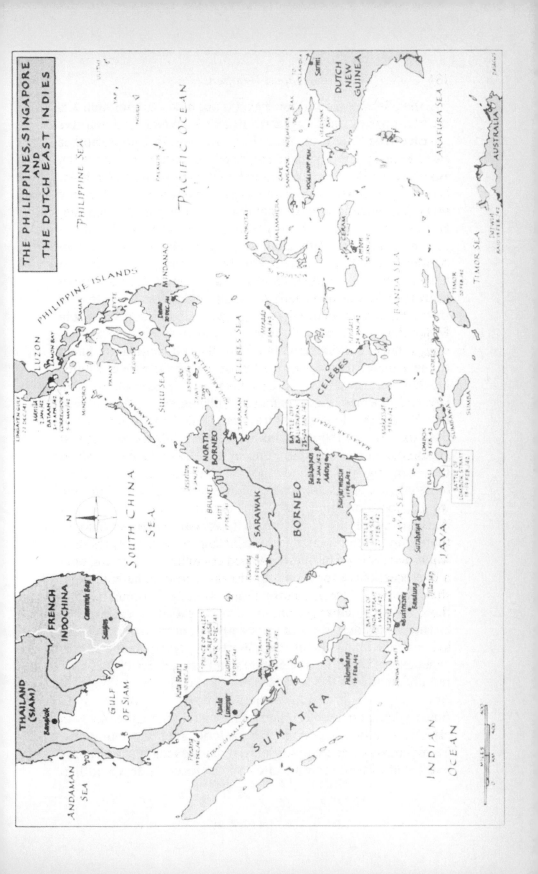

THE PHILIPPINES, SINGAPORE
AND
THE DUTCH EAST INDIES

PHILIPPINE SEA

PACIFIC OCEAN

PHILIPPINE ISLANDS

LUZON

LINGAYEN GULF
22 DEC 41
MANILA
2 JAN 42
BATAAN
1-9 APR 42
CORREGIDOR
5-6 MAY 42

MINDORO

PANAY

PALAWAN

SULU SEA

SAMAR

LEYTE

MINDANAO

Davao
20 DEC 41

MOROTAI

HALMAHERA

CELEBES SEA

NORTH
BORNEO

Tarakan
11 JAN 42

BATTLE OF
BALIKPAPAN
23-24 JAN 42

BRUNEI

SARAWAK

BORNEO

Balikpapan
24 JAN 42

Bandjermasin
16 FEB 42

BATTLE OF
JAVA SEA
27 FEB 42

SOUTH CHINA
SEA

N

CELEBES

Menado
11 JAN 42

Kendari
24 JAN 42

Makassar
8 FEB 42

MAKASSAR STRAIT

BANDA SEA

CERAM

AMBON
30 JAN 42

DUTCH
NEW
GUINEA

Sarmi

ARAFURA SEA

JAVA SEA

Surabaya

JAVA

BATTLE OF
LOMBOK STRAIT
18-19 FEB 42

BALI

LOMBOK
19 FEB 42

SUMBAWA

FLORES

TIMOR
20 FEB 42

TIMOR SEA

BANDUNG

Batavia

Palembang
16 FEB 42

BATTLE OF
SUNDA STRAIT
1 MAR 42

SUNDA STRAIT

THAILAND
(SIAM)

Bangkok

FRENCH
INDOCHINA

Saigon

Cam ranh Bay

GULF
OF SIAM

ANDAMAN
SEA

Kota Bharu
10 DEC 41

Kuantan
10 DEC 41

Kuala
Lumpur

PENANG

STRAIT OF MALACCA

'PRINCE OF WALES'
& 'REPULSE'
SUNK 10 DEC 41

Singapore
15 FEB 42

SUMATRA

INDIAN
OCEAN

MILES
0   400
0   KM

AUSTRALIA

the shocking loss of the two naval dinosaurs was soon inflicted on numerically much superior British Commonwealth forces by Lieutenant General Yamashita Tomoyuki's 25th Army (in fact, a corps of thirty thousand men with two hundred tanks), supported by Lieutenant General Sugawara's Third Army Air Group, in Malaya and Singapore. The Japanese landward drive on the 'impregnable' fortress at the end of the long Indochinese Peninsula (the Imperial Navy being otherwise engaged) began, as we have seen, with landings at Kota Bharu, about four hundred miles to the north. The RAF managed to bomb the command ship, but a beachhead was secured in about three hours; Yamashita began to move his main forces southward from freshly seized bases in neighboring Siam and down the western side of Malaya. Yamashita, known to his men as 'The Tiger,' had as his objectives first Malaya itself, with its abundant rubber, tin and oil; and second Singapore, the British combined headquarters in the Far East. Japanese squadrons based in French Indochina, some six hundred planes, soon overwhelmed the obsolete Anglo-Australian aircraft in northern Malaya – less than half of the 110 on hand could escape to the south. But Lieutenant General Arthur Percival, the British General Officer Commanding Malaya under the C-in-C Far East, Air Chief Marshal Sir Robert Brooke-Popham, had eighty-eight thousand troops, including one Australian, one British and two Indian divisions. Eventually he was to have more than a hundred thousand – but no tanks, even if British artillery outnumbered Japanese by three to one in the larger calibers.

Years of complacent contempt for Japanese soldiers, thought to be incapable of jungle warfare, half blind and generally feeble, now brought a bitter retribution of defeat after defeat. The relentless invaders were inferior only in numbers and average height: they made up for their shortage of guns by the simple expedient of capturing the enemy's and using them against him. In only eight days the Japanese were in Penang, the island and strategic port halfway down the west coast of Malaya. On 11 January 1942, their tanks rolled into Kuala Lumpur, the Malayan capital, with its huge stockpiles of food, fuel and munitions. Yamashita no longer had to worry about the length of his supply line, all the more tenuous after an advance whose speed had astonished the Japanese as much as their stupefied enemies. One of its main secrets was the use of bicycles through the rubber plantations. The Japanese Imperial Guards Division and the 5th and 18th divisions continued their inexorable drive to the south. On 31 January the last retreating Commonwealth troops crossed the Singapore Island causeway, which was then blown up. The equivalent of two divisions had already been killed, wounded or captured by the Japanese, who had by this time lost about two thousand dead and three thousand wounded – 17 percent casualties – and now glowered at the 'fortress' across the narrow Johore Strait. No target had yet presented itself to the camouflaged fifteen- and nine-inch guns of the shore batteries staring silently out to sea on the other side of the island.

Royal Navy officers at the great base were busy destroying the docks, fuel dumps, and other shore facilities even before the last Japanese push began. Setting charges and waiting until the last minute might have served the defenders' crumbling morale better. There was no prospect of an oriental Dunkirk for over one hundred thousand Indian, British, Australian and Malayan soldiers – many of them raw recruits rushed in as reinforcements. Percival, winner of three high awards for bravery in the First World War but known to his men as 'The Rabbit' (because of his buck teeth), now proceeded to bungle the last stand. Despite visible evidence to the contrary across the Johore Strait, he was convinced the attack would come from the northeast, where he concentrated his forces, and not the northwest, where the strait was only half a mile wide even if the ground was swampy. His formidable foes were unlikely to be put off by the prospect of getting their feet wet after advancing five hundred miles through the jungle in two weeks. The Australian 27th Brigade, however, drove off the first Japanese assault on 9 February, inflicting enormous losses – only to retreat unnecessarily in the prevailing confusion. The second wave of small boats swarming with vengeful little infantrymen met no opposition.

The defense was bedeviled by lack of liaison among the three armed services, failure to fortify the northern side, unseasoned troops, friction among the Commonwealth nationalities, secretiveness, hopeless underestimation of the Japanese (they were thought incapable of night flying!) and all-round bad leadership. General Percival had actually refused time and again to fortify the island while there was still time, because it would be 'bad for the morale of troops and civilians' (sic). Exactly the same failure to fortify was exhibited by the British Expeditionary Force in northern France, where Percival had been a corps chief of staff, before the *Blitzkrieg*. Even a personal appeal by General Sir Archibald Wavell, supreme Allied commander in Southeast Asia, on a visit to Singapore in January 1941, to say nothing of a direct and detailed order from Churchill shortly afterward on the defense of the north shore, was ignored by Percival, who was either in shock at the speed of the Japanese advance or otherwise out of touch with reality; at any rate, he abdicated in his mind long before all was irretrievably lost on the ground.

The consequences of this breathtaking display of incompetence, not only by Percival and nearly all his subordinate commanders but also by those who had planned Fortress Singapore and the empty strategy behind it, included one of the best-known and least edifying photographs of the entire war: of General Arthur Percival, ridiculous in the vast, baggy shorts of contemporary British tropical uniform, marching (out of step) with two British standard-bearers to surrender Singapore to Yamashita on Sunday, 15 February 1942. Unlike the Americans now fighting desperately alongside their Philippine comrades on the Bataan Peninsula, the British had even expected significant reinforcements: two divisions of battle-hardened Australians were on their way from the Mediterranean. Proper defenses at the Johore Strait could have held the last redoubt long enough to prepare a determined counterattack, especially since Japanese airpower had been thinned out in support of operations elsewhere.

But in the greatest military defeat in all British history, 138,708 service personnel were lost to the Allied cause in combat or exceptionally brutal captivity which was to endure for nearly four years. The last ship got away, amid ugly scenes of violence and panic in the docks, on 12 February, whereupon the Japanese Navy cut off the maritime escape route; surrender seemed inevitable when the Japanese Army severed the water supply to Singapore City on the 14th. The blow to British prestige, not least among Americans as well as the king's subjects round the world, was one from which the Empire never recovered. Nor was the civilian population spared. All these defeated people were less than nothing to the Japanese soldiery, trained by the contemporary, debased version of the samurai warrior's Bushido code to die rather than surrender or be captured. One-third of the prisoners never came back, but an emaciated Percival was present at the Japanese surrender in 1945. He died in 1966, aged seventy-nine.

I realize how the fall of Singapore has affected you and the British people [Roosevelt wrote to Churchill on 16 February]. It gives the well-known back-seat drivers a field day but no matter how serious our setbacks have been, and I do not for a moment underestimate them, we must constantly look forward to the next moves that need to be made to hit the enemy.

I hope you will be of good heart in these trying weeks because I am very sure that you have the great confidence of the masses of the British people. I want you to know that I think of you often and I know you will not hesitate to ask me if there is anything you think I can do.

I have been giving a good deal of thought during the last few days to the Far East. It seems to me that we must at all costs maintain our two flanks – the right based on Australia and New Zealand and the left in Burma, India, and China.

All in all, February 1942 was one of the low points of the entire war, if not the lowest, for the Allies, and especially for the British. In the week preceding the fall of Singapore, the nadir of the British Army, the three major German warships which had been stuck in Brest, northwestern France, for several months, made a spectacular dash home up the 'English' Channel, despite desperate but uncoordinated attempts to stop them. Not since the Spanish Armada in 1588 had a foreign fleet passed the Straits of Dover in defiance of the Royal Navy. No matter that British mines caused serious damage on the way: news of this consolation came only later, and Hitler was able to make maximum propaganda capital out of his forecast to his admirals that the 'Channel Dash' would be Germany's 'most spectacular naval success of the war' (in fact that distinction belonged to the Wehrmacht's submarine arm). In the meantime, the Russians were making heavy weather of their resistance to Hitler's eastern-front armies, while the British Army in North Africa was under pressure from Rommel.

On the other side of the Atlantic, the United States Navy was being made to learn all over again the painful lessons imposed on the British two years earlier. The U-boats were enjoying their second heyday – or 'Happy Time,' as the German submariners called it – causing havoc on the American East Coast. Ships carefully escorted across the ocean were being allowed to disperse without escort to their American destinations and to make their own unprotected way to convoy assembly points for the eastward voyage. Admiral Ernest J. King, USN, the new COMINCH, knew convoy was the answer but pleaded lack of warships, ignoring well-founded British advice that any escort was better than none. The US Atlantic Fleet had been stripped of every possible ship to reinforce the Pacific after Pearl Harbor. Meanwhile, King seemed prepared to do little or nothing until the fleet of escorts under construction became available, rather than use such resources as the Allied navies could scrape together as escorts in US coastal waters. Millions of tons of shipping and valuable cargoes were needlessly abandoned to the handful of submarines which was all Admiral Dönitz could spare for the job at the time. This great swathe of destruction affected the war in the Pacific as much as the conflict in the West: every ship and every shell was now vital to the global war effort, and every loss, whether of merchantmen or munitions, had to be replaced, mainly by American productive effort. Never had the saying 'War is waste' been more appropriate.

Further, at the beginning of the month the German Submarine Command introduced a new cipher, defeating for the rest of 1942 the British cryptanalysts, who had been so effectively tapping German naval radio traffic over the past nine months. For its part, German Naval Intelligence now completed its penetration of the cipher used by Allied navies to communicate with one another in the Atlantic, which made it even easier to find targets for the U-boats.

*   *   *

The situation on the flanks, which Roosevelt so rightly identified as in need of protection, inspired no confidence. The Japanese Army, after a preliminary attack on 11 December 1941, invaded Burma from Siam on 16 January 1942, and began an advance toward British India, intended to cut the Burma Road, the tenuous and tortuous supply route to Chiang Kai-shek's nationalists in China, and the only overland link left after the Japanese occupation of Indochina. Over to the east, Japanese planes bombed targets on islands north of Australia, such as Rabaul, capital of New Britain, on 4 January, and New Guinea on the 21st. Two days later, Rabaul was seized by Japanese amphibious forces; two days after that, Australia ordered total mobilization. On 3 February the Japanese bombed Port Moresby, capital of Australian Papua New Guinea (now independent). On 19 February Japanese bombers made their first raid on the northern port of Darwin in Australia itself. The town and its naval base were abandoned in the ensuing panic. The entire British Commonwealth and Empire in the Far East, the Pacific and Australasia was threatened by the Rising Sun. Hong Kong, the last British possession in China, had fallen on Christmas Day 1941, six days after Japanese troops advanced on it. In the waters north of Australia the Allied navies endured one defeat after another, and in the Philippines the position of the defenders looked hopeless, as will be described. The Japanese were homing in on their main target, the Dutch East Indies.

On 23 February a Japanese submarine, showing rare initiative for a service that was to be one of the most notable underachievers of the war, bombarded the California coast near Los Angeles. But the Japanese submarines never made a serious attempt to attack American merchant shipping in the Pacific, and none at all to disrupt the traffic through the Panama Canal (itself a tempting target, one would have thought, because of its vastly enhanced importance to the enemy Navy as the link between the Pacific and Atlantic fleets). The Japanese admirals stuck rigidly to the view that submarines were first and foremost an adjunct to the battle fleet, even when they were forced to use the boats to move soldiers and supplies unobserved during the Pacific Campaign. The United States was also singularly fortunate in that the Axis seldom functioned as a military alliance in the Far East: Admiral King's troubles, had he been faced with coordinated submarine campaigns in both oceans simultaneously, hardly bear thinking about. As the war dragged on, from mid-1942 on, the main naval threat to Allied merchant shipping in the Far East, such as it was, came from U-boats operating at huge distances from home, or occasionally from Japanese-held ports, notably Penang in Malaya.

This abdication by the IJN submarine arm is a prime example of an astonishing lack of flexibility in Japanese strategic and tactical planning, so much at odds with their meticulously prepared, stunning succcsses in the opening months of their southward drive. Even the huge harvest of destruction reaped by the German U-boats did not inspire the Imperial Navy to build better boats and attack the long American supply lines. But for the moment, at the beginning of 1942, the Japanese saw no need to think beyond where to strike next for another pushover victory against their disorganized assemblage of enemies.

However triumphantly successful for the Japanese and morally shattering for their enemies these moves against the Western powers may have been, our narrative has not yet brought the aggressors in sight of their main objective, the Dutch East Indies. Here we need to recall the Japanese strategy for southward conquest. Their aim was to become economically self-sufficient at last by seizing the resources of British North Borneo (Sarawak and Brunei) and Malaya, Dutch southern Borneo and the rest of the Indies (Java, Sumatra, Celebes, Ambon, Timor and Bali plus lesser islands). To protect this vast treasure chest, they wanted a perimeter of territory and islands large enough to hold enemy air attacks at bay, from the Kuriles north of Japan via the Marshalls, Wake, the Bismarcks, the Indies, Malaya and round to Burma. All this was included in phase I of the 'Outline Plan for the Execution of the Empire's National Policy,' as agreed at the imperial conference of 6 September 1941, and subsequently translated into operational terms. Five months were allowed for the execution of this unprecedented program of *Blitzkrieg* (a term hard for the

Japanese to pronounce but eagerly borrowed and copied by their staff planners). Hong Kong would need twenty days (it took six), Manila fifty, Singapore one hundred (seventy-two were required) and Java 150, they calculated. These short-term estimates proved unduly pessimistic; the guesswork only began to go wrong in the longer term, most notably over their main logistical requirement, shipping. The further the planners were required to look into the future, the more unreliable their optimism became. Japan proved very good at *Blitzkrieg* but phenomenally bad at the long haul. For all their talk of total war the junta had no inkling of what it meant.

Phase I was executed with extraordinary panache and precision in three stages, of which the first included the occupation of Thailand and the landing in Malaya, the air and ground attacks on the Philippines and moves against British North Borneo. Only in the first week of November 1941 was the Pearl Harbor attack officially incorporated in the plan, which was the subject of a solemn cooperation agreement between the Japanese Army and Navy. Stage two entailed the destruction of Dutch and Allied air strength and landings in Dutch Borneo, Celebes, and other islands including southern Sumatra, culminating in a convergence of forces on the strategic prize of prizes, Java, key to the Indies. The third stage was to be one of stabilization, accompanied by expansion of the Japanese position in Burma on its western flank. Phase II was vaguely described as the consolidation and strengthening of the perimeter thus gained, and phase III required Japan to sit tight on the acquisitions, exploit the sequestered resources, intercept any attacks from outside, and sap the American will to fight by outwaiting them inside the fortified Greater East Asia Co-Prosperity Sphere. Had the junta adhered to this plan, it would have been extremely difficult to halt, let alone reverse.

Phase I looked very risky on paper. Japan, embroiled in a sprawling conflict in China, was about to acquire three new enemies, the United States, and the Dutch and British empires. Its own estimates located up to 70,000 troops and 320 planes in Malaya, 42,000 and 170 in the Philippines, 85,000 plus 300 in the Indies, and 35,000 with sixty in Burma. As most of these forces were 'colonial,' they were, with some justice, regarded as inferior; only the much-outnumbered Royal Air Force, with its new and successful experience in Europe, was feared (unfortunately, neither 'the Few' who had won the Battle of Britain nor their Spitfires could be spared; the RAF in the Far East was largely antiquated in both equipment and outlook). There was also the matter of the US Navy, which globally outnumbered the Japanese by four to three in modern ships. Against all this Japan hurled eleven army divisions and special units (up to 200,000 men), 700 first-line aircraft plus 1,500 others with the Army, 1,600 and 3,300 respectively with the Navy. But these forces never lost their main advantage so long as they were on the attack – the initiative on when and where to strike next, which always belongs to the well-organized aggressor. With all those islands to protect, the Allies were unable to position their forces in the right places at the right times. They could only react, too late and too little, to each Japanese thrust when it came. Even when they knew days in advance where the next blow would fall, the defenders were unable to respond adequately. There was never enough time to assemble and dispatch the ships, ferry the troops and scrape together enough naval and air forces to protect an impromptu defensive position which could be outflanked, bypassed or overwhelmed at the enemy's will. Only a massive accumulation of land-based airpower in advance could have staved off defeat. The British could not spare it; the Dutch could not afford it; the Americans woke up to the need too late.

Japan was therefore wholly confident of acquiring and keeping the initiative in the air from Burma to Wake Island. This belief was based both on quantity of aircraft (and carriers) available in the Pacific region and on quality. The intricate southward advance was planned on the basis of 'hops' commensurate with the best operational range of their fighters; each step was usually no more than about four hundred miles, so that aircraft could soften up the defenses and cover the invasion. The first priority of the landing would be the capture of an airfield, to be used for consolidation of the new gain and then for jumping off to the next one. The key to all this, whether protecting Japanese bombers, ships, and ground units, attacking enemy aircraft, strafing their ground defenses, or beating off their counterattacks from jungle airstrips or the heaving decks of carriers, was fighters: specifically the plane known as the Zero.

The year 1940 was the twenty-six-hundredth since the traditional foundation of the Japanese Empire. It was also the year in which the Mitsubishi corporation put its A6M1 fighter into full production. Since a Japanese military aircraft took its type number from the last digit of the year of its introduction into service, the new fighter was called the Zero-One-One (first air-frame model, first engine design). When the Americans learned to fear it, they code-named it Zeke. Their way of coping with the unmemorable Japanese aircraft designations was to give fighters and seaplanes boys' names (bombers, flying boats, and trainers got girls' names, such as 'Betty' and 'Kate,' while transports were fobbed off with code names starting with 'T'). But 'Zero,' with its undertone of sudden annihilation, seemed more appropriate, and it stuck, so much so that its Japanese sobriquet, Reisen, has been all but forgotten. It was certainly rather more deadly, in performance as well as in name, than its predecessor as carrier-fighter, the Claude.

Designed by Horikoshi Jiro, the Zero proved itself on combat trials in China from August 1940. Armed with two 20mm cannon and two 7.7mm (.303-inch) machine guns, the first model outgunned its opponents (older British, American, and Soviet fighters in Chinese service). The second, at 331 miles per hour, was also rather faster. It could be adapted to carry two 260-pound bombs. It climbed very quickly to a ceiling of 31,000 feet and was extremely maneuverable, one of the secrets of its success being its extremely lightweight metal construction. It immediately achieved a world record for a fighter by comfortably completing a round trip of over 1,000 miles without refueling, more than three hours in the air. The 12th Naval Air Corps was the first to be given one squadron of Zeros; they were so successful that the Navy placed a huge order for more. Meanwhile, Japan dominated Chinese airspace.

Despite this early triumph of the new fighter, hardly anyone in the West took Japan seriously in the air. Alone among the future belligerents on either side, Japan kept its latest weaponry out of military parades. For those prepared to learn, the long-range bombing of China in 1937 from Kyushu in Japan and Formosa (by naval 'Nell' medium bombers flying 1,250 miles in each direction by night in bad weather, albeit with a reduced payload) should have sounded an alarm. The Navy was more alert to the possibilities of both bombers and fighters, insisting on longer range in the former category and identifying the Zero as outstanding in the latter. This was the work of the younger officers, who also regarded the Navy's 5,000 pilots (3,500 front-line) as an inadequate total and called for 15,000. None of this impressed the admirals at the Navy Ministry or IGHQ until it was too late; Japan never caught up, either in pilot training or in aircraft production. Nor was it able to keep up with the swift Western technological developments brought on by the war. Both the West's prewar dismissal of Japanese interest and prowess in aeronautics and Japan's inability to keep up in the Pacific war derived from each side's reckless underestimation of the other. In the opening hours of the war in the Far East, the United States lost two-thirds of its front-line air strength in the Pacific Ocean region.

The Zero played second fiddle to the torpedo-bombers at Pearl Harbor but was crucial in the Philippines and Malaya, and especially in the complex campaign for mastery of the Dutch East Indies. At the outset, the nearest land bases for Japanese airpower against the Philippines were on Formosa, 450 miles from the main US fields at Clark and Iba and 500 miles from Manila. No other fighter could have made such a flight with enough fuel left to operate over enemy territory and then return; but by dint of careful training in fuel economy, improving still further on the design endurance of 1,166 miles, more than 100 were able to accompany the delayed bombers, strafe ground targets, and destroy US fighter strength. Many more Zeros came to cover the conquest of the Philippines and prepare for the first 'hop' south. In the three months from the attack on Pearl to the fall of Java, Zeros claimed 471 enemy aircraft shot down out of 565 and two-thirds of all enemy planes destroyed everywhere (including on the ground). The solitary, initial failure of the Japanese at Wake was largely due to the absence of fighter cover. In the end, a total of 10,938 Zeros of seven main types were built for many roles, including night fighters, seaplanes, dive-bombers, reconnaissance planes, fighter-bombers and, in the end, A6M7 suicide planes. It was the perfect partner in crime of that other Mitsubishi airborne marvel, the G4M 'Betty' land-based naval twin-engined bomber, which could fly 3,745 miles without refueling and deliver a payload of nearly a ton. But even to its crews it was known as the 'Mark I torch,' because, like the earlier Zeros and all except the last Japanese war planes, it lacked self-sealing fuel tanks and tended to burn when hit; some 2,500 were delivered.

Immediately after crippling American airpower in the Philippines, the Japanese made three separate landings on Luzon to acquire the necessary airstrips for covering the main landings from seventy-six ships at Lingayen (48th Division) and Lamon Bay to the east (16th Division). Rightly seeing air superiority as paramount, they readily risked dividing their forces to secure it. The first return on this risky investment was the American abandonment of Manila Bay as a naval base, even for the twenty-nine submarines of the US Asiatic Fleet, by New Year's Eve; all the elderly fleet's major warships got away, as did two hundred thousand tons of Allied merchant shipping, leaving twenty small vessels to do what they could under Rear Admiral Francis W. Rockwell. A further landing was made on Mindanao, to the south of Luzon, where there was a useful fifth column of Japanese workers. The objective was Davao, in the south of the island, the perfect base for operations against North Borneo. It was captured on 20 December 1941, by an invasion covered by two light carriers and three cruisers. A seaplane base was set up at once, giving the Japanese local air superiority.

Two days later, nine ships with four thousand troops left for Jolo in the Sulu Archipelago northeast of Borneo, which was taken on Christmas Day, when Hong Kong also fell and General MacArthur declared Manila an open city (without remembering to inform Admiral Hart of the Asiatic Fleet, who left for Java aboard the submarine USS *Shark* the next day). Another seaplane base was opened immediately off Jolo. Miri in Sarawak had already fallen on the 17th to twenty-five hundred invaders, who swept aside the tiny Anglo-Dutch garrison. Kuching, the Sarawak capital, fell on Christmas Eve. The immediate objective was the local stocks of fuel oil. It was already clear to the Allies that their defense was hopelessly overstretched, spread far too thinly over hundreds of thousands of square miles of island, jungle, and ocean. There was nothing they could do about it, although an old Dutch submarine, *K14*, sank two transports off North Borneo, and American boats sank two more off Lingayen. Allied submarines were little more effective at this stage, when they should have been of most value, in harassing the Japanese, who feared them greatly, than vice versa; the difference was that the Allied side was to learn from these early disappointments and meanwhile made good use of submarines on special missions, notably in the Philippines.

The conference, code-named 'Arcadia,' between Roosevelt and Churchill and their chiefs of staff in Washington over Christmas and New Year's, decided, among many other things, to set up an Allied supreme command for the area under attack by Japan – from Burma in the northwest to southeastern China, and as far south as Australia. It was the American-British-Dutch-Australian command (even though the latter two were not consulted), or ABDA, under the British General Sir Archibald Wavell (ABDACOM). His deputy and his naval chief (Admiral Hart, ABDAFLOAT) were American; his chief of staff and his air commander (ABDAAIR) were British; the heads of ground forces (yes, ABDAARM) and the combined naval striking force were Dutch. ABDA proved much better at producing acronyms than scoring points off the rampaging Japanese, but the new command took over on the last day of 1941 and was active on 15 January from its headquarters in Bandung, Java. Wavell's plan was based on holding Singapore, which he reasonably saw as the anchor of a successful defense of the Indies. There was very little time, as everyone involved already knew.

The Japanese Second Fleet, under Vice-Admiral Kondo Nobutake, was for the time being divided into two main parts for the task of securing Greater East Asia for the emperor. The Eastern Area Force consisted of the Third Fleet, under Vice-Admiral Takahashi, whose job was to gain control of Philippines waters, the Makassar Strait (between Borneo and Celebes), the Java Sea (between that island and Borneo), and points east. A strong detachment under Rear Admiral Hirose was to cover the occupation of Tarakan (off the northeastern coast of Borneo); a smaller one was to take Bali, the 'paradise island' immediately east of Java. The Western Area Force was commanded by Vice-Admiral Ozawa Jisaburo, C-in-C of the First Detachment Fleet, whose area included the South China Sea, Malayan waters and Sumatra. The Second Fleet enjoyed the intermediate support of three battleships and escorting vessels in a Southern Force under Kondo's direct command and with the ultimate backing of Vice-Admiral Nagumo's Carrier Striking Force, now reduced to its original four carriers with two cruisers, eight destroyers, eight tankers and supply ships after its reinforced foray against Pearl Harbor. The Navy had its Special Landing Forces and the world's first fleet of purpose-built landing craft, as well as substantial (more than the Army's) land-based air forces. Neither service possessed a heavy bomber (or was ever to acquire one).

The Japanese Army's separate contribution to the multiple walkover was provided by the Southern Army, commanded by Field Marshal Count Terauchi Hisaichi (1879–1946) since November 1941. At his disposal were eleven divisions plus all manner of extra units, including sixty-six air companies operating his seven hundred front-line aircraft, cavalry units re-equipped with vehicles or bicycles, a special paratroop battalion for use against Sumatra, and another air group flying fifteen hundred support aircraft. He had his own fleet of requisitioned transports and supply ships. Terauchi's command consisted of four hundred thousand men, from generals to laborers. Among the main subordinate commands were the 14th Army (corps, in Western terms), under Lieutenant General Homma, for the Philippines; the 15th, under Lieutenant General Iida Shojiro, for Burma; the 16th, led by Lieutenant General Imamura Hitoshi, the exceptionally able Java commander; and Lieutenant General Yamashita Tomoyuki's 25th Army for Malaya. Army and Navy commands worked independently of one another but were governed by the cooperation agreement negotiated by their respective GHQs in Tokyo, a curiously contractual approach to warfare which was to be repeated many times for the defensive operations later in the war. The concept of the 'combined operation' not only was unknown to the Japanese officer but would have been anathema if it had been known. But in the dark days of 1941–42 the Allied forces, who had to contend with the emperor's soldiers, sailors and airmen coming at them unstoppably from all directions at once, could be forgiven for thinking otherwise.

The Japanese Navy's Western Force completed its conquest of British North Borneo with landings in Brunei Bay on 6 January 1942, and at Jesselton, to the northwest, on the 11th. On the same day, the Eastern Force, using Davao as a base, moved against the Dutch by landing troops on the Menado Peninsula. This is the northernmost arm of the large, crayfish-shaped island of Celebes; Menado is the tail, due east of Borneo. Again on the 11th, paratroops and amphibious forces took Tarakan, an island immediately off the east coast of Netherlands Borneo. The main objective on that enormous island was Borneo's east-coast port of Balikpapan, heart of the local oil industry, which could not be reached (and therefore reinforced) overland. Delighted by the unexpectedly rapid progress of operations so far, Admiral Takahashi decided to bring forward his planned occupation of Balikpapan. Rear Admiral Nishimura's Fourth Destroyer Squadron covered the landing of troops on the evening of the 23rd.

Early that morning the convoy came under attack from the USS *Sturgeon*, a submarine which fired a salvo of torpedoes and was rewarded with some satisfying explosions. These prompted the facetious signal, '*Sturgeon* no longer virgin.' As no Japanese ship was actually penetrated, this claim was premature. So was the ejaculation of large numbers of torpedoes that night by the four old but dashing destroyers of Rear Admiral William A. Glassford's Task Force 5, US Asiatic Fleet, alerted by air reconnaissance from Java. The two light cruisers which were to have accompanied them having in one case broken down and in the other run aground, the First World War destroyers, known as 'cans' to US sailors, caught the invasion convoy off Balikpapan at the perfect moment – in the midst of unloading, with its naval escort haring off in the wrong direction after imaginary submarines. By rushing their attack and firing at unnecessarily long range, the Americans sank only one Japanese patrol boat out of three present and just four transports out of twelve. They were not helped by a disturbingly high proportion of defective torpedoes, a problem that was to recur rather more seriously elsewhere. It could and should have been a complete whitewash; as it was, the US Navy's first surface action of the war in the Pacific region held up the Japanese progress by less than one day. Balikpapan was secured on the 24th.

On the same day, the Eastern Force sent another detachment to seize Kendari with its air base at the end of the south-eastern peninsula of Celebes. This gave the Japanese air forces supremacy over the short route from Java to northern Australia. Surabaya, the Dutch and Allied naval base in northeastern Java, had soon to be evacuated in favor of Tjilatjap, on the south coast, soon renamed Flapjap by American sailors. Six days later, the Japanese sent another formation to consolidate their stranglehold with the capture of Ambon, the original 'spice island,' 350 miles due east of Kendari. This was a much larger affair, backed by Nagumo's carriers, and it began with air raids to soften up the Dutch brigade and Australian battalion defending Ambon Town, which was assaulted on the 31st. By 3 February the Japanese had taken the airfield and were rounding up the defenders. The next day Rear Admiral Karel Doorman, Royal Netherlands Navy, led elements of the ABDA Combined Striking Force – two Dutch and two American cruisers and four destroyers from each navy – in an abortive attempt to disrupt Japanese operations in the Makassar Strait. The only result for the Allies was a severe battering from the air for light cruisers HNMS *De Ruyter* and USS *Marblehead* and heavy cruiser USS *Houston*, which lost the use of her after eight-inch gun turret. The *Marblehead* had to limp home to the US. On 8 February the Japanese Army landed on Singapore Island.

From Ambon the Navy's Eastern Force moved on to make a two-pronged assault on the island of Timor, half Dutch and half Portuguese. Both parts were invaded and occupied on the 20th.

By that time the formation that had set down the occupation forces of Menado and Kendari moved on to Makassar Town, at the end of the southwestern Celebes peninsula, on the 9th, and down to Bali on the 19th. Three days earlier, the occupation of Borneo had been rounded off by the seizure of Adang and Banjarmasin, down the east coast from Balikpapan.

The Western Force had not been idle; on 14 February it arrived off Palembang, the great oil port in the southeast of Sumatra, to back up a parachute drop by seven hundred men, who were to try to seize vital installations before they could be destroyed. This time the Dutch, however briefly, were ahead of the game; not only were they well advanced in wrecking their own and Allied assets, but a battalion of the Royal Dutch East Indies Army was on hand to destroy the airborne intruders. This setback, and a raid the same day by Doorman's Striking Force on the Sunda Strait (between Java and Sumatra), held the Japanese up for a day. But on the 15th (local time) – the day Singapore fell – an air attack from light carrier *Ryujo* dispersed, but did not seriously damage, the Allied ships; and on the 16th the ever-cautious Doorman – no coward, as would soon be shown, but a man who attached at least as much importance to discretion as to valor – called off the planned attack on the seaborne invasion of Palembang, by then in progress. An entire infantry division, which could have been broken in the act of landing, was thus able to get ashore and overwhelm the Dutch garrison in short order.

By this time ABDA was falling to pieces. The Dutch, now with by far the most to lose, wanted their Vice-Admiral Conrad Helfrich, RNN, East Indies naval commander and minister of marine, to take over from Admiral Hart. As the US naval historian Morison bluntly puts it: 'Finally President Roosevelt, [Navy] Secretary Knox and Admiral King, on the ground that Java could not be held much longer anyway, decided it would be better to let a Dutch commander take the rap.' The conspicuously fit Hart therefore left Java pleading 'ill-health,' delegating the tasks, but not the substantive title, of ABDAFLOAT to Helfrich on 16 February. Morale on the battered US ships, residue of the now defunct Asiatic Fleet (later to be replaced by the Seventh Fleet), did not benefit from the

change. Their next setback came when the *Houston* led a convoy
from Darwin, Australia, to reinforce the threatened garrison of
Timor with US artillery and Australian infantry. Realizing that
it would arrive too late, ABDA recalled the convoy on the 18th.
This ensured that it was nicely tucked away in port on the 19th
when the Japanese carriers and extra bombers from Kendari
(Celebes) staged on Darwin the biggest naval air raid since Pearl
Harbor. Twelve vessels, three naval and nine mercantile, and
eighteen aircraft were destroyed, and the port and town badly
damaged. Darwin was temporarily abandoned by its inhabitants,
and ceased to be an important naval base for the duration of
the war.

On the same day, Admiral Doorman, unable to wait for his
ships to assemble in enough strength to achieve local superiority,
opened a three-stage assault on the Bali landing then in progress.
The first wave consisted of two Dutch light cruisers and seven
Dutch plus two American destroyers. One Dutch destroyer
ran aground and another was sunk by a very weak Japanese
presence in the Badung Strait. The second wave, of one Dutch
light cruiser and four US destroyers, went in just after midnight
on the morning of the 20th against stiffer resistance. They took
light damage and inflicted similar. The third wave, of five Dutch
motor torpedo boats, passed through shortly afterward at such
speed that they claimed to have seen, heard and felt nothing at
all. Another chance to wreck a lightly defended invasion convoy
was lost.

Now all the Japanese naval forces in the southwest Pacific –
the Western, the Eastern with its various offshoots, Kondo's
Southern Force and Nagumo's carriers – came together for
the last and biggest operation against the first prize, the
head and heart of the East Indies and also the keystone in
Japan's inverted arch of conquest: Java. The Japanese already
controlled the oil fields of Borneo and Sumatra, but these had
been extensively sabotaged by the defense: Java was not so
important as an oil producer but did have huge stocks, away
from its long coasts. It was also the administrative center of the
Indies; the capital, Batavia, was in the northwest of the island,

which was by a large margin the main center of population in the region. It was entirely appropriate, therefore, that the jewel in the crown of the emerging Greater East Asia should draw the largest amphibious operation of the war thus far: the southward advance was about to yield its ripest fruit. One light and four fleet carriers, four battleships, eight heavy cruisers and a swarm of light cruisers, destroyers and auxiliaries converged on Java. The Western Force brought fifty-six transports and supply ships from the great base at Camranh Bay in Indochina; the Eastern came from Jolo, in the Sulu Archipelago off Borneo, with forty-one more.

Wavell now wanted to abandon Java altogether but was countermanded by London on 20 February. Major General Brereton sent the remaining America B-17s and transport aircraft to India on the 24th. Wavell himself flew to Colombo, Ceylon, on the 25th, at which moment ABDA disintegrated. One Australian and two British light cruisers plus destroyers, which had briefly formed ABDA's Western Striking Force to escort convoys, headed ingloriously in the same direction three days later, reducing the Allies' patchwork navy by a third. The burden and honor of command, such as it was, now fell to the Dutch entirely, at sea, on land and in the air: the phrase 'forlorn hope,' lent by the Dutch language to the English and originally applied to a doomed military outpost, now came into its own. The Dutch could hardly fail to know that the Japanese were coming, and in overbearing strength. On the 25th Admiral Helfrich ordered Doorman to make a sweep along the northern Java coast from east to west to seek the invaders. Lacking air reconnaissance, he found nothing that day or on the 26th.

A final reinforcement was in the meantime still on its way to the beleaguered garrison of Java. America's oldest aircraft carrier, USS *Langley* (Commander R. P. McConnell, USN), a 1912 collier converted in 1922, nicknamed 'the Covered Wagon' for her shape and now in use as an aircraft transport, left Fremantle, Western Australia, with thirty-two US Army P-40 fighters and thirty-three pilots aboard. In company were the freighter *Seawitch*, with twenty-seven more planes still in their crates, and two destroyers. On the morning of the 27th, Japanese land-based air patrols found them seventy-five miles south of Tjilatjap and crippled the *Langley*, which had to be abandoned and sent to the bottom, with the planes, by the destroyers. They managed to escape with the *Seawitch* – but she never got the chance to unload her cargo.

Also on the morning of the 27th, the Japanese air umbrella found Doorman's tired searchers, bombed them, and missed. Even so, Doorman turned back, south toward Surabaya, but was sent out to sea again in response to a definite sighting of Japanese ships to the west, north of Java. The five cruisers turned round once more and sailed into the sunset – Doorman's flagship *De Ruyter* (light), HMS *Exeter* (heavy), USS *Houston* (heavy), HMAS *Perth* (Australian light), and HNMS *Java* (Dutch light), with Dutch, American and British destroyers. This motley old squadron had neither radar nor air cover: the cruisers had even left their spotter-seaplanes behind, in anticipation of a night action. Roughly halfway between the eastern and western ends of Java, and halfway between that island and Borneo to the north, the Combined Striking Force entered upon its last fight, the Battle of the Java Sea.

Late in the afternoon the Japanese heavy cruisers *Nachi* and *Haguro* opened fire on *Houston* and *Exeter* from fourteen miles to the north. The light cruiser *Jintsu* then came up with a squadron of destroyers to attack the Allied escorts. To prevent the Japanese heavy cruisers, with their joint broadside of twenty eight-inch guns, from 'crossing the T' – the classic fleet maneuver which gave the side that did it the chance to concentrate all its fire on the head of the enemy line – Doorman turned aside to run parallel. The light cruisers, in which department he had his only advantage, were now able to join in the full-speed running battle. Another Japanese light cruiser, *Naka*, and her seven destroyers charged across the Allies' bows, joining in a long-range torpedo attack, which had no effect. No significant damage was inflicted by either side for almost an hour – until a plunging shell from the *Haguro* hit the *Exeter* (Captain O. L. Gordon, RN) at extreme range, passed into a magazine and exploded. She swung out of the line; all the other cruisers deduced that Doorman wanted a southward turn and followed suit. The line was reduced to a shambles. One Dutch destroyer was sunk by a torpedo. The Japanese commander, Rear Admiral

Takagi, attacked again, soon sinking a British destroyer. The cruiser action was resumed, the *Exeter* managing to keep going at fifteen knots, wreathed in smoke and still firing, but gradually losing touch with the rest of the battle. The *Houston* (Captain A. H. Rooks, USN), still without the three guns of her after turret, swung to and fro to make maximum use of the six forward guns until the ammunition for them ran out: more was manhandled from aft.

The four speeding cruisers still in the Allied line, having turned south, then east, then north, were once again running west, the Japanese forces still to their north – which is to say on their starboard side. The Japanese invasion convoy, Doorman's target, was thirty miles farther still to the north. Doorman turned northeast at 6:30 P.M. in search of it. More by luck than judgment, he was heading straight toward it, completely unaware because he had no better means of detection than the eyes of his lookouts. The Japanese cruisers were on a converging course from the east. The cruiser battle resumed after dark under Japanese star shells; Doorman reversed course for Java, still hoping to stumble across the convoy he had set out to destroy. He ran into a brand-new Dutch mine field nobody had told him of, and another British destroyer was lost. What with the confusion, losses and diversions, the Allied cruisers were without destroyer escort when the Japanese heavies appeared on their port beam, also heading south in pursuit. The gunfire was renewed to little discernible effect until the Japanese launched their Long Lance torpedoes, the deadliest in the world, at eight thousand yards. *De Ruyter* and *Java* were hit fair and square and soon sank. Doorman ordered the other two, *Houston* and *Perth*, to retire to Batavia, and chose to go down with his flagship seven hours after the battle began. One Japanese destroyer had been badly damaged, a small price for the loss of half the Allied force. The invasion convoys were held up by two days at most.

The punishment of the residual Allied forces in the Indies continued but was rapidly if bloodily drawing to a close. The British Western Striking Force, as we have seen, made off to Ceylon. Remaining American and British officers saw no point in a useless fight for someone else's collapsing empire and told Helfrich as much. The crippled *Exeter* was in Surabaya with a British and an American destroyer, *Houston* and *Perth* in Batavia with a Dutch one. Helfrich ordered both groups to head east for the Sunda Strait and then round to Tjilatjap for a last combined foray from there. Unfortunately, the Japanese Navy's Western Force already bestrode the narrows, covering one of the main thrusts of its invasion of Java, at the northwestern tip of the island. *Perth* and *Houston* arrived, saw the convoy unloading and opened fire at the moment of maximum embarrassment to the invaders. But there were four undamaged enemy heavy cruisers, a light carrier and sundry other warships present. The two Allied cruisers sank four transports out of fifty-six, including that of the commanding general, Imamura, who thus arrived ashore soaked to the skin. He did not treat this as a bad omen.

The Japanese surrounded the two cruisers, and in a fierce exchange which could end in only one way sank the *Perth* (Captain H. M. L. Waller, RAN) at five minutes past midnight on 1 March. Forty minutes later, every round of ammunition aboard, from eight-inch shells to rifle bullets, having been fired, USS *Houston* succumbed to torpedoes and shellfire, turned turtle and sank. One-third of her crew, 368 men, survived. The lone destroyer with them was also lost in this one-sided Battle of the Sunda Strait.

I was able to slide down the flaring bow without any difficulty [Lieutenant (j. G.) H. S. Hamlin, USN, of the *Houston*, remembered after his release from Japanese imprisonment in 1945] because we were listing so far that it was no more than vertical. When I reached the waterline I found myself able to stand on the ship's bottom, and as I walked down the ship's bottom toward the keel, I found myself rising out of the water to get over the bulge that the *Houston*'s bow had. I hit the water on the other side of that bulge and gave the best imitation of a torpedo that I could, trying to get away from suction.
I got a couple of hundred feet away and I turned back to take a look at her, and she was full of holes all through the side . . . Her guns were askew, one turret pointing one way and another the other, the five-inch guns pointing in all directions. There was a big, bright flame coming up just about the mainmast and she was listing way over. I couldn't help think[ing] what she looked

like when I first joined her, when she was the President's yacht. She shone from end to end. I think I will always remember that last look, though. And as I watched her she just lay down to die, she just rolled over on her side and the fire went out with a big hiss.

At 10:20 A.M. on the 1st the charred *Exeter* was caught by four Japanese heavy cruisers as she tried to creep down the Sunda Strait. More than an hour later, having told her two destroyers to run for it, the *Exeter* too was abandoned, then rolled over and disappeared. Both destroyers were caught and sunk soon afterward. Only six American destroyers, including four of Squadron 58, managed to escape the fall of Java; the squadron left Surabaya on 28 February, and two got away from Tjilatjap. All made it to Australia, despite being sighted and briefly fired upon by Japanese destroyers.

First thing in the morning of 1 March, the governor-general of the Dutch East Indies, Mr A. W. L. T. van Starkenborgh-Stachouwer, announced the dissolution of the ABDA naval command. Eight days later, this stiff and conservative ruler of seventy million Indonesians, fifty-four years of age but managing to look a decade younger after more than five and a half years in his job, refused to proclaim the dissolution of the Netherlands empire in the Far East. The Japanese Army, fanning out from its three beachheads on Java, took Batavia on 6 March and soon reached Buitenzorg, the governor-general's palace fifty miles south of the capital, and Bandung, the main Dutch East Indies Army base. At the Bandung military airfield the now impeccably turned-out General Imamura confronted the equally immaculate governor-general and his military commander, General Hein ter Poorten.

The contemporary diary, found later in the war, of an unnamed Japanese staff lieutenant colonel describes the scene. Imamura's demand for the unconditional surrender of Java and the Indies was met with a blank refusal by Starkenborgh and an offer from ter Poorten to surrender Bandung only. The governor-general bureaucratically explained that he was in charge of both the civil government and the Army but the Navy came directly under the queen and he was therefore not its C-in-C. Barely suppressing his rage, Imamura snapped: 'Why did you come? There is no use for further questioning. If you do not surrender unconditionally, there isn't any other way but to attack continuously . . . I will give you ten minutes from now to make the final decision.' Starkenborgh ignored the threat. Instead, he twice asked for the official Japanese photographer, present to record the humiliation of Dutch power in the Far East after three hundred years, to be ordered out of the room! Finally, ter Poorten stepped in and broke the deadlock by agreeing to a total surrender. The anonymous colonel noted:

[The governor-general's] splendid attitude cannot be compared with that of [the British General] Percival, whom I saw in Singapore. There is no race in the world that bargain as the Dutch do. Even today, at this very moment, they came to negotiate a bargain. In this manner they probably pestered the [abortive Japanese] economic mission of a year ago. This is unlike the time we were in Singapore . . .
   At 1030 hours on the 9th I heard a broadcast made by the [Dutch] army commander. His voice shook [as he ordered his men to cease fire. The next day brought the formal surrender of the Dutch East Indies at Bandung.] It surrendered absolutely, like Singapore. The faces of the [Japanese] officers and men were beaming with joy.

Starkenborgh was held by the Japanese for the duration and survived to support the postwar effort by the Dutch to re-establish themselves in Indonesia. After a singularly vicious colonial war, independence was granted at the end of 1949.

The Japanese had knocked out the Dutch and completed the southward advance but for a few loose ends. The fighting was still going on in Bataan, to the astonishment and annoyance of IGHQ and Terauchi, and also in Burma. Wavell was back in India as C-in-C, but on 5 March Burma was hived off his fief as a separate command under General Sir Harold Alexander. Three days later, the capital, Rangoon, fell to the Japanese Army. On 23 March the obvious jumping-off point for a counterattack on Burma, the Andaman Islands, was pre-emptively seized by a small Japanese task force. The British retreated over the Irrawaddy River in Burma at the end of April, and Mandalay finally fell on 20 May, the high point of the northwesternmost

Japanese advance. Allied overland supplies to China via the corkscrew Burma Road were cut off.

The British meanwhile were trying to re-establish a naval presence in the Far East after the loss of their deficient deterrent off Kuantan. The new base was Ceylon, the great island off the southern tip of India in the middle of the Indian Ocean. Admiral Sir James Somerville, formerly of 'Force H' at Gibraltar, became C-in-C of the British Eastern Fleet on 24 March, which was new in name but not so much in composition. Two new fleet carriers, *Indomitable* and *Formidable*, were its most promising assets; with the light carrier *Hermes* the British could launch fifty-seven obsolescent strike planes and thirty-six mediocre fighters, hardly enough to deprive even the nervous Nagumo of much sleep. The five battleships were all veterans of the First World War, third-class in Japanese eyes; there were two heavy and five light cruisers, sixteen destroyers, and seven submarines. There was very little shore-based airpower, and the BEF's three bases at Colombo and Trincomalee in Ceylon, and at Addu Atoll, six hundred miles to the southwest – so secret that the Japanese never got to know of its role – were spacious but deficient in almost every other respect.

Ultra intelligence revealed just in time that Admiral Kondo of the Second Fleet had ordered a double assault on the Indian Ocean at the end of March. Nagumo was to attack the Ceylon bases and Ozawa the shipping in the Bay of Bengal. Somerville was under orders to avoid a fleet action with the much superior Japanese Navy; the Royal Navy had beggared itself in assembling his mostly superannuated fleet and had no more big ships in reserve anywhere in the world. He organized his capital ships into a fast division (the two large carriers and battleship *Warspite*) and a slow (the other four battleships and light carrier *Hermes*), each with cruiser and destroyer support, and put to sea. He was safely tucked away at Addu on the afternoon of 4 April, when an RAF Catalina flying-boat patrol sighted Nagumo 360 miles southeast of Ceylon. Somerville was in the midst of refueling and unready for sea; he stayed put, hoping to be able to organize a night counterattack by air.

On the morning of the 5th ninety-one bombers and thirty-six fighters subjected Colombo, on the western side of the island,

1. Emperor Hirohito receives homage from War Minister Tojo on the 2,600th "birthday" of the Japanese Empire in October 1940.

2. Prime Minister Tojo Hideki reviewing troops in Bangkok.

3. Admiral Yamamoto Isoroku, commander-in-chief of the Imperial Japanese Navy's Combined Fleet and architect of the Peral Harbor attack.

4. "Tenacious Tanaka": Rear Admiral Tanaka Raizo, stubborn commander of the Tokyo Express in the Solomon Islands.

5. Sent abroad to lie for their country: Ambassador Nomura Kichisaburo (right) and Special Envoy Kurusu Saburo waiting with Japan's ultimatum at the US State Departament even as Pearl Harbor is attacked.

6. Japan's brightest naval commander: Vice Admiral Ozawa Jisaburo, who fooled Halsey at the Battle of Leyte Gulf.

7. & 8. Japan's leading naval air aces: Captains Fuchida Mitsuo (left, postwar) and Genda Minoru (right, prewar).

9. "Even-tempered... always in a rage": Fleet Admiral Ernest J. King, USN, commander-in-chief, US Fleet, and Chief of Naval Operations.

10. The US Navy's finest: Admirals Chester W. Nimitz, Pacific commander-in-chief (left) and Raymond A. Spruance, Fifth Fleet commander, in the Marshalls.

11. William F. "Bull" Halsey, the sailors' admiral and commander, Third Fleet.

12. "I have returned": General Douglas MacArthur, Supreme Allied Commander, South-West Pacific, wading ashore at Leyte in the Philippines in October 1944.

13. The flight deck of a Japanese carrier just before the launch of the Pearl Harbor attack.

14. Pearl Harbor in flames.

15. Battleships *West Virginia* and *Tennessee* burning.

16. Carrier *Hornet* entering Pearl Harbor six months after the raid.

17. Carrier *Wasp* with all planes on deck.

18. Jumping ship: a man in transit from one US destroyer to another in the Solomons.

19. Death throes: USS *Wasp* in September 1942.

to the Pearl Harbor treatment. Forty-two RAF fighters took them on, losing nineteen planes. The Japanese lost seven. Most shipping had been sent out to sea, but the attackers sank one destroyer and one armed merchant cruiser (converted liner). They also damaged shore facilities, though they did not put the port out of action for long. But among the dispersed ships were heavy cruisers *Dorsetshire* and *Cornwall*, which were found at sea that same morning by more than fifty of Nagumo's aircraft. In a speeded-up reprise of the Kuantan disaster, both these valuable ships sank in a few minutes with the loss of 424 men; 1,122 were saved. Nagumo, meanwhile, having failed to locate Somerville, was moving slowly to a position east of Ceylon for his second strike, against Trincomalee, on the eastern side of the island. Shipping was again dispersed and the RAF put up twenty-two fighters and nine Blenheim bombers against an attack of similar strength to the first. Five of the Blenheims were shot down and one merchantman sunk; the port was battered but not paralyzed. Once again the sting in the Japanese tail did the worst damage. Nagumo's pilots found the first of Somerville's ships as they were returning on the morning of the 10th. They sank the *Hermes* with all her aircraft, plus a destroyer, a corvette and two tankers. The remaining RAF fighters in Ceylon did not have the range to come to their aid.

Japanese losses on this spectacular Indian Ocean raid amounted to a mere seventeen aircraft. A total of twenty-three Allied merchantmen of 112,300 tons were sunk in Ozawa's parallel operation, supported by a rare display of aggressive war on commerce by Japanese submarines, off the west coast of India, which netted five more (32,400 tons). The BEF, not knowing whether the assault was a prelude to more spectacular Japanese landings and unable to cope with overwhelming enemy maritime air superiority, prudently if ignominiously withdrew all the way to East Africa for a while.

Unbeknown to them, Nagumo's epic voyage of destruction to and fro across the North Pacific and through the South China and Java Seas to the Indian Ocean was at last at an end. The western and eastern extremes of his travels were eight thousand miles apart, four thousand miles in each direction from Japan's Inland Sea, where they had begun and ended. His bag included five battleships, one carrier, two cruisers, seven destroyers, dozens of merchantmen, transports and smaller vessels, hundreds of aircraft from six nations and all manner of damage to ships and shore facilities – all this without so much as a scratch· on any ship in the Mobile Force from enemy action. Losses were limited to a few dozen pilots (of disproportionately high value) and planes to enemy action and accidents, enough to make replenishment a matter of moderate urgency. But that seemed a very small price for the successful completion of an intricate multiple mission of strikes and support actions across millions of square miles of sea and land. The seemingly irresistible iron fist of the Japanese Empire had certainly earned and could well afford a rest; Greater East Asia was secure, despite the running sore in China and the irritation in the Philippines. Nobody in the world at that time had any reason to imagine in the wildest of dreams that Nagumo's next voyage would have such a different outcome, not even the neurotic admiral himself, who returned a national hero to Tokyo Bay.

The 'China Incident' was to last the entire war; the fighting withdrawal of the bulk of the Philippine garrison (including twenty thousand Americans) down the Bataan Peninsula to the offshore fortress of Corregidor was slowly but surely drawing to a close as Nagumo went home. His Army colleague Lieutenant General Homma found the weather a greater obstacle to his aim of taking Manila than the two Philippine divisions of the North Luzon Force that initially opposed him on Lingayen Gulf. The defense had metaphorically shot itself in the foot by a change of strategy before the Japanese arrived. MacArthur's original plan for the defense of Luzon was to meet the enemy at the beach and throw him back into the sea. Food dumps were therefore sited with this in mind. The decision to withdraw into the Bataan Peninsula, which surprised the Japanese, left too little time to recover most of the rations, adding malnutrition and starvation to the burdens of disease and enemy action which soon fell upon the defense. General Jonathan Mayhew Wainwright's plan was to stand and fight on a series of five defense lines; at the third, he was to hold out long enough to be joined by Brigadier

General Albert M. Jones and his South Luzon Force, pulling back from the subsidiary Japanese advance from Lamon Bay in the southeast of the peninsula.

After a week the defenders were on half-rations, or two thousand calories a day. The Filipino 26th Cavalry finally shot its horses and ate them. The only fresh meat after that was water buffalo, which was usually of shoe-leather consistency. The dismal diet (barely one thousand calories a day at the end), the unspeakable humidity, the lack of suitable clothing and protection against insects caused epidemics of malaria and intestinal and nutritional diseases. Eighty percent became ill. MacArthur, his command staff and the Philippine government withdrew from Manila to Corregidor, with its network of tunnels, at the turn of the year. He paid his one visit to the front line by PT (motor torpedo) boat on 9 January 1942. It lasted ten hours. This failed to dissuade the fighting men, who styled themselves 'the Battling Bastards of Bataan' (not to be confused with the 'Bataan Gang' of staff cronies who swarmed round MacArthur for the rest of the war and after), from calling him 'Dugout Doug.' The American press, ever hungry for heroes and importunately urged on by his skilled and uniquely overbearing personal publicity machine, named him 'the Lion of Luzon.' At this time Homma's northern front was manned by one division, the other having been pulled out for operations against the Dutch, and one third-rate brigade, conclusive evidence that the Japanese high command saw Bataan as a sideshow. The thrust from Lamon consisted of seven thousand Japanese troops, or about two brigades. Homma had already occupied an undefended Manila on 2 January; Philippine airspace was completely secure, and so, most of the time, were the local seaways.

But Homma could not eradicate the last resistance unaided. He was heavily outnumbered, even if the opposing troops were inferior in quality of command, training, equipment, supplies, experience and almost everything else of use except for individual bravery. The Japanese were no more immune to the climate and disease than their enemies, and Homma suffered a personally disastrous loss of face by asking Terauchi for reinforcements at the beginning of February. There was a slackening in the fighting until General Yamashita, the conqueror of Malaya and Singapore, was sent to the aid of Homma with fresh troops from the 21st and the whole of the 14th divisions, bringing to fifty-seven thousand the total of Japanese troops involved in the capture of the Philippines. Extra bombers were transferred from Burma, and the final Japanese push in Bataan moved irresistibly forward on 3 April. MacArthur, doubtless heartened by the gift, sanctioned by Roosevelt, of $500,000 from an unduly grateful (or hopeful) President Quezon of the Philippines, stayed put. On 22 February General Marshall, Army chief of staff, acting on Roosevelt's behalf, directly ordered MacArthur to withdraw – but left the choice of date to him. He departed in a PT boat and another blaze of publicity on 12 March 1942, promising, 'I shall return.' We too shall return to the subject of MacArthur, in the next chapter. General Wainwright was left in command on the ground. Roosevelt and Marshall did not flinch from following to its conclusion the logic of their almost frantic exploitation, for morale-boosting purposes, of MacArthur's extraordinary gift for self-promotion: they gave him America's highest award for gallantry, the Congressional Medal of Honor. This is a taste of what it was like for those he left behind, after Major General Edward P. King surrendered Bataan on 9 April 1942:

We drove along through the very congested road [said Lieutenant Michael Dobervich, USMC, after his liberation; he was driving a captured truck on Japanese orders]. We saw the beginning of the looting, bayoneting, face-slapping . . . It was hard to take. The stragglers were either bayoneted or shot . . . Americans from general to private had to salute every and any Jap or suffer a blow with the rifle or a slap . . . I arrived at camp on 11 April 1942 . . . [We had to] stand for sixteen hours in the terrific heat; many passed out from lack of water . . . The Japanese were great at mass punishment. They tried it at the least provocation . . . I saw several soldiers come back from a working party that were dead . . . I had ten of my men die in my presence after coming back from working parties, too sick and beyond recovery . . . The situation was very bad and there were not enough people to bury the dead. Our camp was a camp of the living dead. People moved about in a daze, gaunt and stary-eyed [sic] from lack of nourishment. Words cannot describe the conditions . . .
    At this particular burial they piled about thirty bodies into one large pit . . . like so much cordwood. Before the covering started, one of

the dead bodies began to move; it was a feeble effort . . . to raise its head. The Jap guard ordered this Marine of mine to strike the head with a shovel. He hesitated and that enraged the guard so that the bayonet was thrust at him so he was forced to obey . . . The odor was terrific, the graves were shallow and a prevailing wind kept our camp well supplied with the stench.

Such was the sixty-mile 'Bataan Death March' and its aftermath. Of the seventy thousand sick and starving Filipinos and Americans who took part in it, fourteen thousand died on the march alone, and untold thousands more in the camps. The last eleven thousand defenders held out for another month, until 5 May, when Wainwright formally surrendered Corregidor and the Philippines after the final assault by two thousand Japanese. Only then was the Japanese Navy free to use Manila Bay – by which time its availability was all but academic. Meanwhile, America had some real heroes to celebrate – and to mourn.

Even before he knew he was to lead the mighty counterattack against the Japanese tidal wave of conquest, Rear Admiral Chester W. Nimitz, USN, noted laconically in his diary on 10 December 1941: 'We must not again underestimate the Japanese.' Some four months later, as he braced the Pacific Fleet for an expected Japanese drive against New Guinea, Admiral Nimitz, now commander-in-chief, Pacific Fleet (CINCPAC), set down a more considered appreciation of the enemy in his 'Estimate of the Situation' on 22 April 1942:

The Japanese are flushed with victory. Their morale is high . . . Their planning, training and experience is excellent and must not be underestimated. The attack on Pearl Harbor was a workmanlike job in every respect. That operation may be taken as an example of what they can do in the way of planning and execution of the plan. Timing, the objective, the surprise, driving home and attack – all show that we can expect excellent work. Their army has shown ability and the will to reach objectives even in the face of considerable opposition. They can travel over very bad terrain. Their food supply is very simple – a few handsfull [sic] of rice plus what the country can offer seems to keep them going in fighting trim. In landings they are perfectly willing to accept large losses from enemy fire, and drownings, and they do reach objectives in spite of losses.

Admiral Yamamoto, C-in-C of the victorious Combined Fleet, was aware of the danger of hubris, that overweening pride which comes before a fall. Four months into the war in the Pacific, he wrote to a colleague: 'The mindless rejoicing at home is really appalling; it makes me fear that the first blow against Tokyo will make them wilt at once . . . I only wish that [the Americans] had also had, say, three carriers at Hawaii . . .' His emperor also felt uneasy. On his forty-second birthday (by Japanese reckoning; in Western terms his forty-first), on 29 April 1942, Hirohito remarked privately to an aide: 'The fruits of victory are tumbling into our mouths too quickly.' They had in fact virtually ceased to do so: Japan was about to undergo a shocking change of diet.

4

The Giant Awakes

The temporary flagship aboard which Chester William Nimitz took over command of what was left of the United States Pacific Fleet on 31 December 1941, was appropriate not only in unusual modesty of size, there being no battleship available, but also in type of vessel. USS *Grayling* was a portent not easy to make out among the wrecked warships and general damage in Pearl Harbor except from close quarters, even though she was 308 feet long. She then boasted only one three-inch gun and four machine guns and had a long, narrow deck just a few feet clear of the water. She was a submarine, number *SS 209*, one of the twelve in the T-class, the most modern in US service. Her main armament, invisible below the waterline, was ten twenty-one-inch torpedo tubes, six forward and four aft.

Six hours after Pearl Harbor, Admiral Harold R. Stark, chief of naval operations, had issued a terse order to the fleet: 'Execute unrestricted air and submarine warfare against Japan.' This eight-word command contains, as briefly as it can be summarized, the formula by which the United States Navy rounded on Japan and made possible its destruction as a military power – a process to which the first atomic bombs were a dramatic and convenient but strategically irrelevant embellishment.

We can note here for chronological reasons that the administration decided less than twenty-four hours before the Pearl raid to switch from research to development of the Bomb. At first hindsight, this looks like a most portentous event in the history of the Second World War. Look back over the war again and a different view emerges: the real portent was for the postwar future of mankind as a whole. Had the Japanese junta chosen to recognize when it was beaten by a much superior enemy (honorably, in anyone else's terms) instead of putting the salvage of its collective face before the fate of its country, it would have acknowledged that the submarine and the plane had well and truly sunk Japan's imperial ambition together with its merchant fleet. Only the stubborn stupidity of the Japanese leadership ensured the Bomb a role in the last multilateral war of the prenuclear age. We shall return to these themes at the proper time.

Stark's order was obeyed with surprising speed. It was the submarines that set out on the first American counterattack against Japan, the carriers being otherwise occupied. Lieutenant Commander Elton W. Grenfell, USN, took USS *Gudgeon* out from Pearl Harbor on 11 December all the way to the Japanese Inland Sea for a cautious patrol, during which they were to sink the Japanese submarine *I173* on 24 January 1942, 220 miles west of Midway Island. It was the first sinking of a warship by a US submarine (the first American sinking of such a vessel in the war was achieved by two SBD dive-bombers from USS *Enterprise*, which caught submarine *I170* north of Hawaii on 10 December). But before we go any further into the strategically decisive exploits of US submarines and air forces, we need to consider the men who now became responsible for the deployment of the US Navy – including Nimitz as CINCPAC, and his immediate superior, the commander-in-chief of the United States Fleet, Admiral Ernest J. King – as well as other key factors in the administration's uncharacteristically decisive initial response to the Japanese onslaught. To do that, we need to go back to Pearl Harbor on the morning after the beginning.

The forces left to Admiral Kimmel as (outgoing) CINCPAC were not inconsiderable. Ships in his command undamaged or easily reparable included three fleet carriers (*Enterprise*, *Lexington*, *Saratoga*); battleship *Colorado* in a California yard; a dozen heavy and eight light cruisers, a good six dozen destroyers and related types; fourteen submarines at sea or in Pacific ports. Admiral King, still commanding the Atlantic Fleet, was ordered to send carrier *Yorktown*, three battleships and a squadron of destroyers from his Atlantic command through the Panama Canal as reinforcements. All along both US coastlines shipyards were busy building many more of all types, for the British as well as the US, including the fleet carriers *Hornet* and *Wasp*, which were all but ready for duty with the Pacific Fleet.

In the last week of America's peace, Colonel Robert McCormick's Chicago *Tribune* had published details of Washington's plan for war, known to the initiated as WPL 46 or 'Rainbow 5' (small wonder that Nimitz more than once describes American difficulty in keeping secrets as 'a national weakness' in his war journal). It was the last throw of the isolationists. Thanks to the Japanese successes in the opening phase, the leaked plan had to be so radically modified as to be unrecognizable, even as the vast majority of the US population swung behind the president in his new role as active – not just constitutional – commander-in-chief. Instead of sweeping majestically across the Pacific against the Japanese-mandated islands with the Royal Navy sturdily looking after the area south of the Equator, the US Navy was now told to hold Hawaii and the Western hemisphere to 180 degrees longitude and protect the strategic routes in the Pacific. These ran from the West Coast to Hawaii, from Hawaii to Australia, and from Panama to New Zealand. The US Navy, in the person of Vice-Admiral Herbert F. Leary, took charge of the ANZAC command, adjoining ABDA's southeastern limits and including Australia, New Zealand, eastern New Guinea, the Solomon Islands and Fiji (the Australasian segment of the British Commonwealth). As a sea command, ANZAC lasted from 26 January to 22 April, when a new Allied strategic arrangement came into force, as will be seen. Meanwhile, the Pacific Fleet was not to get involved in fighting for Guam or the Philippines at this strategically defensive stage; tactically, the submarines and carriers would go on the attack as soon as possible.

Rightly anticipating his imminent selection as a principal scapegoat for the Japanese surprise attack, Kimmel, short on imagination but long on energy and resolution, nevertheless reacted with speed, sagacity and decisiveness to the emergency thrust upon him by the events of 7 December. The situation on Oahu was nothing like as bad as it seemed at first, despite the shocking loss of eighteen ships to a sneak attack. Most of the battleships were recoverable in shallow water, which would not have been the case had they been caught by Japan's superior airpower at sea; the shore facilities were only superficially damaged; the carriers were intact. The main worry was that nobody knew where the Japanese were or what they would do next, which might include more moves against Hawaii, Wake, Midway and/or the Aleutian Islands in the North Pacific. Kimmel therefore ordered battleships to base themselves at San Francisco as they became available, out of harm's way and clear of congested Pearl Harbor. He set up a convoy system, air reconnaissance, offensive submarine patrols and carrier task forces, while urgently reinforcing the defenses of Pearl itself in cooperation with the Army. Nobody could have done more or acted faster with what was available.

His reward was to be barred from active service on the order of the president on 17 December. Kimmel (1882–1968), though more sinned against than sinning, was made to retire on 1 March 1942, in his substantive rank of rear admiral and spent the rest of the war (and beyond) testifying at one inquiry after another. His offense was to be caught nearly as unprepared as his country, of which he had been a true and faithful servant. His Army opposite number, General Short, suffered a similar fate. Neither hapless commander was granted his day in court: a good lawyer would have made too much capital of Washington's greater sins, which it heaped upon the two Hawaiian commanders' shoulders. The replacement of both was inevitable and in itself reasonable because they had, whatever the cause, been caught napping; denying them the chance to redeem themselves was not. But then political buck-passing never was a pretty sight.

Kimmel was temporarily relieved by his deputy, Vice-Admiral William S. Pye, who took over just in time to preside over the on-off Wake Island relief fiasco described above. No sign of Nagumo's fleet had been found, thanks largely to an *Enterprise* pilot's false report of Japanese ships south of Hawaii (in fact the heavy cruiser *Minneapolis*, with three light cruisers and a dozen destroyers).

Apart from the large detachment providing deep cover for the seizure of Wake, Nagumo's men had gone home briefly, to collect their victors' laurels amid jubilant scenes and to replenish, before moving out again by the turn of the year to back up the other Japanese operations. Meanwhile the imperial submarine arm, unfazed by its complete failure to contribute to the Pearl attack, in which one fleet submarine out of twenty deployed and all five miniature boats were lost for no return, was already enjoying its most profitable month of the war. In the twenty-five days of hostilities in the Pacific up to the end of December 1941, eleven boats out of two dozen on patrol in the Western hemisphere sank some 40,700 tons of shipping (eight vessels, of which six were American) and damaged seven totaling 47,500 tons. No attempt seems to have been made to interfere with the obvious and inevitable move to reinforce the Pacific Fleet from the Atlantic via Panama. A small handful of brief bombardments by individual boats against targets on the West Coast and isolated US Pacific islands, and a couple of seaplane sorties against Oahu, were all that the Japanese Navy was disposed to deliver against its new enemy when he was in his most vulnerable and least organized condition. The Americans were prudent to fear rather more, and justifiably surprised when it did not come. For the immediate defense of Pearl, the Army could muster 114 and the Navy ninety-four aircraft; including urgent reinforcements.

Admiral Ernest Joseph King was 'the most even-tempered man in the Navy,' one of his six daughters once said: 'he is always in a rage.' He hated, in no particular order, incompetence, paperwork, Englishmen and the media, and loved himself, the US Navy, women and liquor. He was also the owner of the highest intellect to be found among the wartime US Joint Chiefs of Staff (JCS).

But in 1939, when the war began in Europe and Harold Stark, rather than he, was appointed chief of naval operations, the insatiably ambitious King, rising sixty-one and facing mandatory retirement at sixty-four, concluded that his career was over. Stark, one of the few King-admirers who also liked the man, tried to get him the Pacific Fleet instead of Kimmel. Had the CNO succeeded, King would probably have disappeared into the footnotes of history. In the event, that fate was reserved for Kimmel, not 'the Bald Eagle,' as King was known for his lack of hair, lofty height and manner.

He was born in Lorain, Ohio, in 1878, shortly after his Scottish father had migrated there to become a foreman in a railroad repair shop. He graduated fourth in the class of 1901 at Annapolis Naval Academy. By then he had already had a taste of action at sea in the Spanish-American War of 1898. In 1909, as a lieutenant of thirty-one, he got a prize for a paper on shipboard organization from the US Naval Institute. In 1917 he was on the staff of the US Atlantic Fleet, with which he served alongside, and presumably learned to dislike, the British (more precisely, the English ruling class, so heavily overrepresented in the Royal Navy). He finished the war as a relatively youthful captain of thirty-nine, with the Navy Cross.

Between the wars King helped to design a gunnery range-finder, commanded a submarine division, experimented with tactical innovations and wrote a paper on the training and development of officers, still in use sixty years later. In 1925 he won fame by salvaging a sunken submarine off Rhode Island. As naval aviation began to develop in the late 1920s, King, still a captain, saw a window of opportunity and jumped through it. The law wisely required commanders of carriers and naval air stations to be qualified as air crew. Naturally, none of the new breed of naval pilots was senior enough for such a vacancy; King was, provided only that he could win his 'wings,' which he did in 1928, just short of fifty. He then took command of the *Lexington* (and did his mock attack on Pearl Harbor). He achieved flag rank in 1932.

Rear Admiral King became chief of the naval Bureau of Aeronautics in 1933, commander of all naval patrol-plane squadrons in 1936 and, as a vice-admiral, chief of all carriers and aircraft in the fleet in 1938. Like Yamamoto in Japan, King was the prime mover behind the development of US maritime airpower in the crucial prewar years. Only congressional budgetary restraints on the Navy prevented him from doing more. Pipped at the post by Stark, King accepted a position on the General Board of the Navy with ill grace until his enormous experience was recognized and he was allowed to go to sea again. He gladly took one step down to rear admiral to become chief of the Atlantic

Squadron, US Fleet, in December 1940. Two months later he was a vice-admiral again and commander of the newly re-created US Atlantic Fleet, which soon brought him his fourth star as a full admiral. As such he led the unofficial but rapidly growing naval support for the British as they fought to keep open the transatlantic lifeline against the German submarines in the most important strategic struggle of the entire war. He did not allow his distrust of the British to get in the way of his professionalism or his readiness to face the operational and logistical challenges of Roosevelt's policy of 'all aid short of war.'

The most important opportunity in King's life arose in August 1941, when he personally escorted Roosevelt to his first meeting with Churchill at Argentia, Newfoundland, in his flagship, heavy cruiser *Augusta*. He was therefore on hand when Roosevelt and Frank Knox, the Navy secretary, needed naval advice in the talks with the British, who had the practical advantages of two years' experience of war and the backing of a finely tuned, supple staff apparatus. King's phenomenal grasp of strategy, knowledge of detail, breadth of experience and forthright incisiveness made an abiding impression. Luck placed King in port on the East Coast on 7 December 1941, enabling him to get to Washington the following day. A week later, Roosevelt and Knox decided to ease the burden of naval command by removing from the responsibilities of the CNO the post of commander-in-chief, US Fleet. Stark stayed in place, but in their view the new 'CINCUS' could only be Ernest King. (He liked the job but not the bad pun of the acronymic title, which was changed to COMINCH in March 1942.) At that time Roosevelt sent Stark to London to command US naval forces in Europe and gave King his final wish: the addition to his duties of the role of CNO.

King's dispositions for his new and heavy responsibilities included forswearing hard liquor on 31 October 1941, the day a German submarine sank the US destroyer *Reuben James* on convoy duty in the Atlantic – a great shock to the American public, which had no idea how deeply 'neutral' America had become involved in backing Britain. As COMINCH, King took up residence on the converted yacht *Dauntless* in Washington Navy Yard, whence he would visit his family, lodged in the Naval Observatory, on occasional weekends. The always well-groomed

King worked extremely hard at a surprisingly cluttered desk until 4:00 P.M., when he would walk out to points unknown, to reappear for dinner aboard the *Dauntless*. Whenever he needed a longer break, he would simply disappear for two or three days, to the acute embarrassment of his small staff, who had to cover, even if it was the president who wanted him. He was wont to stay with friends on one of two farms, in Maryland or in northern Virginia, to recharge his batteries.

Even Roosevelt and Knox learned to put up with this typical manifestation of King's autocratic behavior. The self-consciously tough naval chief, for whom the five-star rank of fleet admiral was specially created in 1944, had no doubt whatsoever that he was the cleverest man in the US Navy (with the possible exception, as he once uncharacteristically admitted, of Raymond Spruance, whom we shall meet at Midway). This truth being self-evident, he saw no need for modesty, restraint, tact or even holding in check his fearsome temper. He never hesitated to bawl out defenseless subordinates in public and doled out praise in small quantities in private. Only consistent competence could stave off a thunderbolt from the hardest of taskmasters, who believed unshakably that he could do every job in the Navy better than anyone else. He had to delegate, but there was no law to say that he should enjoy it. Yet for those few who gained his trust and respect King could never do enough.

This brilliant, self-indulgent bully was a superb administrator who nevertheless made at least one serious mistake in his conduct of the Navy's war, during which his service took over from the British the domination of the world's oceans which it enjoys to this day. He refused to learn from the Royal Navy in an area where the older service had unique practical knowledge: antisubmarine warfare. The most obvious result was a six-month hemorrhage of Allied shipping on the US Atlantic seaboard because King would not accept that any convoy escort was better than none. He insisted that a weak escort was worse than nothing, the exact opposite of British experience, and left the field free for the Germans to torpedo a vast array of rich and undefended targets. Only when his Army opposite number, General Marshall, demanded action did King develop the requisite sense of urgency and plug the gaping hole in Allied defenses. It was a manifestation of profound incompetence all the more remarkable for being so rare in King's outstanding career. For all his breadth and depth of professional knowledge, Admiral King was a bigot.

Without openly challenging the 'Germany-first' policy of the administration and the alliance, which in naval terms meant 'Atlantic first,' King fought long and hard, and with success, for a greater allocation to the Pacific of resources within the joint war effort from the very beginning. The result was a huge preponderance of American naval commitment in the Pacific throughout the war from 1942. The 'Battle of the Atlantic' against the U-boats was very nearly lost for the lack of a couple of squadrons of long-range US Liberator bombers, the only land-based planes capable of closing the North Atlantic air gap which had become the Germans' favored operational area. For an ex-submarine commander, King once again proved remarkably obtuse about the danger to the entire, worldwide war effort from the attrition by U-boat of the Allied shipping stock, as important in the Pacific as in the Atlantic theater. Or perhaps it was because it was once again the British who wanted him to change his mind, as they had the most to lose from his obduracy.

Be that as it may, King masterminded the Washington end of the Pacific Campaign with intellectual vigor and political skill. By the time he came into his full inheritance in March 1942 as COMINCH and CNO, he had developed the strategy that was to win the war against Japan: hold Hawaii, sustain Australasia, and, as soon as possible, start driving northwestward from the Australia-Fiji line against the Japanese-held islands. He took the view that the Marines should take the islands and be relieved by Army garrisons, an attitude that prompted General Dwight D. Eisenhower, then the Army's chief Pacific planner, to confide to his diary: 'One thing that might help win this war is to get someone to shoot King'! To him it seemed King just wanted the US Army to make the Pacific safe for the Navy. Even so, Marshall went along with King's strategy on the JCS and usually backed him in his victorious fight for the Pacific theater in the Combined Chiefs and other Allied councils. Nimitz, of course, could not manage without the considerable and growing help of the US Army in the Pacific, its assault divisions and air forces as well as its garrisons.

By refusing to concede unity of command in the Pacific unless to an admiral, King bore a major share of the responsibility for the waste of time and effort caused by the two-pronged counterattack against Japan, despite his best efforts to pass the impasse off as a 'whipsaw strategy.'

Unfortunately, the cooperation between King and Marshall at the very top and the backing of the president for both – a triangular relationship based on mutual respect – did not extend uniformly down the chains of command, where interservice rivalry too often loomed larger than the enemy in the thinking of too many officers, if not quite so viciously as in Japan. The Army had to have a supremo of its own, separate from Nimitz, against Japan, in the person of MacArthur, as Supreme Allied Commander, South-West Pacific. But he could hardly manage without considerable naval assistance. And even when the two separate offensives which thus developed were supposed to merge for the final thrust, neither service would give way, so that MacArthur became overall Allied supremo only at the moment the war ended. The United States alone could afford such a fundamental division of forces against such a formidable enemy and still win. Paradoxically, nothing shows up more clearly the folly of the Japanese junta in going to war against America. But it was lucky for the Americans that the Japanese spread themselves so broadly and thinly across the Pacific and Far East. They thus made exactly the same mistake as the Western powers in 1941–42, failing to concentrate their forces to defend their most vital strategic assets and laying themselves open to piecemeal defeat.

One of King's first anxieties as fleet C-in-C was the replacement of Kimmel as Pacific commander. The appointment was, however, taken out of his hands by the C-in-C of the United States, Roosevelt, in a role he was beginning to take increasingly seriously, if not positively to enjoy. Backed by Secretary Knox, the president relied on the same instinct that had led him to choose King, and on the same personal experience of the candidate's qualities at close quarters. It was Nimitz's good luck to be head of the misnamed Bureau of Navigation (i.e. Personnel) at the appropriate moment. As such, it was his responsibility to advise the president on senior naval appointments, a task in which Roosevelt took particular interest from his days as assistant Navy secretary. Nimitz had shown his shrewdness and wisdom to such effect that Roosevelt decided he was the ideal man for the Pacific job. It was a typical 'hunch' appointment and one of his very best. King, it need hardly be said, was unhappy about the choice but would have been no less so about any other man picked for such an exciting and important post. He 'knew' he would do it better himself, if only that were possible. Besides, the naval personnel chief was too ready to listen to others and had to be something of an office politician. But it was these very qualities of tact, flexibility and diplomacy, so conspicuously lacking in King, which helped to make Nimitz the right man to take over and rebuild a shattered formation whose morale was in disarray.

Nimitz was a German by descent on both sides, a second-generation American and a Texan, born in Fredericksburg in 1885. Lacking the money to go to college, the young Chester, like many another in the same position, applied to the Military Academy at West Point. There being no vacancy, he tried for the Navy at Annapolis and was accepted, graduating seventh out of 114 in 1905. Promoted ensign in 1907, he got a very early taste of command on a gunboat in the Philippines. He was only twenty-two and still an ensign on being given command of an old destroyer: his career almost sank without trace when he ran her aground the next year. Though he was reprimanded at his court-martial, his overall record was so remarkable that he still made lieutenant that same year, and never looked back. He moved into submarines, went to Germany (he spoke the language well) to study them, and at twenty-six commanded a division of boats, becoming a technical expert on diesel engines and inventing a system of refueling on the move. He also invented the circular formation for the battle group at sea and even developed ways of integrating the new carriers into the fleet. In the First World War he was engineering staff officer with the Submarine Force of the US Atlantic Fleet in

British and Mediterranean waters. Afterward, among various posts, he oversaw the building of the submarine base at Pearl Harbor in 1922. Rising as required through alternating staff and sea berths, he made rear admiral in 1937 and commanded the First Battleship Division for two years until his appointment to Navigation.

Nimitz had already turned down a tentative proposal to make him C-in-C, US Fleet, early in 1941 on the grounds that he was much too junior. Hindsight reveals this becoming modesty as an unconscious favor to King, but the latter probably saw it as another reason to distrust his gifted subordinate. Nimitz himself wisely deduced that such a huge jump to the top command would make him many enemies, and he was not a man who liked to have any. Even so, there were twenty-eight admirals senior to him when he was given the Pacific. There was also widespread agreement that he had been well chosen.

He could hardly have seemed more different from his irascible boor of a chief. Uninterested in personal publicity, he was pleasant and friendly, always calm, unpretentious, even humble, and led by example. The quiet manner concealed excellent judgment of both men and issues, a first-rate intelligence and an unassuming self-confidence which enabled him to take sometimes startling decisions and stick to them. It did not take him long to appreciate the worth of the decidedly eccentric and casually insubordinate cryptanalyst, Rochefort, now Commander, an assessment which was soon to change the course of history. He worked extremely hard but was able to relax in even the greatest crises by shooting at targets with a .45 automatic while awaiting news from thousands of miles away, or indulging in the old Texan pastime of pitching horseshoes in competition with his staff. Nor was he the kind of boss whom it was best to let win. He took up residence on Makalapa Hill, overlooking Pearl Harbor, in a house with a bunker beneath, and made a point of getting to know every commanding officer under his flag. He lived simply but kept a good table – and a photograph on the wall of MacArthur, his fellow Supreme Allied Commander in the Pacific, as an awful warning against making bombastic statements!

The new commander-in-chief, Pacific Fleet, took the view that 'it was God's mercy that our fleet was in Pearl Harbor on 7 December.' The damage was nothing like as bad as it had appeared in the time of shock. He decided at once that no retreat to California would therefore be necessary, that he would have to keep his headquarters ashore at Pearl because of the unprecedented size of his command, and that there was no need to purge Kimmel's staff. Morale rose at once as it became clear that a cool and coherent intelligence had taken over. Staff conferences were short, crisp and to the point; the admiral was prepared to listen and even to accept expert advice. Nimitz then dug in against King's overanxiety to go on the offensive and raid Japanese Pacific islands with all available forces. He got the plan cut down to a couple of cautious hit-and-run carrier strikes. This was only the first of many clashes with King in which, in his quiet way, he gave as good as he got. King was no respecter of persons, but Nimitz was too tough-minded, in the best sense, to be bullied. When King became CNO as well as COMINCH in the great American and Allied redistribution of commands in March and April 1942, Nimitz added the operational brief of C-in-C, Pacific Ocean Areas (North, Central, and South), or CINCPOA, to his responsibilities.

He also possessed that rare quality, magnanimity in victory. He probably saved the life of Grand-Admiral Karl Dönitz at the Nuremberg trials by sending a statement to the court confirming that the Pacific Fleet had waged unrestricted submarine warfare against Japan from day one of hostilities, just as the Germans did in the Atlantic. All in all, Nimitz has a claim to be considered the most important military leader in the war against Japan, the greatest admiral in American history, and the outstanding naval officer of the Second World War. It seems clear in retrospect that, whereas Nimitz could have done King's job, King could not have done Nimitz's so well. (King retired after the war and declined in health, though he lived on until 1956. Nimitz, also promoted to fleet admiral at the end of 1944, succeeded him as CNO after signing the Japanese instrument of surrender on behalf of the United States. He went on in 1949 to work as a 'goodwill ambassador' for the

United Nations before retiring from public life in 1952. He died in 1966.)

Nimitz was like the personally modest General Marshall in having no obvious character trait or physical peculiarity for journalists and cartoonists to use as a handle (or a stick to beat him with). The same can hardly be said of the third US naval commander to be honored with a fleet admiral's fifth star (in December 1945) – William Frederick Halsey, Jr., born at Elizabeth, New Jersey, in 1882, and nicknamed 'Bull' by the press for the fighting spirit which was his most obvious and valuable quality. Comfortably placed in the middle of his class on graduation at Annapolis in 1904, Halsey served as a commander in destroyers in the Atlantic in the First World War (Navy Cross) and in various posts at sea and ashore thereafter.

The turning point in his career came in 1935 when, at the age of fifty-two, he qualified as a pilot and took command of the carrier *Saratoga*. Halsey may not have been an intellectual luminary, but he knew instinctively where the future of naval power lay. He stayed with naval aviation and in the spring of 1940 headed the Pacific Fleet's carriers in the rank of vice-admiral. It was in this capacity that he came to be at sea with the *Enterprise* when Pearl Harbor was under attack. On arrival in the battered port he said of the perpetrators: 'Before we're through with 'em, the Japanese language will be spoken only in Hell!' This demonstration of the admiral's use of English as a blunt rather than a precision instrument was just what the sailors, who loved him, and the press and the public, who soon learned to do so, wanted to hear. It was the first of a series of caustic and pugnacious remarks which contributed almost as much to morale at home as his dashing, though not always felicitous, exploits at sea. Nimitz could not have chosen a more appropriate leader, a real fighting admiral, to carry out King's order to counterattack as soon as possible.

Halsey was, however, more of a bull at the gate than the kind that wrecks the proverbial china shop. He made his fair share of mistakes, from the linguistic to the strategic, but not out of recklessness or timidity. Impatient of paperwork and of keeping up with essential reading, he was only too ready to delegate staffwork to staff officers, taking the reasonable view that this was what they were there for. He had the self-protective knack of choosing excellent subordinates. The informality of his approach to command and his carelessness worried his senior staff and led to serious errors, but the air crews and the lower deck would do anything for him and probably gave him more than they gave to any other commander. He was always on their side, the very model of a 'sailor's admiral.' (Halsey retired in 1947 and died in 1959.)

Well led though the Pacific Fleet might be as it struggled to get back on its feet, disaster still lay in wait just around the corner – or, on 11 January 1942, under the waves, five hundred miles south of Oahu. Commander Inada Hiroshi, IJN, in the aircraft-carrying submarine *I6*, concluded one hour after sunset that evening that he had the American carrier *Lexington* in his sights and fired a salvo of Type 93 torpedoes at very long range. In fact it was her sister ship *Saratoga* (Captain A. H. Douglas, USN) on patrol in Task Force 14, under the flag of Vice-Admiral Herbert F. Leary. One missile hit the thirty-three-thousand-ton carrier on the port side well below the waterline, killing six men and injuring five, flooding three compartments. No sign of the submarine or the torpedo (its part-oxygen propellant left no bubbles) was seen. The stricken ship was lucky to have been built on the armored hull of a battle cruiser in 1925. The 'thin skin' of later carriers would not have saved her. *Saratoga* took three hours to get back on an even keel by counterflooding and then managed to raise sixteen knots to return to Pearl. TF 14 was broken up, Leary was sent to command ANZAC, and *Saratoga* went to California for repair and modernization.

Vice-Admiral Wilson Brown, commanding TF 11 from *Lexington*, was forced to abandon an air raid on Japanese-occupied Wake Island on 23 January, when Lieutenant Commander

Togami in *I172* sank his oiler, the *Neches*, in the same area.

Having covered a troop convoy to reinforce Samoa, northeast of Fiji, Halsey's Task Force 8 (built round *Enterprise*) and Rear Admiral Frank Jack Fletcher's TF 17 (*Yorktown*) were ordered by Nimitz to attack the Marshall Islands, in the western central Pacific, mandated to Japan after the First World War. It was from there that a Japanese attack on Samoa was most likely to come, if ever. The Americans left Samoa at speed on 31 January and struck half a dozen islands, including Kwajalein, by cruiser and destroyer bombardment and air raids by fighters and torpedo- and dive-bombers. The Japanese attempted to counterattack the *Enterprise* with land-based bombers from island airfields.

Very little damage was done by either side, but the news of the first surface attacks – and the wildly exaggerated damage claims, an easily understandable weakness on both sides – were good for naval and public morale. Halsey became one of the war's earliest American naval heroes and went on to raid Wake on the 24th, again with little measurable effect. He then pushed his luck by launching an air attack on Marcus Island, a thousand miles southeast of Tokyo and two thousand miles west of Hawaii, on 4 March.

Late in February, Wilson Brown and *Lexington* were temporarily attached to ANZAC command to attack the Japanese strategic base at Rabaul, New Britain, in the Bismarck Archipelago east of New Guinea. Approaching from the east on the morning of the 20th, three Japanese patrol aircraft sighted the group 350 miles out. One got away, and in the afternoon the first large 'dogfight' of the war between Japanese and American maritime aircraft took place. Lieutenant Edward 'Butch' O'Hare distinguished himself by shooting down five Kate bombers in his Wildcat fighter; the Americans lost two planes and one pilot. Surprise having been lost, and it being clear that Rabaul would be heavily defended, Brown canceled the raid.

He was given Fletcher's *Yorktown* group for a second

attempt in March, for which King himself issued the superfluous and therefore insulting order, 'Attack enemy.' The target was changed on 8 March in the light of news that the Japanese were attempting to establish themselves in northeastern New Guinea (Papua), on the Huon Gulf, at Lae and Salamaua, where the air bases were. Brown adopted the clever stratagem of attacking from south of New Guinea, over the Owen Stanley mountains, fourteen thousand feet high, and from only fifty miles offshore. The 'over-the-top' raid by 104 planes began at breakfast-time on 10 March. For the loss of one aircraft they sank three Japanese ships, a small result that only briefly incommoded the Japanese operations on the ground.

These first carrier sorties by the Americans were tactical pinpricks but had psychologically important consequences. For the Americans, they boosted morale and provided some much-needed combat practice. They did not inconvenience the Japanese unduly but caused loss of face to some commanders, who had boasted that the Americans would be unable to strike back for months. But there was one more American air strike to come during this period of greatest Japanese success in the first six months of the Eastern war, and for this naval undertaking the help of the US Army Air Force was indispensable.

As the days after America's humiliating baptism of fire turned into weeks, King and his advisers were very conscious of the pressure of public opinion: Where is the Navy? Why don't they *do* something? The war with Japan was barely a month old when COMINCH ordered Captain Donald B. Duncan, his air-operations officer, to examine the possibility of bombing Tokyo in specific retaliation for Pearl Harbor. This could only be done by carrier, but the Navy had no aircraft capable of doing the job, which required longer ranges and heavier payloads than those that carrier aircraft could manage. The raid would have to be launched at least five hundred miles out from the Japanese capital, outside the three-hundred-mile scope of enemy coastal air patrols and the picket line of guard boats known to be on station five hundred miles out. Duncan took three days to draft a plan for a raid by two carriers, one carrying the bombers, the other providing protective cover at sea. The brand-new USS *Hornet* would embark the raiders and the *Enterprise* would cover; Admiral Halsey would command this specially created Task Force 16. (Let it be noted here that the Navy created, dissolved, and merged groups, task forces, and even fleets of ships throughout the war; this admirable flexibility – and the sheer size of the expanded fleet – led to a plethora of designations which the reader is best advised to take on trust!) The most suitable bomber would be the twin-engined Mitchell B-25 flown by the Army. King therefore enlisted the support of General Henry H. 'Hap' Arnold, chief of the Army Air Forces.

It was a hair-raising plan: 'suicidal' is not too strong a word. The planes were too big to be stored below, so would have to stand on the flight deck throughout the approach to Japan. And though launching the B-25, suitably adapted, with catapult assistance might seem hazardous enough, landing one on a heaving deck was out of the question. The aircraft would therefore not return to the *Hornet* at all but would have to fly on to China and hope to land at a field eleven hundred miles from Tokyo, controlled by Chiang Kai-shek's prickly nationalists, who could not be told the reason. The alternative was to ditch. For the aircraft, if not for the two hundred pilots, air crew and maintenance men, the voyage with the *Hornet* would have to be on a one-way ticket. Sixteen B-25s, the most the *Hornet* could carry, were 'borrowed' from the Army's 17th Bombardment Group for an initially unspecified 'special mission,' for which seventy officers and 130 enlisted men nevertheless volunteered.

The commander of the raid itself was Lieutenant Colonel James H. Doolittle, aged forty-five, a businessman who had already been called to active duty from the reserve in 1940, to advise on converting car plants to aircraft production. He was given a month to get his men trained. Lieutenant Henry L. Miller, USN, taught the Army pilots how to take off from a carrier with a load of thirty-one thousand pounds, two thousand over specification, at sixty-five miles per hour, just over half the specified takeoff speed. He did it without going near the sea, measuring out the *Hornet*'s flight deck on the ground at Eglin

Field in Florida. He then installed a catapult and launched them. 'Landing-on' was not taught. Only one plane stalled in training and nobody was hurt. Then all sixteen flew across to San Francisco Bay for loading on the *Hornet*, on Fools' Day 1942. Only well after Captain Marc A. Mitscher took the carrier to sea the next day were the men told why they were there. Talking to the crews by public-address hookup, Halsey said: 'Our destination is Tokyo. We are going to attack Tokyo.' The Army contingent and the entire ship's company erupted in cheers; Navy pilots board were acutely jealous.

TF 16 also included *Enterprise*, with twenty-seven fighters, eighteen torpedo-bombers and thirty-six scouting dive-bombers aboard, as well as four cruisers, eight destroyers and two oilers. On 17 April, a thousand miles from Tokyo, the task force refueled and detached oilers and destroyers for the high-speed run to the launching point, about five hundred miles out. Each Mitchell was loaded with 1,140 gallons of fuel and four five-hundred-pound bombs (except for Doolittle's and three others which took incendiaries instead). On this basis the bombers would just be able to land at the chosen field (which, unbeknown to Doolittle, had not been readied, because of bad weather and Chinese evasiveness: radio silence prevented this rather important fact from reaching the task force).

Halsey planned to launch on the afternoon of Saturday the 18th, so that the B-25s would arrive over Tokyo after dark. Doolittle would go first, to drop his incendiaries on Tokyo as markers for twelve other bombers; the remaining three would drop incendiaries on Nagoya, Osaka, and Kobe to confuse the defense. But in the early hours of the 18th, over seven hundred miles out, the American radar picked up two Japanese ships, on picket duty much farther forward than expected. TF 16 sheered off, only to be sighted at 5:00 A.M. from forty miles by another picket, number twenty-three, the *Nitto Maru*, which got off a clear warning before the cruiser *Nashville* sank her. Halsey, having lost surprise and concerned for his carriers, decided to launch at once, despite the rough weather – from 650 miles out. It would have to be a daylight raid.

Commander Mott, gunnery officer on the *Enterprise*, was among those who looked on in amazement as the B-25s lumbered down the deck of the *Hornet* and staggered into the air at what should have been stalling speed. As the carriers withdrew,

We turned on Radio Tokyo and then a very amusing thing happened. Somehow or other, inadvertently, the Reuters news agency had released the broadcast the day before [sic] that Tokyo had been bombed. I don't know to this day how they ever got this information but they were a little previous with it, and the Japanese broadcaster was talking on at great length about the fact that this was the most laughable joke to the Japanese people and that no foreign bomb or shell had ever landed on Japanese soil and that as he looked around him that day in Tokyo he saw nothing but calm serenity and cherry blossoms. About one half-hour later he suddenly went off the air. It doesn't take much imagination to follow through as to what had happened.

The Japanese had also been partly alerted by signals traffic analysis (they never broke the important US ciphers) that enemy carriers were at sea, despite the American radio silence once the mission had begun. Admiral Yamamoto had therefore ordered the pickets to be specially vigilant and reinforced the very light air defenses around Tokyo. An anti-aircraft exercise took place over the city that morning. The *Nitto Maru* had reported three carriers approaching; it was calculated in Tokyo that the single-engined planes they carried would have to be launched from as close as two hundred miles, implying a raid after dark on Saturday by the time the carriers were close enough. So, by launching two-engined planes from so far out, Halsey achieved tactical, if not strategic, surprise after all: the Mitchells needed less than two hours to make up the distance it would have taken the ships over twelve hours to cover. No sooner was the exercise over than Tokyo was confronted with the real thing, at noon local time, at least eight hours earlier than expected. The Americans had flown in low over the sea to come in at treetop height, climbing only at the last moment to evade blast from their own bombs. A few fighters tried to interfere, to little effect; such anti-aircraft fire as there was did no damage to the raiders.

What with the exercise and the spread of targets, the ordinary population of the city hardly noticed the effect of the bombing. In official circles it was a different story, however. Inevitably, there was loss of face among the braggarts for whom a setback, let alone a defeat, at the hands of a foreign enemy was unimaginable (the underlying reason why Japan was never ready for a war of attrition). Many thought that the bombers must have come from Midway Island as they were 'too big for carriers'; some who knew of the carriers' approach assumed they had a purely diversionary role. As Radio Tokyo's English-language service recovered its voice and made the predictable play on Colonel Doolittle's name, the city's air defenses were hurriedly reorganized. The Army was put in charge of a new anti-aircraft command, and the Navy was ordered to cooperate. Correctly deducing that the raid was, for the time being, an isolated act of defiance and that the Americans were unable permanently to divert ships or planes from their many other commitments, staff officers never did plan a proper air defense system and were thus caught napping a second time, irretrievably, when the Americans took up bombing on a much bigger scale. The most immediate consequence for the Japanese was the temporary diversion of four Army fighter groups urgently needed elsewhere.

Sister Jackson, the naval nurse captured on Guam (see pp. 30–31 twenty-seven), had been in internment in Kobe for about a month when a lone American bomber dropped its incendiaries over the city.

The effect of this raid, whatever it might have been from a military point of view, had quite an effect [sic] on Japanese morale, because just about two weeks previous to that, one of their military spokesmen had gotten up and sounded off to the effect that Japan could not be raided by an outside power, and here was an outside power showering them that day with bombs and showing them it could be done . . . The bombing was certainly a welcome visitor as far as we were concerned.

One of Doolittle's sixteen planes had to make an emergency landing at Soviet Vladivostok on running short of fuel. Allies or no, the crew of five were interned for more than a year until they were able to escape to Iran. The rest got to China.

Four B-25s made crash landings at or near the designated field, where they were identified as hostile, so that the landing lights were switched off. The remaining eleven crews bailed out; five men were killed in the process. Three men from one crew and all five from another were captured on the ground in Japanese-held areas and put on trial for bombing nonmilitary targets (the result of unintended error). They were condemned to death, but the sentence was commuted in five cases; three were executed and one died later in camp. This brought the final death toll from the Doolittle Raid to nine out of eighty – in the circumstances, astonishingly low. Doolittle was awarded the Congressional Medal of Honor.

A fact that is usually overlooked in this context is that the Japanese, suspecting rather more active Chinese collaboration in the raid than the reluctant minimum actually provided, killed 250,000 Chinese people in reprisal – a worse atrocity than Nanking. But this was not the Americans' fault: as ever when dealing with terrorism, it is essential to remember that such crimes are the responsibility of the person who fires the gun.

Task Force 16, having sunk or driven off the sixteen Japanese patrol craft which were the only enemy forces to make contact, returned to Pearl on 25 April, only to be ordered back to sea as quickly as possible to oppose more Japanese threats (see chapter five). But the news of the raid unleashed euphoria in the United States, as it was meant to do, even though the strictest secrecy was (for once, we may interpolate) maintained on the details. Roosevelt, who was not told of the execution of the raid until Doolittle had taken off from *Hornet*, teasingly announced that the bombers had flown from Shangri-La. Halsey described the Army airmen's exploit as 'one of the most courageous deeds in all military history.'

It was in fact a pinprick, albeit a spectacular one. Its most important consequence by far was to remove all serious doubts among the junta about the wisdom of extending the Japanese defense perimeter even farther, instead of settling for what phase I had brought them and consolidating it, as their original strategy demanded. Now they planned to seize the Aleutians to their northeast, Midway to the east, and the Solomon Islands to the southeast. But first they needed to build up their position in eastern New Guinea, as a threat to Australia.

The other weapon immediately to hand when America was taken unawares, as we have briefly seen, was the submarine. The order to 'execute unrestricted warfare' went against American undertakings enshrined in article 22 of the London Naval Treaty of 1930, which imposed Prize Rules on submarines. Unless a potential victim was visibly armed, the boat had to stop and search – which meant throwing away its one advantage of temporary invisibility – and also to ensure that passengers and crew were evacuated with the means to reach safety, all this before a merchant ship belonging to the enemy or carrying goods for him could be sunk. American submarines, like Japanese, were therefore originally envisaged as adjuncts of the battle fleet to reconnoiter and to attack warships from underwater. Prewar developments in depth charges, aircraft and sonar (active sound-detection, known in Britain as Asdic) led to extremes of caution in the American submarine arm and a preference for attack from deep water by sonar. There was no experience of periscope attack by night or by day, still less of the night surface attack so devastatingly perfected by the Germans in two years of destruction in the Atlantic. In reasonable weather in the Pacific, an aircraft could see a submarine down to 125 feet below the surface.

As an example of the cautious mood which prevailed in the US service at the outset, there was a standing order to stay submerged in all areas within five hundred miles of any known enemy air base. Even in the Pacific that was rather inhibiting; among the islands of the southwestern Pacific it was stifling, literally and metaphorically, especially on the older S-class boats, which had no air conditioning. This premium on caution, together with slow peacetime promotions, ensured that many American boats had relatively elderly and decidedly unadventurous commanders in December 1941. Against this, the aggressive call for unrestricted warfare was based on the fact that no Japanese merchant ships were innocent in wartime, because they were all ferrying troops, military supplies or materials looted from captured territory for war purposes. Further, Japan's Nazi ally was waging unrestricted war on the Allied shipping stock, which was strategically indivisible. Finally, of course, Japan had made an unprovoked attack without warning on Pearl Harbor and had thus punched a hole the size of several aircraft-carriers through international law.

Another factor which made American submarines markedly less useful than they might have been, especially at the beginning, was their armament. The S-class boats, of which the Pacific and Asiatic fleets began the war with six each, used the Mark X torpedo of First World War vintage which carried a 500-pound warhead all of 3,500 yards at 36 knots, barely fast enough to catch most Japanese warships. Later submarines used the Mark XIV, developed in 1931. This too was a 21-inch torpedo, but with a 640-pound charge and a range of 4,500 yards at 46 knots (and twice as far at 31, a setting that was seldom used). Compare this with the performance of the Japanese Long Lance already referred to, and it soon becomes clear that the American submariners had a problem on their hands. In fact they had by far the worst torpedoes of any important belligerent, the result of complacency and lack of initiative between the wars. No American torpedo had been tested in real or simulated battle conditions; nor had their exploders (detonators, or 'pistols' as the Europeans called them). The top-secret Mark VI double exploder offering a choice of settings, of magnetic or contact detonation, had never been properly tested either, not even when it became known what troubles the Germans and the British had been having with their detonators in combat. The Japanese, by contrast, had taken the expensive precaution of sinking real ships with live torpedoes during exhaustive testing. However, the size, construction and performance of the US boats themselves, and the comfort of the post-S-class submarines, made them ideal for the kind of submerged cruiser-warfare on merchant shipping practiced by Dönitz in the Atlantic. From about 1940 the American submarines outclassed even the formidable U-boats in size and speed above and below water and in diving speed, maneuverability, and comfort. They would have done better from the beginning had their torpedoes also not run deep – the result of testing them with dummy warheads in peacetime! The extent of the great American torpedo crisis only became clear in mid-1942. How this arose and what was done about it will be described later.

A strict and heavy training schedule in 1940–41 in preparation for combat caught most Pacific Fleet boats unready for sea on 7 December 1941; only three of fourteen were able to set out as soon as they could be made ready in the postraid chaos of Pearl Harbor, *Gudgeon* first, then *Pollack* on 13 December and *Plunger* the next day, all to the Inland Sea. The first boat to fire a shot in anger was *Tautog* (all later US boats bore the names of ever more obscure fish) during the Pearl raid: she claimed to have downed a Japanese plane by machine-gun fire from her dry dock. Soon eight submarines from the West Coast and elsewhere joined the force led by COMSUBPAC (commander, submarines, Pacific Fleet). The Atlantic Fleet sent eleven more S-boats, while the first of thirty-six oceangoing submarines to be added to the Pacific Fleet during 1942 began to arrive from the mainland. By New Year's Day seven boats were on patrol between Hawaii and Japan.

But the honor of being the first US submarine to sink a merchantman (as distinct from a man-of-war or another submarine) went to a boat attached to Hart's Asiatic Fleet. On 15 December 1941, USS *Swordfish* (Lieutenant Commander C. C. Smith, USN) sank the *Atsutasan Maru*, 8,663 tons, off the island of Hainan, south of China and west of the Philippines (all Japanese merchantmen bore the generic title *maru*, meaning, roughly, a self-contained entity). *S38* sank another *maru* in Lingayen Gulf on the 22nd, one of the few noteworthy submarine blows against the Japanese invasions.

The Asiatic Fleet had twenty-nine boats at its disposal on the outbreak of hostilities, including six S-boats. It had more than twice as many submarines as the Pacific Fleet, thanks to the belated policy of strengthening the defenses of the Philippines begun in the summer of 1941. Sadly, they had been unable to contribute much in this defensive role because of the weaknesses already outlined. After Manila was abandoned at Christmas 1941 (together with 230 precious torpedoes) the boats withdrew and, although collectively styled Task Force 3, Asiatic Fleet, were soon divided into two commands, one based on Surabaya in the Dutch East Indies and the other at Darwin, Australia. When those harbors were abandoned in February 1942, the two groups went respectively to Fremantle, Western Australia, and Brisbane in Queensland, on the other coast; they remained there under command of the Seventh Fleet, formed to support MacArthur as Supreme Allied Commander, South-West Pacific, when this post was created for him.

But there was constant interchange between the boats of COMSUBSOWESPAC and the Pacific Fleet, under which all submarines in the entire ocean were grouped for administrative purposes. Submarines from both commands began running clandestine missions to and from the Philippines from January 1942 until the fall of Corregidor in May, delivering ammunition and vital equipment and taking off people and assets. Soon a routine of sixty-day oceangoing patrols was established, with two weeks' shore leave between patrols. Those fortunate enough to spend the latter period in Hawaii had the run of the Royal Hawaiian Hotel, at the famous Waikiki Beach, to relax in: the entire luxury establishment was rented for the benefit of submariners, echoing the treatment given by the Germans to theirs in occupied France.

Yet there is no getting away from the fact that the opening moves by American submarines were ineffectual in the Pacific command and incompetent in the Asiatic/South-West Pacific, where little tactical intelligence and no air reconnaissance were available to help the boats in their defensive role. Shallow and constricted waters gave them little room for maneuver and added considerably to their difficulties. The main reason for the overall failure was inexperience; the US Navy learned to sail submarines in the First World War but not to fight in them: they saw virtually no combat and did not practice war on commerce. The naval Bureau of Ordnance, one of whose responsibilities was developing torpedoes, was an organization of breathtaking narrowness of vision which saw nothing odd in producing weapons without testing them. It also distrusted new ideas and refused to own up to errors even when confronted with them (loss of face was never an exclusively oriental fear). The known hesitancy of some skippers led commanders in distant bases to ascribe the mounting reports of torpedo failures to the men rather than to the weak weapon with which they had been sent, themselves untried, into one of the most stressful forms of warfare.

Small wonder, perhaps, that Commander Morton C. Mumma, Jr., cracked up six days into the war, when his *Sailfish* was depth-charged off the Philippines by two Japanese destroyers. The boat was regarded as jinxed by her crew; originally named *Squalus*, she had sunk with the loss of two dozen men in 1939 and had been salvaged and refitted. Mumma himself was a strict disciplinarian, but under enemy attack he handed over command to his executive officer and had himself locked in his own stateroom. The boat got away unscathed. As so often happens in these cases, Mumma was credited (wrongly) with sinking a destroyer – and was awarded the Navy Cross, the highest decoration available short of the Congressional Medal of Honor. This sad story does not appear in the official US-submarine operational history but was discovered by Clay Blair, author of the definitive and exhaustive history, *Silent Victory*. Mumma's breakdown was one of the most dramatic of a series of failures of command which crucially hampered the effectiveness of American submarines in the months when they were most needed for defensive purposes. In the case of the Asiatic Fleet submarines, Blair takes the bleak view that training, maintenance, strategy and tactics were all as poor as the torpedoes before the war, even after it became clear that conflict was imminent. And although the operational history objectively concludes that the service became 'the world's greatest submarine force' (shades of Karl Dönitz!), it does not cover up the weaknesses – on the contrary – and therefore constitutes a usable historical source, including lengthy sections on torpedoes and 'fatal and near-fatal errors.' We shall have to draw upon it again.

To return to events on the surface, the overwhelming success of phase I of the Japanese plan of southward expansion tempted the junta to defer the scheduled second phase, of consolidation, and take the expansion a significant stage further. The loss of face caused by the Doolittle Raid, which a moment's reflection would have identified as a gallantly desperate throw by an enemy still trying to get back on his feet, only stoked up the spirit of untrammeled, mindless aggression behind the Japanese war effort. It was as if their leaders – including Yamamoto, who was at least as responsible as anyone else for the imminent attempt by the Japanese to overreach themselves – did not exist unless they were on the attack: I strike, therefore I am.

The Imperial Japanese Navy in the spring of 1942 was markedly superior in numbers and more often than not in quality of individual ships to the US Pacific Fleet: ten carriers (six large), ten battleships, thirty-five cruisers, and 110 destroyers against four, four, twenty-four, and ninety respectively, with about forty-four oceangoing submarines each. So far, the Navy had carried everything before it; so had the Army; and between them they had established air superiority over the entire sprawling eastern theater of the worldwide conflict. Was this not the moment to strike again, to profit from an advantage which would otherwise become a wasting asset as the United States built up its strength – before the 'sleeping giant' was fully awake?

Not really, because Japan was already overextended. Psychologically it was in the position of the child whose eyes are bigger than its stomach – wanting to swallow more out of sheer greed but unable to digest it. 'That was a fatal error by a nation which, unlike Germany, fully understood the role of seapower but had underestimated the war potential and fighting spirit of her enemy,' was the reaction of Admiral Thomas C. Kinkaid, the cruiser commander who was to become chief of the Seventh Fleet, in an unpublished memoir. The notorious Japanese opportunism was on the point of undoing the stupendous gains of the first few months by offering the Americans a whole series of chances to counterattack which would not have arisen had the conquerors stayed within the strong perimeter gained by their opening conquests.

No increase in the overall number of troops of their Southern

Army group was either contemplated or possible, given the dragging commitment in China. Even so, IGHQ changed its plans and identified four more objectives. Troops were to take Port Moresby in Papua New Guinea, to expand the Japanese position there and in the Bismarck Archipelago to the east; the Navy was to capture Midway, the US outpost in the middle of the Pacific, to extend the perimeter and force a decisive battle with the US Fleet; the Aleutian Islands, off Alaska, were to be invaded to establish a position in the North Pacific; and the New Caledonia, Fiji, and Samoa groups of islands were to be occupied to cut off Australia as the rearward base for an Allied counterattack. Only then would the promised consolidation take place. Moves subsidiary to this hastily spatchcocked extra program of conquest included the bombing of Darwin on 19 February, the raid on the Indian Ocean early in April, seaplane raids on Hawaii planned for May, and the dashing miniature-submarine raid on Sydney harbor in Australia at the end of that month. Australia, mobilizing as fast as she could, reasonably took the view that she was also on the menu of conquest – an event the Commonwealth was determined would occur only over the dead bodies of her stubborn soldiers in New Guinea.

Of all the territories in South and East Asia ruled by Europeans until the end of 1941, only India, heart of the British Empire, remained out of Japanese hands at the beginning of March 1942. Noting the discomfiture of the British, Indian nationalists were on the verge of mounting a 'Quit India' campaign, despite London's abortive attempts to buy them off with promises of political jam tomorrow. Britain had not foreseen a Japanese threat to the 'jewel in the crown' of the Empire and had been using the volunteer Indian Army as a source of trained reinforcements for the Middle East and North Africa. Wavell returned from ABDA to Delhi as C-in-C, India. General Alexander, commander of the last stage of the Dunkirk retreat in 1940, arrived in March to take command in Burma – just in time to supervise another withdrawal, by the British Army to India, conducted with stubborn brilliance on the ground (nine hundred miles of appalling jungle) by Lieutenant General Sir William Slim, who would eventually be widely recognized as Britain's best field-commander of the century. Like MacArthur, but without boasting about it, he would return.

The American and British governments, specifically Roosevelt and Churchill, realized that a more radical, worldwide shakeup of commands was urgently required against the Axis, which was calling the shots on almost all fronts. Chiang Kai-shek became Supreme Allied Commander in China, partly because he was the sitting tenant of an area from which Allied forces (Chinese manpower and US officers plus American airpower) might strike back at Japan, and partly because Washington was concerned to build up China as a great power (for after as well as during the war, to supplant Japan and balance the Soviet Union in Asia). General Joseph W. Stilwell, USA, was appointed chief of the Allied staff advising Chiang. 'Vinegar Joe' was also to wear the hats of commander of US forces in the China-Burma-India (CBI) theater, regional administrator of Lend-Lease aid and operational chief of the planned Sino-American intervention in Burma from the north, in collaboration with the British from the west. Personality clashes and shortcomings on both sides ensured that the collaboration was never easy.

On 9 March, the day Java surrendered, Roosevelt put forward, in a long letter to Churchill, a proposal to divide the world into three areas: the US would take sole charge of the Pacific, the British would take on the Indian Ocean and the Mediterranean, while the two allies shared the Atlantic on existing lines, shoring up China and Russia as best they could. This was broadly how it was done, without consulting Australia and New Zealand, which quite reasonably took the view that they should have a say in a strategy on which their very survival depended. Washington's answer to this objection was the Pacific War Council, on which the two Australasian states, Canada, China and the Dutch and French governments-in-exile joined American and British representatives. It was a face-saver with no power to influence events. A parallel council was set up in London but did not last long: it soon proved as irrelevant as Britain's reservation of the right to return to the Pacific when it could (which depended almost wholly on Lend-Lease supplies from the United States). Even before the proposal went to Churchill, the US presumed acceptance and made arrangements to go it alone in the Pacific.

Unable to resist the US Army's demand for a supreme command alongside the Navy's obvious lien on one, and unable also to agree on a single supremo for its western front against Japan, Washington, as we have noted, set up two supreme 'Allied' commands. Nimitz in the Pacific (soon subdivided into the North, Central, and South Pacific Ocean areas) and MacArthur in what was specially termed for the occasion the South-West Pacific: the Dutch East Indies (minus Sumatra, left in the British Indian Ocean sphere) plus the Philippines, Borneo, New Guinea and Australia. By the end of March 1942 the US Army had already diverted or specially sent eighty thousand troops to Australia and hundreds of aircraft to Australasia to protect this important investment, much larger at this stage than its commitment in Europe – too large to be left in the care of an admiral. The 159th eastern line of longitude parallel, which separated the two commands, passed through the middle of the Solomon Islands, an Anglo-Australian archipelago. French New Caledonia, British Fiji and Samoa and the Dominion of New Zealand all fell within the US Navy's South Pacific area.

A key factor in the creation of SOWESPAC was the personality of MacArthur himself, who thus made sure to benefit from his vicarious last stand in the Philippines, the only area on the map of conquest that had not fallen into Japanese hands within weeks, days, or hours of being attacked. Since the Japanese did not provide for setbacks of any kind in their war planning, the Philippines sideshow was much more of an irritant and a loss of face to them than it needed to be, an attitude that played into American hands: if they were so upset, surely it must hurt, and if it hurt it must be important. MacArthur's public-relations skills were such as to make his defeat by numerically inferior forces an even bigger moral victory for the US than the extraordinary escape from a superior enemy at Dunkirk had been for the merely perverse British. MacArthur could hardly be awarded the Medal of Honor and then put on the scrap heap. The administration's need, and 'public opinion's' (meaning the media's) demand, for heroes in the first major war readily accessible to electronics, combined with MacArthur's insatiable ambition to make his appointment inevitable. Having made their hero, they were stuck with him. There were also party-political considerations: a senior general with so many influential Republican friends could not be sidelined at a time when president and people required national unity against the enemy.

The appointment also brought the advantage of indefinitely deferring a settlement of the struggle between Army and Navy for ultimate supremacy in commanding the war against Japan. Sheer weight of numbers would give the Army dominance within America's role in the European theater, where the scale of the US contribution in turn gave Washington control over the West's entire campaign against Germany, the chief enemy. The geography of the war against Japan would favor the Navy, for which MacArthur's campaign was to be little more than a sometimes useful, sometimes distracting diversion on the left flank. So who and what was this MacArthur factor which already loomed so large in the administration's calculations?

As is so often the case, notably but not exclusively in the United States, yesterday's unquestioned hero is today's discredited reject. To be drawn too deeply into the cont- inuing debate on MacArthur in a general account of the Pacific Campaign would be to fall into a trap that could have been laid by the supreme Allied confidence man himself: it is a controversy perhaps second only in fascination to the Pearl Harbor debate – but it is much less important. On the other hand, it is suspiciously easy to dismiss MacArthur as a loud noise with a clay pipe at one end and feet of the same substance at the other – if only because the pipe was a corncob.

Douglas MacArthur was born at the Arsenal base at Little Rock, Arkansas, on 26 January 1880. His father, Arthur

MacArthur, was an Army officer who won the Congressional Medal of Honor for leading a charge up Missionary Ridge at Chattanooga in the Civil War and further distinction against the Indians; he went on to lead a brigade in the Philippines in the Spanish-American War, became governor general of the islands in 1900, and the US Army's most senior general from 1906 to 1909. Douglas, urged on by his vicariously ambitious mother, lost no time in seeking to outbid such an eminent example in every particular. He came out top in the West Point entrance examination and was top of his year there for three years out of four, until he joined the Corps of Engineers as a second lieutenant in 1903. He was allowed to serve as his father's aide in 1904–05, when MacArthur senior was sent to observe the Russo-Japanese War and afterward to tour the Far East. After various postings to the Philippines, Central America, and the US, MacArthur junior was promoted to colonel soon after the US entered the First World War and went to France on the staff of the 42nd 'Rainbow' Division in October 1917. Even though a Medal of Honor eluded him at this stage, he showed much personal bravery and dash, winning the Silver Star (the first of his seven in this war, a record), the French Croix de Guerre, and other decorations: he was eventually to receive the Distinguished Service Cross twice, the Distinguished Service Medal, and countless foreign awards. He won the Purple Heart twice for wounds and rose to the rank of acting major general on 10 November 1918 – the day before the fighting stopped. But he had been the youngest American brigadier general of the conflict despite annoying General Pershing by removing the wire stiffening from his officer's cap – subsequently one of his most famous trademarks – sporting a yellow scarf and a bright sweater in the front line. In this he was no worse than the American General Patton with his pearl-handled pistols or the British Montgomery with his two-badge beret in the Second World War: generals should be known to their men, and this is an easy way to do it. MacArthur was also endowed with a strong populist streak which never ceased to surprise those who knew him in his earlier days.

Nobody questioned his courage, but there was criticism of his expensive methods, even in the general slaughterhouse of the western front, where uninspired generals hurled the flower of their countries' young men into the mud in frontal assaults on fortified positions because they could think of nothing better to do. Pershing, the American Expeditionary Force commander, stifled an inquiry, describing MacArthur as his best leader of troops, whom he intended to promote. Unfortunately for MacArthur's ambition the Armistice supervened. He had to settle for reverting to colonel and becoming the youngest-ever superintendent of West Point – which he transformed in what may have been his most solid achievement as an officer. Then he went back to the Philippines as Manila district commandant, and later as commander of the entire archipelago, being promoted to the substantive rank of major general in 1925. He sat that year on the court-martial of Brigadier General William 'Billy' Mitchell, America's prophet of strategic airpower (he sank a battleship in trials as early as 1921), and concurred in the aviator's conviction for flagrant and persistent insubordination. Typically, MacArthur was dead-set against military aviation until he realized the potential of the B-17 for his own position in the Philippines in 1941. In 1928 MacArthur made an unusual departure for a serving officer by accepting the post of head of the American Olympic Committee, which brought his promotional talents into play and his name before the public. President Herbert Hoover made him America's youngest-ever chief of staff of the Army, in the acting rank of full general, in 1930.

The top job was a bed of nails in the Depression, when most of MacArthur's energies were absorbed in a desperate struggle, decisively lost, to maintain the minimal strength of the Army through endless budgetary cutbacks. Faced with the choice between paying the men and buying modern equipment like tanks and aircraft, he chose the former, but even so was forced to preside over the nadir of the United States Army as a standing force.

Two incidents stand out from what should have been the climax of MacArthur's already considerable career, in which he had surely matched if not surpassed his father.

The first occurred shortly after his appointment, in July 1930. A protest movement of unemployed veterans called the

'Bonus Expeditionary Force' was formed to demand immediate payment of a war bonus promised by Congress for many years later. Some fifteen thousand people, including families, camped out peaceably enough on the Anacostia Flats, on the far side of the Eleventh Street Bridge from the government district of Washington. Left-wing agitators succeeded in provoking a skirmish with the police, which gave the embarrassed administration an excuse to send in troops to restore order. MacArthur responded with indecent haste and gusto, calling up infantry, cavalry and even light tanks, and leading them, on horseback, against the three thousand or so who had marched unarmed over the river. He chased them along Pennsylvania Avenue between the White House and the Capitol. Ignoring two explicit orders from the president, MacArthur pursued the demonstrators back over the river and cleared their encampment, subsequently destroyed in a fire which lit up the night sky all over Washington. Then he called a press conference at which he proudly announced that there had been no serious injuries. This was two dozen cases short of the truth, including a baby who died from tear gas.

The second incident could have been terminal for a career already publicly called into question by the first. When Roosevelt succeeded Hoover, he decided the 1934 military budget had to be cut even further in favor of projects to revive the economy. MacArthur admits in his own *Reminiscences* – the only source – that he succumbed to a temper tantrum and seriously insulted the president, who bawled him out. His offer to resign was unaccountably rejected, and he was kept on for an extra year until 1935. Yet Roosevelt had long since formed the opinion that MacArthur, with his outsize ego, was one of the two most dangerous men in the United States (the other being the ultrapopulist Louisiana state politician Huey Long). MacArthur had the presence and the aura to play – had already played at Anacostia – the role of 'the man on horseback.' This was a euphemism for the strong leader yearned for by a lot of authoritarian Americans in all walks of life, from Congress to commerce and from the communications industry to the Catholic church. It was an age when so many, left and right and worldwide, were attracted by totalitarian solutions to intractable problems. Roosevelt concluded that it was safer to have MacArthur on the inside helping out than on the outside horning in. The general even went along with the use of the Army in various New Deal relief and training schemes. It is not hard to imagine the positive reaction in Washington when MacArthur, whom even the preternaturally tolerant Marshall described as 'conspicuously temperamental,' went back to his beloved Philippines for the next six years as military adviser to President Manuel Quezon, his friend since 1903.

In March 1941 MacArthur, already sixty-one and claiming to have trained twelve Philippine divisions, offered his services to the administration. In July, when Washington prematurely concluded that Japan was on the point of going to war, he was restored to the active list as a major general (later lieutenant general and, in December 1941, full general) and US Army commander in the Philippines and Far East. By this time he was field marshal in the Philippine Army, a position reflected by the gold braid on his unstiffened cap – and the cool half-million dollars he accepted from Quezon on 3 January 1942. This mind-boggling fact was suppressed by all concerned, including Roosevelt, until 1979, when Ms. Carol M. Petillo unearthed it while researching a doctorate. The circumstances have never been explained, any more than has MacArthur's supine conduct between hearing of the raid on Pearl Harbor and losing his air force on the ground many hours later.

He left Corregidor with his second wife, their four-year-old son, Arthur MacArthur (the absorbing delight of his old age), a nurse and seventeen staff aboard four PT boats on 12 March 1942, for Mindanao, and reached Australia by B-17 five days later, having promised the forlorn Philippines garrison help he knew would not be coming. At a press conference on arrival, he promised dramatically, 'I shall return,' an undertaking that was to prove spectacularly expensive to honor. He also announced that he had been chosen to lead the American offensive against Japan, which was news to the Navy and Roosevelt alike: the president had decided on 22 February, the day he ordered MacArthur to leave the Philippines, to appoint him to defend Australia. Marshall nevertheless went ahead and personally drafted the citation for his Medal of Honor, formally approved on 25 March. When, amid a shower of more and more outlandish adulation back home, the National Father's Day Committee anointed him 'number one father of 1942,' MacArthur expressed the hope that when he was dead little Arthur would remember him by saying, 'Our Father who art in Heaven.' As Roosevelt had feared nine years earlier, there were calls for him to be made C-in-C of all US forces, president, even dictator. In April 1942 he was confirmed as Supreme Allied Commander, South-West Pacific. A long war lay ahead of him; but he had already passed into American mythology.

5
The Turn of the Tide

Even though Japanese war-plan 'scenarios' were habitually 'best-case' and not 'worst-case,' and even though the brave troglodytes of Corregidor were still holding out in their tunnels, the junta was so surprised by the scale and speed of its successes after four months of war in the Pacific that it was reduced, early in April 1942, to wondering what to do for an encore. Anticipating warship casualties of a good 25 percent, for example, the triumphant admirals had lost barely one-tenth of that (26,400 tons, twenty-three vessels, nothing bigger than a destroyer); merchant shipping losses of over three hundred thousand tons (under 6 percent) were offset by shipping captured in overrun Allied harbors. What was thought in advance to require six months had only needed three, and the forecast that it would need another six months to get the captured oil fields back on stream after the inevitable enemy sabotage was also proving unduly pessimistic.

As usual, Navy and Army pulled in opposite directions in the debate, which began early in February; also as usual, within each service there were two passionately differing schools of thought. In the Army the ambitions of many opportunists were once again directed at the Soviet Union, which had almost gone under to the German invasion of June 1941; the Russians had managed to stem the tide in their terrible winter, but surely the Germans would launch the final blow in the spring. Given the existence of the Axis, which for reasons of mutual mistrust and racial disdain never developed into a true military alliance, this would undoubtedly have been the correct strategy. Cooler heads at IGHQ (Army) pointed out that there was a not inconsiderable item of unfinished business, still known to all concerned as the 'China Incident,' to resolve before hurling Army groups into Soviet Asia.

The Navy, having advanced southward with such success,

was torn between turning northwest across the Indian Ocean, perhaps even to link up with the Germans somewhere in the Middle East, or southeast to neutralize or even invade Australia, the obvious enemy springboard for a counterattack. Overall, the Army came down on the side of caution and the consolidation of existing gains laid down in phase II of the original war plan. The Navy wanted to improve on the perfection of phase I by extending the perimeter of occupation even farther – the Aleutians, Midway/Hawaii, Fiji, Samoa, New Caledonia, New Guinea, Australia and the Indian Ocean – then to omit phase II as superfluous and go straight into the aggressive defense laid down in phase III until the US lost the will to fight.

The result was an irresolute, multifaceted compromise. There was to be no advance across the Indian Ocean, but there would be a Pearl-style raid to keep the British on the defensive, duly done in April by Nagumo's carriers. There would be no invasion of Australia, because it was too big and too far away, but it could be cut off by the capture of the New Caledonian, Fijian, and Samoan archipelagoes. The defensive perimeter could reasonably be extended by adding the western Aleutians and Midway to the chain of occupied islands; if this brought the US Fleet out for a decisive battle, the Combined Fleet would be happy to oblige, said its chief, Admiral Yamamoto. And finally there was Port Moresby in (Australian-administered) eastern New Guinea, defended by just one Australian Army brigade: from there and from Tulagi, at the southern end of the Solomon Islands, northern and eastern Australia could be attacked by land-based bombers as often as necessary. The Japanese landed on 8 March at Lae and Salamaua on Huon Gulf, on the northern side of the Papuan peninsula and separated from Port Moresby on the south side by the Owen Stanley range. This was intended to back up their occupation of the strategic base of Rabaul on the island of New Britain east of Huon. Admiral Wilson Brown's two carriers bombed but did not seriously incommode them from over the mountains; neither the Australians nor MacArthur felt able to spare the resources necessary to support a force significantly larger than the brigade in place in Moresby since January, even though intelligence made it clear that the Japanese had designs on it.

The design drawn up by Yamamoto was code-named Operation MO. The main objectives were Moresby and Tulagi, in order to threaten Australia; subsidiary targets were the phosphate-rich Nauru and Ocean Islands, on the Equator between the Marshalls and the Solomons, for the benefit of Japanese agriculture. Because Nagumo's four carriers were resting while at the same time Operations MI (Midway) and AL (Aleutians) were in preparation (spurred on by Doolittle), MO would be executed by a relatively modest naval force. Vice-Admiral Inouye Shigeyoshi of the Fourth Fleet, at Rabaul, was in charge, supported by some 150 land-based aircraft and seaplanes plus the carriers *Zuikaku* and *Shokaku* with two cruisers and six destroyers. These units were to cover separate invasion forces for Moresby and Tulagi, the former, from Rabaul, rather larger than the latter, which set out from Truk in the Caroline Islands. The invasion groups had their own escorting forces, including the light carrier *Shoho*; the plan as a whole reflected the Japanese taste for intricacy of design and depended too heavily on complicated timetabling, leaving too little margin for the unfathomable effects of enemy countermoves.

There were bound to be some. American Army and Royal Australian Air Force planes were constantly patrolling the entire Coral Sea area, between the east coast of Queensland and New Guinea and beyond. Old-fashioned spies in New Britain reported the gathering of forces at Rabaul; Commander Rochefort was reading the electronic runes and predicting a thrust from there; Nimitz concluded by 29 April that the blow would fall on or about 3 May. To counter it, he assigned Task Forces 11 (*Lexington*, Rear Admiral Aubrey W. Fitch having relieved Wilson Brown) and 17 (*Yorktown*, Rear Admiral Frank Jack Fletcher) from Hawaii and New Caledonia respectively. Vice-Admiral Herbert F. Leary, now commanding MacArthur's naval forces (assembled from Asiatic Fleet and ANZAC elements), sent Task Force 44, of one American and two Australian cruisers led by the seconded British Rear Admiral J. C. Crace, RN, from Sydney and Nouméa, New Caledonia. The coming clash would occur on the boundary between MacArthur's South-West and Nimitz's South Pacific Ocean areas. Admiral Halsey's double

Task Force 16 had returned from the Doolittle Raid to Pearl on 25 April; he would be sent with his two carriers as soon as possible, even though the Coral Sea was thirty-five hundred miles away, in case the Japanese were delayed. Fletcher would command unless Halsey arrived in time: the orders were simply to check any 'further advance by the enemy in the New Guinea-Solomons area.'

Fitch joined Fletcher at the eastern end of the Coral Sea, about 250 miles west of the New Hebrides, at dawn on 1 May. The Americans refueled from tankers at their leisure, Fletcher's force first. Fitch did not think he would be ready until the 4th, so Fletcher sailed northwestward on the evening of the 2nd. He was topping up his fuel five hundred miles south of the Solomons on the morning of the 3rd when, unbeknown to him, the Japanese started landing on Tulagi. Under radio silence, Fitch was also topping up a hundred miles from Fletcher but was unaware of the latter's precise position or the Japanese operation. Only at 7:00 P.M. did intelligence from MacArthur's air patrols about the Tulagi invasion reach Fletcher by radio. He raced north with the *Yorktown* group without waiting for Fitch, sending oiler *Neosho* to meet him on the 4th at the rendezvous chosen on the 2nd, with orders to join up three hundred miles south of Tulagi at dawn on 5 May.

But Fletcher continued at high speed, intent on attacking the Japanese before they could settle in. He reached a point about a hundred miles southwest of Tulagi in the early hours of 4 May. The Japanese, overconfident after all their easy landings up to now, withdrew most of their covering forces as soon as their troops were ashore and prepared to send the Moresby invasion convoy to sea on the evening of the 4th. That morning Fitch found the *Neosho* and was joined by Crace's cruisers, whereupon *Lexington* and the rest put about for the new rendezvous. This had the effect of sending them farther away from Fletcher instead of toward him. Fletcher, meanwhile, crept closer to Tulagi through bad weather and launched twenty-eight dive-bombers and twelve torpedo-bombers against the Japanese landing force, holding back his eighteen fighters to protect the carrier. The three attacking squadrons went in independently after 8:00 A.M., damaging a destroyer and sinking three minesweepers. Their second attack destroyed two seaplanes and did a little damage to shipping. One American plane was lost, and two were forced down. A third attack by the dive-bombers sank four landing craft. Fletcher, believing he had done rather more damage than was actually the case, thanks to the usual inflationary reports from the pilots, withdrew toward Fitch.

The Japanese now knew that at least one US carrier was in the vicinity; their own had been lurking north of the Solomons, out of reach of US patrols. Vice-Admiral Takagi Takeo brought his two fleet carriers racing south in response – belatedly, fortunately for Fletcher, because they too had been refueling – to the call for help from Tulagi.

The Americans linked up and reorganized themselves, shooting down a land-based Japanese reconnaissance plane from Rabaul on the morning of the 5th while the Japanese bombers from there staged a 'softening-up' raid on Port Moresby: the invasion fleet was already on its way there. B-17s from Moresby found and attacked the light carrier *Shoho* in the northern Solomons but missed. Task Force 11 was subsumed into Task Force 17 so that the two carriers could operate as a single group under the tactical command of Fitch, the more experienced airman. Each side was looking for the other, but with only limited and fleeting success. Neither got enough information from air patrols to establish a clear picture of the other's whereabouts. The Japanese pressed on regardless; the Americans waited for something to happen. The opposed carrier forces passed within seventy miles of each other unawares as the US garrison on Corregidor finally surrendered on 6 May.

Contact came at last on the morning of the 7th, when Takagi's patrol-planes found the flat-topped *Neosho* and the destroyer USS *Sims* ten miles from the Japanese ships and reported them as a carrier and a cruiser. The opposed forces were now some seven hundred miles south of the Solomons. Rear Admiral Hara Tadaichi, in tactical command of the two Japanese carriers, launched a strike. High-level bombers missed, but dive-bombers soon blew up the *Sims*, killing all but sixteen hands. Within minutes the *Neosho* was hit seven times by dive-bombers but somehow stayed afloat, dead in the water but blown westward by the wind, her position wrongly reported by her navigator. She was found and dispatched by a US destroyer on the 11th, and 127 men were rescued.

Fletcher, meanwhile, at dawn on the 7th, sent Crace's three cruisers and two US destroyers ahead toward the eastern end of New Guinea to attack the Port Moresby invasion convoy as it rounded the point. This was unwise as it weakened Fletcher's screen while exposing Crace to attack without air cover. This is precisely what happened after lunch on the 7th, when waves of Japanese planes came in. Sharp and energetic maneuvering and anti-aircraft fire prevented the forty-two attackers from obtaining a single hit. Shortly afterward the destroyer USS *Farragut* came under attack from three bombers – of the US Army Air Force based in Queensland. Fortunately they too missed. Crace sailed on until he heard at midnight that the Japanese invaders had turned back, whereupon he did the same and went back to Australia for a well-earned rest.

Fletcher himself launched search planes which reported 'two carriers and four cruisers' to the north at breakfast-time. Both his carriers launched two-thirds of their aircraft for a strike two hundred miles off. When the scouts returned, they corrected their report to two cruisers and two destroyers; even that was wrong. What they had glimpsed was two old light cruisers and a few armed merchantmen. The Japanese had meanwhile located Fletcher; at 9:00 A.M. on the 7th Admiral Inouye was prompted by Crace's swift advance to rein in the Moresby assault force until the divided opposing force had been seen off by his own carriers and supporting forces. That was as far as the invaders ever got.

On the way to the wrong target, the American pilots sighted a carrier, four cruisers and destroyers off the Louisiades, the scatter of islands immediately southeast of New Guinea. This was light carrier *Shoho*, which promptly came under attack from ninety-three planes and sank shortly after 11:30 A.M. Only three US planes were lost. The rest of Rear Admiral Goto's covering group withdrew to the northeast. Fletcher still had no idea where the Japanese heavy carriers were; however, they had not yet pinpointed him either.

Even so, the carrier planes found one another and engaged

in a series of dogfights in poor weather; nine Japanese and two Americans were downed as darkness fell. So close to each other were the two carrier groups that the *Yorktown* was caught unawares after dark as she turned on her landing lights to help returning aircraft. Three planes came in from starboard, flashing their signal lamps, to which the *Yorktown* cheerfully flashed back. Then a voice came over the Tannoy system of an accompanying destroyer: 'Have any of our planes got rounded wingtips?' Someone on the *Yorktown*'s bridge answered, 'Wait.' On the destroyer the voice said, 'Damned if those are our planes,' and a hail of anti-aircraft fire passed across the *Yorktown*'s bow. The Japanese pilots sheered off; the leader was only 250 yards short of the flight deck. The *Yorktown* was ready when three more Japanese pilots made the same error shortly afterward and shot one down.

Fletcher and Inouye, cruising up and down about a hundred miles apart, the Americans to the west, each contemplated and rejected a night attack, preferring to await a new day and sailing north during the night. On the morning of the 8th the opposing carriers at last pinpointed each other by reconnaissance and launched attacks of equal strength in opposite directions: 122 Americans, 121 Japanese. The Americans were in clear weather and open water, the Japanese in a convenient patch of bad weather. Thirty-nine planes from the *Yorktown* attacked the *Shokaku* in the first American strike against a Japanese fleet carrier. The torpedoes, launched prematurely, missed completely, and the bombs scored two hits, starting a fire, preventing the Japanese vessel from launching but not from recovering aircraft. Only twenty-one planes from the *Lexington* found the *Shokaku*, scoring one hit; altogether forty-three US planes were shot down by Zero fighters and anti-aircraft fire. The Japanese carrier put out her own fire and made for home with middling damage, having lost 108 men.

Experience tells; and the Japanese attack scored an early hit on the *Yorktown* with an eight-hundred-pound bomb and two more on the *Lexington* (Captain Frederick C. Sherman, USN), which also came under a heavy torpedo attack. The converted battle cruiser fought on as damage control tried to put out the fires and repair the damage caused by internal blast. But leaking fuel fumes eventually ignited and set off an uncontrollable fire and chain of explosions on the unlucky 'Lady Lex,' crippling her beyond salvage; American destroyers finished her that evening after a ten-hour struggle to save her. Only 216 officers and men were lost from a total complement of 2,951. The *Yorktown* (Captain Elliott Buckmaster) had sixty-six fatal or serious casualties; the one enemy bomb she took had passed through three decks. The resulting fire was quickly extinguished, and flight operations were not affected. Fighting ceased before noon.

So the final score in the Battle of the Coral Sea looked bad for the Americans. One large carrier lost and one damaged out of four available in the Pacific, one oiler and one destroyer sunk to one Japanese small carrier, one destroyer and several small craft sunk, and one fleet carrier out of six damaged. But the Japanese not only left the scene of battle, assuming from pilots' reports that Task Force 17 was destroyed; Admiral Inouye also cautiously put off the invasion of Port Moresby until 3 July, 'another day' that never came. The Japanese Navy never did 'come round the corner' of eastern New Guinea. Its southward advance had been halted. Yamamoto bypassed Inouye to order Takagi directly to finish off the remaining US ships, which had withdrawn in the evening of the 8th on orders from Nimitz.

Nimitz was already thinking ahead to the next battle. The Japanese plan to take Nauru and Ocean Islands was abandoned when a US submarine sank the flagship and Japanese air reconnaissance sighted Halsey's carriers coming west from Pearl Harbor (they arrived too late to take part in the Coral Sea clash and were immediately ordered to return at top speed). Both sides made ludicrously inflated claims. What lends the outcome of the fight for the Coral Sea such great interest psychologically is that such light losses for such an apparently aggressive belligerent as the Imperial Japanese Navy caused it to recoil so sharply. Indeed, in most people's terms it had won. One is driven to conclude that the admirals were hoist on the myth of their own invincibility, to such an extent that the loss, after so much unrelieved triumph, of one fairly important ship – *Shoho* – was not an acceptable misfortune of war but a serious loss of face – despite their having budgeted for ship casualties of 20 to 30 percent!

Coral Sea was a tactical defeat for the Americans, because as the weaker side they lost more men, ships, and aircraft. But it was a strategic defeat for the Japanese, who felt obliged, for the first time since they had opened their southern front, to halt a major advance and to change their plans. That such a drastic reaction was not called for, given the scale of action, is immaterial: the victors behaved as if they had been beaten, and therefore they were beaten. In this sense the first confrontation between carriers at sea, in which neither fleet hove in sight (let alone gun range) of the other, bears an uncanny resemblance to the Battle of Jutland, twenty-six years earlier, when the British and German battle fleets clashed, amid blunders on both sides, in their only major action of the First World War. On that occasion the British lost numerically and the Germans claimed a victory; but the German battle fleet never seriously challenged the Royal Navy again in the war, effectively surrendering to the British blockade, which was the key to victory on the western front. There is thus a powerful case for arguing that the Coral Sea, not the much bigger and more dramatic clash exactly four weeks later at Midway, was the real turning point of the Pacific Campaign.

Admiral Inouye and others may have been overaffected by the loss of the *Shoho* in the first carrier-battle with the Americans, but Yamamoto reasonably regarded the Coral Sea as no more than a setback to be corrected later. He saw no reason to change or delay by a single moment his intricate plans for Operations MI and AL just because MO had not been a complete success. One of his objectives was to remedy the most obvious shortcoming of the Pearl Harbor attack, the failure to catch the American carriers. That this was a serious omission in urgent need of rectification was proved by the Doolittle Raid. The sinking of the *Lexington* in the Coral Sea was an important step in the right direction. With the *Saratoga* still out of action because of torpedo damage, the Americans were for the time being reduced to what they had had on 7 December – three fleet carriers in the Pacific, one of which, *Yorktown*, was damaged. There was no prospect this time of immediate reinforcement. Yamamoto wanted an all-out fleet action in which Japan's superiority in carriers and battleships would complete the discomfiture of the US Navy and leave Japan secure behind its defensive perimeter of islands.

To get his 'decisive battle,' he felt that a serious attack on one of the few American possessions left unconquered in the Pacific was necessary. This would be sure to bring the US Fleet racing to the rescue, straight into a trap to be laid by the bulk of the Combined Fleet. Although he thought the capture of Midway and the Aleutians would usefully extend the perimeter, this was probably secondary in Yamamoto's mind to the fleet action he hoped a move against them would provoke. The Combined Fleet would assemble from bases all over the Japanese-dominated western Pacific. Operation AL was to be a diversion for MI, to which by far the greater forces would be committed. If the simultaneous attack on the Aleutians persuaded the Americans to divide their inferior forces, they would be even weaker at Midway; if it did not, they would still be weaker than the Combined Fleet. Originally Yamamoto planned to use all six fleet carriers for the MI operation, but as a concession to the general Army-Navy compromise he detached two to cover MO. Even if *Shokaku* had not been damaged, the Battle of the Coral Sea would have prevented both of these from being ready in time for Midway, since they had first to get home, replenish, and sail halfway across the Pacific.

The first straw in the wind indicating that the Japanese were thinking of a strike at the Hawaiian Archipelago, which includes Midway, was noticed in the first week of March. Two Japanese 'Emily' four-engined flying boats made an ineffectual bombing attack on Honolulu. A week later, another was caught on its way to repeat the job by Marine fighters from Midway and shot down. There was something of a mystery for the Americans about how these aircraft could have managed the long flight from the nearest Japanese seaplane base at Wotje in the Marshall Islands. They deduced that the big planes must have been refueling from submarines in the sheltered waters provided by the French Frigate Shoal, a rocky outcrop five hundred miles northwest of Pearl. US patrol boats were sent to mine the waters and deter submarines; the Japanese airborne probes, which were more concerned with reconnoitering the whereabouts of the US Fleet than bombing, duly stopped. Midway itself, an atoll six miles across with a lagoon and two islets used as an air-and-seaplane base, was briefly bombarded on 7 December 1941, and was reinforced by the Americans a fortnight later. Fighters drove off three Japanese submarines which briefly attempted to bombard the atoll in the ensuing weeks. American submarines meanwhile took to refueling at Midway.

To work as intended, Yamamoto's plan needed at least tactical and preferably strategic surprise. This was to be denied to him by the alertness and efficiency of American Naval Intelligence, which picked up the first whisper of new, large-scale fleet operations across the Pacific before the end of March. Early in April the Americans concluded that the Japanese were preparing something very large and complex indeed. The cryptanalysts at Op-20-G (Pacific) in Washington and Joseph Rochefort's listening post in Hawaii – which was officially an outstation of Op-20-G not answerable to Nimitz – agreed on this but differed on the likely Japanese objectives. Admiral King backed the judgment of the Washington team; Nimitz and his fleet intelligence officer, Commander Edwin T. Layton, USN, had learned to rely on the casually brilliant, unruly Rochefort. King could not bring himself to trust his most important subordinate, Nimitz – the man who could 'lose the war in an afternoon,' as the British Admiral Jellicoe could have done at Jutland – so it is not surprising that he would not accept Nimitz's evaluation of the available intelligence. It was fairly clear by late April that Japanese naval attacks on Port Moresby, possibly the southern Solomons, and an unidentified third objective (Nauru and Ocean Islands) were imminent. On 29 April, therefore, Nimitz bravely committed his immediately available strategic forces – the two carriers under Fletcher – to frustrate the first Japanese moves, but at the worrying cost of one sunk and one damaged. And the Japanese had taken one of their three MO objectives, Tulagi, a place the Americans had been eyeing since March but had not felt ready to seize and hold.

Sigint, however, revealed that some weeks would pass before the continuing Japanese naval preparations led to another, even bigger move on the vast Pacific board. The operational order was issued as early as 5 May, three days before the climax of the haphazard Coral Sea clash. It was intercepted at once by the Americans, who had mastered no more than a third of the Japanese codes, even though they had penetrated the current JN25 cipher. It took more than a week to unravel the text, but the coded references within it were still not clear in mid-May, when Nimitz warned King that the Japanese could be preparing to attack Hawaii again, or even the West Coast, at the beginning of June (Washington thought the blow would fall in the middle of June). Particularly puzzling was the significance of the letters 'AF,' obviously, from the context, referring to a place – but which? References to bad weather and other clues enabled the Americans to guess that 'AO' stood for the Aleutians, which were to be the target of a much smaller attack at the same time as AF. Two days later, CINCPAC was confidently forecasting a major attack on Midway. Those forty-eight hours marked the apotheosis of Joseph Rochefort as the inspiration behind the greatest single strategic triumph in the history of both signals intelligence and the United States Navy.

Admiral King disliked Rochefort personally, which comes as no surprise. Unbiddable and irrepressible, the intuitive cryptanalyst was a permanent challenge to all naval orthodoxy. Despite Rochefort's known record in educated guesswork, King asked him only once for his advice, early in April 1942. What, he asked, would the Japanese Navy do in the medium term on the basis of existing communications intelligence? Rochefort confidently replied that the attack on the Indian Ocean was the end, not the beginning, of Japanese naval activity there; that there would be no invasion of Australia; that a large operation was planned from Rabaul (MO); and that there would be some kind of major strike in the Pacific somewhat later (MI). Rochefort would soon cap even this magnificent feat and blow away the last shreds of mistrust which had attached to signals intelligence since the Pearl Harbor 'failure' unfairly attributed to it.

He worked out a scheme to settle the disputed identity of 'AF' once and for all, developed from an idea put forward by one of his brightest assistants, Lieutenant Commander Jasper Holmes, USN. Holmes had been to Midway and knew that there was no natural freshwater supply on the island, eleven hundred miles northwest of Pearl Harbor, at the other end of the Hawaiian chain. The garrison had to get its water by evaporating and recondensing seawater to get rid of the salt. A call for extra water would therefore ring true to a Japanese eavesdropper. There was a completely secure under-water telephone-cable between Pearl Harbor and Midway. This could be used to order the garrison to broadcast by radio, in plain English, a spurious request for emergency water supplies by tanker. More cunningly still, a detailed message about an alleged explosion in the water-filtration plant was to be sent as a follow-up, in a low-grade cipher which the Americans knew the Japanese had broken with the help of material they captured on Wake. Midway duly sent out the false messages, and the Japanese listening posts picked them up and retransmitted them to their own sigint people. They reported shortly afterward to IGHQ and the fleet that 'AF is short of water' or words to that effect, making sure that the invasion force took extra supplies. Eavesdropping on the enemy eavesdroppers, the Americans now knew beyond any possibility of doubt that AF stood for Midway. The American ruse had proved to be a masterstroke.

This was more than enough for Nimitz, who boldly decided to back Rochefort's team to the hilt by staking the entire available American carrier strength of three on his judgment. Halsey was summoned back from the South Pacific, where he had been since missing the Coral Sea engagement (the only important carrier-action of the entire war in which his flagship, USS *Enterprise*, did not take part). Fletcher was also called home with the damaged but full-powered *Yorktown*. Because he knew almost the entire Japanese order of battle from the intercepted operational directive, Nimitz deduced that the simultaneous attack on the Aleutians was intended wholly or partly as a diversion. He therefore sent a modest North Pacific Force of five cruisers and ten destroyers – the most he could spare – under Rear Admiral Robert A. Theobald to cover them. Nimitz even warned him to expect invasion of the outlying islands of Attu and Kiska, but Theobald ignored his master's voice and stuck to his belief that the Japanese would go for Dutch Harbor, just off the southwestern tip of Alaska. He was therefore caught in the wrong place when the Japanese arrived on 3 June.

Yamamoto's overall plan was to establish a new forward-defense line from the Aleutians through Midway, Wake, the Marshalls, Gilberts and southern Solomons to Port Moresby. Proper air patrols would ensure that no American force passed this line unseen. From mid-May the Japanese were also hard at work planning the seizure of New Caledonia, Fiji, and Samoa to cut off Australia, but not to such an extent that this was allowed to weaken the fleet, which was made up of the following components. Admiral Nagumo was back in harness with his Mobile Force of four fleet carriers (his flagship *Akagi*, *Kaga*, *Shiryu* and *Hiryu*) protected by two battleships, two heavy cruisers and a dozen destroyers. The Midway occupation force consisted of twelve transports, minesweepers, heavy cruisers and a covering group of two battleships, four cruisers and destroyers, all commanded by Vice-Admiral Kondo Nobutake. The First Fleet, which was to act as distant cover for both operations, MI and AL, was commanded by Yamamoto himself in the monster *Yamato*, with six other battleships, one light and two seaplane carriers, two light cruisers and a dozen destroyers. An 'advance expeditionary force' of sixteen submarines was to scout ahead of the attackers. Finally there was the Northern Area Force for the Aleutian operation, a double convoy (one segment for each island to be taken) covered by two light carriers, two heavy cruisers and destroyers. Yamamoto hoped to achieve surprise by assembling his massive force from a proliferation of ports according to a timetable complicated enough to have been a challenge to the German General Staff on the western front in 1914.

The end result was the greatest concentration of naval tonnage since the British battle fleet at Jutland, the largest assemblage of seapower ever to sail under the Japanese flag, the biggest yet seen in the Pacific Ocean and the most powerful in all history. For Nimitz to think in terms of 'setting a trap' for this most modern of the world's navies can be seen as impertinent, but that is what he set out to do. *Yorktown* (nucleus of Rear Admiral Frank Fletcher's Task Force 17) was diagnosed as needing three weeks for repair but was turned round in three days – thanks to herculean dockyard labor at Pearl Harbor, inspired by the frequent presence of Nimitz himself. To his eternal disappointment, Halsey was diagnosed on his return as having severe dermatitis. He was obliged to relinquish the seagoing command to Rear Admiral Raymond Spruance of Task Force 16 (*Enterprise* and *Hornet*). Task Force 17 was protected by two cruisers and six destroyers, Task Force 16 by six and nine respectively. For the first and only time in the wartime Pacific command, United States submarines were assigned a defensive role, with twelve screening Midway, four Oahu and three others on call. On Midway Island, Nimitz could and would call upon thirty-two naval Catalina flying boats and six torpedo-bombers plus twenty-seven fighters and twenty-seven dive-bombers of the Second Marine Air Wing. The atoll was America's immovable but unsinkable fourth aircraft-carrier for the coming battle. Spruance assembled the three mobile 'flat-tops' with their escorts some 325 miles northeast of the island on 1 June, Task Force 16 having left Pearl on 28 May and Fletcher two days later with the patched-up *Yorktown*. The trap was set.

Fittingly enough the Midway air garrison got the first sight of the fleet that was intending to overwhelm the island. Ensign Jack Reid, USN, was flying his lumbering Catalina about seven hundred miles west of Midway on the morning of 3 June, 1942, when he saw a formation of eleven ships heading east at nearly twenty knots. He tracked it for several hours. It was a major part of the Japanese Occupation Force. Captain Cyril T. Simard, USN, commanding Midway, immediately mobilized part of his heaviest armament, the nineteen B-17 Flying Fortresses 'on loan' from VII Army Air Force (supported by four B-26 Marauder medium bombers). Nine B-17s found the convoy 570 miles off and bombed it from twelve thousand feet, claiming hits on four ships but missing them all. Four Catalinas, each carrying one torpedo, came in next, flying low and hitting an oiler which nevertheless managed to sail on into the gathering dusk. At this moment the American carriers were over three hundred miles east of Midway, and the Japanese four hundred miles west of the Americans. Admiral Fletcher, in tactical command because senior to Spruance, planned to be two hundred miles north of Midway at dawn to attack the enemy, who was expected by intelligence to launch a raid on the atoll from the northwest at that time.

The unbearable itch which had removed the dashing and impetuous Halsey from his admiral's bridge put the best tactician and perhaps the best mind of the US Navy in command of the fate of the United States on the 180th day of the war in the Pacific. As will shortly be shown, he was called upon to take charge for the decisive phase of the battle. Raymond Ames Spruance, sensitive, honest and self-effacing and one of the fastest thinkers in uniform, was born in Baltimore, Maryland, in 1886 and rose in the Navy by sheer ability and dedication, despite a most unhelpful tendency to seasickness. He had no private money, no personal enemies, no friends in high places – and no need to burn the midnight oil because no task detained him long. Endowed in equal measure with physical and moral courage, he had a pronounced fatalism that helped him to stay supernaturally cool and calm in any crisis under fire. He could delegate and did so willingly, spent as little time as possible at work, and took special care of his health at all times, making sure he got all the sleep he needed. He was no less concerned to avoid casualties. Yet he missed nothing, being always ready to intervene with one of his famously concise commands at the crucial moment. Hating publicity, speechmaking and paperwork, he was nevertheless a past master of 'clear thought, clearly expressed,' and had no equal in the service for staffwork. As we have seen, even Admiral King recognized that Spruance was more than a match for himself. Spruance did not shrink from criticizing his own country's war policy, notably the internment of ethnic Japanese

US citizens in concentration camps and the indiscriminate bombing of Japan.

But for the moment Fletcher was giving the orders from *Yorktown*, calling for airborne sweep-searches from each carrier in turn before daybreak. Once again it was the Catalina fliers from Midway who located the approaching enemy, at 5:34 A.M. They saw two carriers and a swarm of planes already heading for their island. Fletcher told Spruance to take *Enterprise* and *Hornet* southwestward and launch an attack, promising to follow when *Yorktown* had recovered her scouts. In fact all four Japanese carriers had put up a total of 108 planes to attack Midway – thirty-six torpedo-bombers and similar numbers of dive-bombers and protecting Zero fighters. By 6:30 A.M. the bombs were falling and the hopelessly outnumbered and outclassed Marine fighters were involved in dogfights all over the sky.

Among the mass of preparation and detail Nimitz had to contend with in planning the trap for the Japanese at Midway, the Pacific C-in-C remembered to send a film unit to the atoll. It was directed by a certain Commander John Ford, USNR, who had made his name before the war with the Western *Stagecoach*.

My eyes were sort of distracted by the . . . leader of the [Japanese] squadron, who dove down to about five thousand feet, did some maneuver and then dove for the airport [he recalled afterward]. We have all heard stories about this fellow who flew up the [seaplane] ramp on his back, but it was actually true. He dove down about one hundred feet from the ground, turned over on his back and proceeded leisurely flying upside-down over the ramp. Everybody was amazed, nobody fired at him, until suddenly some Marine said, 'What the hell,' let go at him and then shot him down. He slid off into the sea.

I was close to the hangar and I was lined up on it with my camera, figuring it would be one of the first things they got . . . The whole thing went up. I was knocked unconscious . . . I did manage to get the picture. You may have seen it in the *Battle of Midway* . . . You can see one big chunk coming for the camera.

Of the twenty-six Marine planes that went up, seventeen were shot down and seven badly damaged. It was all over by seven-fifteen, and Midway was not attacked again. How many planes the Japanese lost over the island is unknown because of the result of the battle as a whole; probably few. But they did little serious damage to the ground installations. Nimitz, meanwhile, from his command post at Pearl, had already ordered Captain Simard to send his six naval Avenger and four Army Marauder torpedo-bombers against the enemy carriers. It was the first American strike, and it was an almost unmitigated disaster.

I believe about ten torpedoes were actually fired against the *Hiryu* [said Commander Kawaguchi Susumi, the carrier's air officer]. We received no hits from any of them . . . Those torpedoes were very slow; they seemed to surface, go down again, surface, go down again. For some reason those torpedoes didn't seem to have any speed at all. There was one occasion when a torpedo came toward us on the surface. We hit it with a machine gun and blew it up.

The small American wave of torpedo-bombers lacked not only an effective main armament but also any form of fighter protection; they were shredded by the Japanese fighters and anti-aircraft barrage. One Avenger and two Marauders, all damaged, managed to get back to Midway. They scored no hit but unwittingly contributed to the fateful moment upon which the battle hinged. Lieutenant Tomonaga of flagship *Akagi*, who led the attack on Midway, had called in to recommend a second strike on the atoll. The ill-fated torpedo attack ten minutes later supported his judgment, as the twin-engined Marauders could only have come from Midway, not a carrier. Admiral Nagumo decided to follow the lieutenant's advice. His carriers had planes on deck armed with torpedoes, ready to attack any enemy surface ships; he cleared the decks to receive the aircraft returning from Midway and ordered the waiting planes to swap their torpedoes for bombs for a second strike at the island base. As this work, which normally took a full hour, was in hand, a Japanese reconnaissance aircraft sent in the first sighting report of the American fleet – ten surface ships heading southeast from the north. Halfway through the rearming, therefore, Nagumo ordered the process stopped. Those planes still loaded with torpedoes were to keep them while he waited impatiently for more information from the scout. Total confusion prevailed below; the Japanese decks remained empty, awaiting the arrival of Tomonaga's pilots. It was only then that the scout reported the enemy force as including at least one carrier.

Meanwhile Midway had sent in another wave, this time Major Lofton Henderson's sixteen Marine dive-bombers, forcing the carriers to maneuver sharply. Most of the American pilots were new to the job, and the major decided on a gliding descent on the targets rather than a powered dive. The Japanese fighter cover soon shot down eight and badly damaged six; only eight managed to get back to Midway, where just two proved salvageable. The gallant Henderson was among the dead. Once again there were no hits by the Americans.

Their third attack was mounted by Lieutenant Colonel Walter C. Sweeney's formation of fifteen B-17s, sent to make a second strike on the transport convoy but diverted by Captain Simard. They dropped more than sixty tons of bombs from the safe (but also completely ineffectual) height of twenty thousand feet, hit nothing, were hit by nothing, and returned to make seriously exaggerated claims. It was not yet obvious that there could be no such thing as precision-bombing from such an altitude (a fact that was still true in the 1970s over Indochina and in 1982 over the Falkland Islands). The top-secret Norden bombsight (one of the few major technological advances the Americans refused to share with the British in 1940) was an excellent piece of equipment with a built-in precomputer, but it was not capable of the kind of miracle achieved in the Gulf War in 1991.

Strike number four was delivered by the Marines from Midway in eleven lumbering Vindicator bombers commanded by Major Benjamin Norris. Despite being pounced on by a cloud of Zeros, ten managed to lob their bombs at the battleship *Haruna* and nine got back to base. But the United States forces had still not managed to scratch the paint of a single Japanese ship despite serious air losses. Half the Midway air garrison had been destroyed. The submarine USS *Nautilus* had managed to fire one of her substandard torpedoes at a battleship despite heavy depth-charging by the Japanese. She too missed.

*   *   *

Spruance, meanwhile, with *Enterprise* and *Hornet*, was planning just after dawn on 4 June to launch an attack on the Japanese carriers, which he was approaching from the northeast. Reports of the attack on Midway led him to bring forward the launch by two hours, which meant dangerously overtaxing the limited fuel carried by his aviators. But it should also have meant that his planes would catch the Japanese while they were refueling. What it actually meant was that the American carrier pilots struck during the chaos below decks when Nagumo's other aircraft were having their armament changed. As the American raiders were forming up over their own task force, the Japanese seaplane that had been tracking the American carriers saw what was going on and warned Nagumo. Spruance was aware that he was under enemy observation but nevertheless decided to press ahead with an attack by every aircraft he could spare. Fletcher in *Yorktown*, on his way to rejoin Spruance, decided to launch most of his, which included some planes and pilots transferred from the damaged *Saratoga*. The weather was calm and clear with a light wind. The Japanese carriers were still pressing forward toward Midway to recover their first wave and launch a second, sailing in a box formation, *Akagi* leading *Kaga* by a mile, *Hiryu* leading *Soryu* similarly three miles to port, all inside a circular screen formed by two battleships, three heavy cruisers and eleven destroyers.

The first wave was recovered by about 9:15 A.M., and the Japanese strike force turned northeast toward the US carriers, replenishing the returned aircraft with fuel and weapons as it went. The change of course caused Lieutenant Commander Samuel Mitchell's thirty-five SBD dive-bombers and twenty-seven attendant Wildcat fighters from *Hornet* to miss the enemy carriers and the battle. All the fighters were lost for lack of fuel and their pilots had to be fished out of the sea; thirteen of the bombers reached Midway (with the loss on landing of two more of the original twenty-two). Lieutenant Commander John Waldron's fifteen low-flying TBD-1 torpedo-bombers were the only element of *Hornet*'s air group left in the battle. They had no fighter protection, having lost touch with Mitchell's high-flying formation because of a large patch of cloud. If they pressed on now and found the enemy, theirs would be tantamount to a suicide mission. They did, at 9:30 A.M., and it was. Not one came back. Zeros and a hail of anti-aircraft fire from the Japanese screen brought them all down. The Japanese fleet was still 100 percent intact. Of the thirty air crew, only Ensign George Gay, of Torpedo Squadron 8, aged twenty-five, from Waco, Texas, survived.

We had had no previous combat flying . . . The Zeros . . . turned out against us in full strength. It's been a very general opinion that the anti-aircraft fire shot our boys down, and that's not true . . . The Zeros shot down every one of them . . . I think we made a couple of grave mistakes. In the first place if we'd only had one fighter with us I think our troubles would have been very much less. We picked up, on the way in, a [Japanese scouting] cruiser plane . . . It tracked us and I know gave away our position and course and speed . . . I know that they knew we were coming. [Commander Waldron] put us in a long scouting line, which I thought was a mistake at the time.

I was just lucky. I'll never understand why I was the only one that came back but it turned out that way, and I want to be sure that the men that didn't come back get the credit for the work that they did. They followed Commander Waldron without batting an eye. I trusted him very well . . .

So I flew right down the gun barrels, pulled up on the port side [of the Japanese flagship *Akagi*], did a flipper-turn right by the [carrier's superstructure] island. I could see the little Jap captain up there jumping up and down raising hell and I thought about wishing I had a .45 so that I could take a pot-shot at him . . . but then I dropped right back down on the deck and flew aft looking at those airplanes . . . I had a thought right in a split second there, to crash into those planes.

But Ensign Gay was attacked by five Zeros and forced to ditch, lightly wounded in an arm and a leg and unable to rescue his rear gunner, manning the aircraft's single medium machine gun. Aboard the *Akagi* his abortive attack was closely watched by Rear Admiral Kusaka Ryunosuke, Nagumo's chief of staff from before the Pearl Harbor raid.

I was very busy on the bridge, taking evasive action from the torpedo tracks approaching the ship itself from other planes, when it was suddenly brought to my attention [that] this one torpedo plane was approaching very close without having dropped its torpedo. I watched it and . . . it appeared to me that it couldn't avoid hitting the bridge, and all those standing on the bridge thought they were done for. I am not certain why the plane missed the bridge; it may have been the movement of the ship or it may have been the change of course of the plane, but in any case it went by very close to the bridge and very low, and I thought that it had probably flown on past the ship, but the lookout said that it had crashed on the other side.

Admiral Kusaka also survived, to take up the quieter pursuit of farming after the war.

The fourteen Devastator (TBD-1) torpedo-bombers from *Enterprise* led by Lieutenant Commander Eugene F. Lindsey had no better luck attacking the *Kaga*. Only four escaped being shot down, few launched their torpedoes, and none scored a hit. Lieutenant Commander Lance Massey's thirteen Devastators were next to appear in the shooting gallery over and around the Japanese carriers; their attack on the *Soryu* also had no effect, with seven being shot down before they could launch their missile and three of the five that managed to do so also being destroyed. The total score for the forty-one old TBDs from the three US carriers was forty-one attacks, thirty-five losses – and no hits. But they had kept the Japanese carriers twisting and turning, preventing them from launching any more planes to join those already in the air. They also commanded the almost total and deadly attention of the Japanese fighter 'cap' (combat air patrol). This was of radical advantage tactically to the American Dauntless SBD dive-bombers when their turn came, a few minutes after the last tragic failure of the torpedo planes.

Lieutenant Commander Clarence McClusky, leader of the *Enterprise*'s air group, arrived over the four Japanese carriers at the head of two squadrons of thirty-six Dauntless dive-bombers shortly after 10:00 A.M. Lieutenant James Gray's Wildcat fighter squadron from the same carrier had already got there but had not intervened in the previous torpedo-bomber fiasco. It had latched on to the *Hornet*'s torpedo planes by mistake and therefore never got the call for support it was expecting from its *Enterprise* colleagues, who were, as we saw, almost wiped out. Gray broke radio silence at about 10:00 A.M. to report the position of the Japanese carriers, thus giving Fletcher and Spruance their first hard news of the main enemy. Gray's twenty-seven planes were also running out of fuel. The neat box of four carriers had long since been broken up by the constant, if thus far ineffectual, American attacks, leaving the *Hiryu* four miles clear to the north of the other three, which formed an inverted triangle two miles across. *Akagi* and *Soryu* were in line abreast with the *Kaga* behind them to the south.

The Japanese fighter 'cap' was still in disarray and at low altitude after its murderous attack on the torpedo-bombers, leaving the sky clear for the *Enterprise* dive-bombers to divide and attack *Akagi* and *Kaga* simultaneously – just as Lieutenant Commander Maxwell Leslie led seventeen SBDs from *Yorktown* in a diving attack on the *Soryu*.

The *Akagi* took two bomb hits at 10:26, one of which caused havoc and huge explosions on the flight deck amid forty reloading planes. The other bored into the hangar below, where the torpedo magazine blew up. Rear Admiral Kusaka had virtually to overpower his distraught chief, Admiral Nagumo, and chase him off the blazing carrier onto a destroyer for transfer to the light cruiser *Nagara*, where he spent the rest of the battle. A desperate fight to save the Mobile Force's flagship went on for nearly nine hours, until Captain Aoki Taijiro, befittingly the last to leave, abandoned her at 7:15 P.M.

The dive-bombers attacked my ship [he said later] while we were still taking evasive action from the torpedo-bombers. We were unable to avoid the dive-bombers because we were so occupied in avoiding the torpedo attacks . . . At one time there were so many splashes round the *Soryu* that I couldn't see the ship at all and thought, maybe it has been sunk. . . . We only received two hits . . . *Akagi* was sunk by torpedoes from a Japanese destroyer early next morning because, as a result of the two hits, the whole ship was on fire . . . We never saw the *Hiryu* again after the engagement.

The *Kaga* was hit four times fair and square. One hit destroyed the bridge; the others set off refueling aircraft on deck and more below, where gasoline and bombs exploded, turning the carrier into an inferno within a few seconds.

The *Kaga* was attacked by about thirty dive-bombers which approached the ship from the port side [said Flight Commander Amagi Takahira]. Four hits were scored, the first one landing just forward of the island – and that caused so much damage to the island itself, and there was so much debris flying around, that I didn't see clearly where the other hits fell . . . Huge fires were started on the flight deck and in the hangar, so that it became impossible to see other ships in the formation. For about twenty minutes we tried to put out the fire but it became impossible, so I abandoned ship, and once I was in the water it seemed to me that I could see all four carriers burning . . .

While I was swimming in the water, very near the burning *Kaga,* I saw a submarine periscope come up above the surface, and I saw a torpedo fired at the *Kaga* . . . [It] bounced off the side of the ship and circled slightly, after which the head of the torpedo dropped off and sank, although the body of the torpedo remained floating near me . . . Several of our sailors clung to the floating after-part . . . The *Kaga* was burning fiercely from stem to stern, and the anti-aircraft batteries were firing from induced explosion of their magazines. Even the paint on the side of the hull was burning . . .

It sank in the evening, after sunset . . . As I was watching there was a tremendous explosion . . . probably from the gasoline tanks or the magazine.

Commander Ohara, the *Soryu*'s executive officer, recollected how *Yorktown* pilots scored three direct, thousand-pound bomb hits on his ship in three minutes.

The *Soryu* was attacked first from the starboard bow by three dive-bombers and by a second wave of four dive-bombers off the starboard quarter and by a third wave of perhaps five dive-bombers off the port bow. These waves came at intervals of thirty seconds to one minute and three hits were scored. The first hit was scored amidships; the second hit came just forward of the bridge on the starboard side; and the third hit was on the . . . aftersection of the flight deck . . . Our planes were being made ready for a second sortie [against Midway Island] and were all lined up on the flight deck ready to take off. The planes in the hangar below decks were loaded with bombs and fuel, ready to be brought to the flight deck, so that the first bomb started all these planes burning. The bombs loaded aboard the planes went off one by one by induced explosion . . . Due to the large fires aboard ship, the large gasoline storage area in the stern of the ship exploded, leaving only the bow afloat, after which the gasoline storage tanks in the forward part of the ship exploded and sank the ship.

Three and a half hours after the dive-bomber attack, when

she was burning less fiercely, *Soryu* came under fire from USS *Nautilus* (Lieutenant Commander W. H. Brookman, Jr., USN). This boat, from the bulkiest class of submarines in the Navy, fired three torpedoes into the blazing hulk and was severely but unsuccessfully depth-charged for her contribution to the big carrier's death-agony, which ended when she broke in two shortly after 7:15 P.M., a few minutes before the *Kaga* went down.

Watching all this from the water was Ensign Gay of the *Hornet*, kept afloat by a cushion from his lost torpedo-bomber and hanging on for dear life to the rubber dinghy he would inflate after dark to await rescue by flying boat.

The [Japanese] carriers during the day resembled a very large oil-field fire . . . The fire coming out of the forward and after end looked like a blowtorch, just roaring white flame and the oil burning . . . Billowing big red flames belched out of this black smoke . . . and I was sitting in the water hollering Hooray, hooray!

With all the luck in the world on their side, the American dive-bombers had knocked out three Japanese fleet carriers in as many minutes. Only the *Hiryu* remained in contention, a few miles to the north. The US carriers were untouched but had lost many planes: the *Hornet* thirty-two, *Enterprise* twenty-five, *Yorktown* seventeen. Lieutenant Commander John S. Thach, leading *Yorktown*'s VF3 squadron of twenty-five Wildcat fighters, was able to give Admiral Fletcher the first eyewitness report of the stupendous damage to the enemy when he returned to his carrier. He had been able to keep down losses to the superior Japanese Zeros by employing the 'Thach Weave' he invented – a system whereby pairs of Wildcats would cover each other by flying a synchronized crisscross pattern, so that if one came under attack the other would swing in and counterattack.

Tactical command of the Mobile Force, including all Japanese aircraft still in play, had been passed by Nagumo to Rear Admiral Abe Hiroaki of the protective screen. He in turn ordered Rear Admiral Yamaguchi Tamon into the attack with the latter's flagship, the separated *Hiryu*, whose air group, sixty-three strong, was as intact as her hull. The target was the *Yorktown* task group, some 240 miles north of Midway just before 11:00 A.M., when the first Japanese wave made for it. A second left two and a half hours later, all forty attackers being homed in on the Americans by spotter planes from a cruiser. Admiral Kondo, commanding the force covering the intended invasion of Midway, was meanwhile bringing his two battleships, four cruisers, and seven destroyers northeastward at top speed to reinforce the battered Mobile Force. In the deep background – so deep that the Americans were unaware of his presence and position – Admiral Yamamoto, the Combined Fleet C-in-C, diverted the light carriers *Ryujo* and *Junyo* from covering the Aleutians operation to the aid of Abe, who told him at 10:50 A.M. that he was going to attack. The situation was serious, but the day was not yet lost.

A dozen Wildcats were flying 'cap' over *Yorktown* and her victorious dive-bombers were overhead waiting to land-on, when the carrier's radar (the Japanese had none in this battle) revealed up to forty 'bogeys' (hostile aircraft) coming in from the southwest as noon approached. The dive-bombers were told to get clear. Eighteen 'Val' bombers came in, and although Thach's fighters shot down or drove off more than half and anti-aircraft fire from the screen got two, six pressed home their attacks and three scored hits. One bomb went off on the flight deck, one plunged down the carrier's single smokestack and blew up below; a third passed through three decks before exploding. Admiral Fletcher, robbed of command communications, gave Spruance a free hand to carry on the fight and moved to the cruiser *Astoria* while damage-control parties fought to save the *Yorktown*, already weakened by her patched-up damage at the Coral Sea. They seemed to be making headway, and the carrier was actually gaining speed to twenty knots, when *Hiryu*'s second wave came in. Ten fast 'Kate' torpedo-bombers covered by six Zeros attacked from four different directions and very low altitude at about 2:45 P.M. The four heavy cruisers round *Yorktown* put up a screen of water by firing high-explosive shells into the sea, but four Japanese bombers got through, and two of their deadly torpedoes struck home on the port side. The *Yorktown* went dead in the water and listed alarmingly to port. By 3:00 P.M. the crew were abandoning the carrier on the orders of Captain Buckmaster, who feared his ship was about to turn turtle. Four destroyers came to the rescue.

Buckmaster was unduly pessimistic. The shattered *Yorktown* was tougher than she looked and, though drifting helplessly, refused to sink. Nimitz ordered an all-out effort to save her. Before dawn on 6 June the captain led a repair party of 170 men back aboard to make a second attempt. The destroyer USS *Hammann* made fast alongside at the starboard bow to provide pumping and electrical power. At noon, by dint of pumping fuel out of the port tanks while pumping seawater into the starboard tanks, the carrier was approaching even keel and all fires were under control.

But in mid-afternoon one of the screening destroyers reported four torpedo tracks zipping toward the *Yorktown*'s blackened starboard side, which loomed over the *Hammann* like a steel cliff. Lieutenant Commander Tanabe on the Japanese submarine *I168*, sent in by Yamamoto on the strength of a sighting by a seaplane patrol from a cruiser, was about to do for the *Yorktown* what Brookman had done for the *Soryu* in identical circumstances. The first and third missiles hit the *Yorktown* and the second the *Hammann*, amidships. The back of the destroyer was broken and her depth charges ignited in a series of shattering sympathetic explosions. This was too much for both stricken ships. The *Hammann* sank in less than four minutes. Twice bitten, Buckmaster did not abandon hope; even though he had abandoned his ship for the second time, he planned to return aboard at first light on the 7th. But just before dawn the *Yorktown*'s renewed list passed the critical point. She turned turtle, disappearing into six thousand feet of water as the crews of the attendant destroyers lowered their colors to half-mast and stood, hats off, to attention.

The last attacking stroke of the *Yorktown* before the Japanese air attack arrived had been to send out ten Dauntless dive-bombers 'on loan' from the *Saratoga*'s VB5 Squadron under Lieutenant Wallace Short, USN, to look for the one remaining enemy carrier. After three hours, by which time the *Yorktown* was *hors de combat* and they could not have landed back on her, they sighted the *Hiryu*, accompanied by two battleships, three cruisers and four destroyers. The *Enterprise* sent Commander McClusky and two dozen SBDs, which caught up with the *Hiryu* at about 5:00 P.M. Four direct hits damaged the island and started a series of raging fires below. Three US dive-bombers were lost; the rest got back to the two surviving American carriers, which now dominated the battlefield. The *Hiryu* was abandoned by her crew but did not sink, despite a salvo of torpedoes from two destroyers, until 9:00 A.M. A last foray by the Midway B-17s did no further damage; nor did one by the eleven serviceable Marine bombers from the island.

When it was ascertained that the ship was in a sinking condition, Admiral Yamaguchi and Captain Kaku [Tomeo] decided that they would go down with the ship [Admiral Kusaka recalled]. They all shared some naval biscuits and drank a glass of water in a last ceremony. Admiral Yamaguchi gave his hat to one of his staff officers and asked him to give it to his family; then there was some joking among them – the captain and the admiral – that their duties were finished when the ship sank.

The scene surely matches anything on record about the last hours of the *Titanic*. An admiral of great promise (who would have served his country better by saving himself, but that was not the Japanese way) and his flag captain bade farewell to their staff at a surreal drinks-party with cocktails of water and canapés of hard tack, in the last act of a naval *Götterdämmerung* without precedent. Nagumo's Mobile Force was now a hollow shell, a screen with nothing left to protect, a carrier striking-force with no carriers. It had been a truly glorious day for the United States Navy, which had extracted maximum profit from the irresolution of Admiral Nagumo. On this occasion he chose to live.

Admiral Yamamoto, aboard the *Yamato*, five hundred miles west of Nagumo, stoically absorbed the dreadful news from the Mobile Fleet. The only outward sign of emotion was a request for news of Genda Minoru, his air-tactics genius, who had been on the *Akagi* (he survived). Yamamoto was suffering from severe stomach pains (not nerves but worms) as the strategy for which he bore total responsibility collapsed in the flaming shambles of the Battle of Midway. Knowing that the Americans had two carriers intact, Yamamoto abandoned his race to the aid of Admiral Kondo, whom he had ordered to go to the relief of Nagumo, and just before 3:00 A.M. on 5 June ordered a general withdrawal.

Rear Admiral Spruance knew that four Japanese carriers had been sunk or driven blazing from the field but could not be sure that there were no others. By no means all of the Combined Fleet was plotted by the Americans, who knew their enemy still had five more carriers at his disposal somewhere, any or all of which could be involved in the enormous Japanese deployment. Like Yamamoto, he recoiled from the idea of a night encounter with forces of unknown quality and quantity, so he sailed east, away from the enemy, with *Enterprise* and *Hornet* for nearly five hours, until midnight. His carriers were not equipped or trained for combat night-flying, and for all he knew the entire enemy battle fleet, which would have an enormous advantage in a night action, for which it was thoroughly trained, lay in wait to the west. The scale of the American victory was far from clear. Commendably, COMINCH and CINCPAC commands were very cautious in their early claims, hinting at victory but not yet claiming the triumph which was their due.

The nemesis of the Japanese fleet had, however, not quite run its full course. The submarine USS *Tambor* saw the four heavy cruisers and two destroyers of Rear Admiral Kurita's group, which was to have directly covered the Midway landing, start to pull back before dawn on 5 June, in response to Yamamoto's orders. Sighting the submarine, Kurita ordered an emergency turn, and his fourth cruiser, *Mogami*, collided with the third in line, *Mikuma*. *Mogami* crumpled her bow and caught fire. The damaged pair limped westward nursed by the two destroyers and were sighted by a Catalina patrol from Midway, which thought they were battleships. The B-17s were mobilized but failed to find them, so Captain Simard for the last time called in the Marines, who mustered a dozen aircraft. One, hit by the heavy anti-aircraft fire, deliberately crashed onto the *Mikuma*; the rest scored no hits. Spruance, sailing west since midnight, diverted toward Midway in response to the *Tambor*'s sighting – only to receive reports of another Japanese formation two hundred miles to the northwest. He chose to attack the latter in the afternoon of the 5th, but his carrier aircraft found nothing. So he turned his attention to the damaged pair of cruisers and began to look for them after dark. Yamamoto by this time had assembled two light carriers, eleven battleships, eight cruisers and dozens of destroyers from the various forces he had deployed for Operations MI and AL, but he was still retiring slowly homeward out of range of all American aircraft except the Midway Catalinas.

An *Enterprise* scouting flight found the damaged cruisers on the morning of the 6th, and Spruance sent three strikes totaling 112 aircraft after them between breakfast and lunch, before the Japanese could get within range of their own aircraft on Wake Island. *Mogami* was hit at least five times but still managed to stagger to safety at Truk in the Caroline Islands, where she took two years to recover. *Mikuma* was abandoned and sank after dark. Spruance was now four hundred miles west of Midway. As darkness fell on the evening of 6 June 1942, he closed the book on the Battle of Midway by turning east to refuel from his tankers. Yamamoto spent another day hoping that the Americans would come chasing after him, but on the 8th he cut his losses and made for home.

Checked and forced to change their plans for the first time in their history at the Battle of the Coral Sea, imperial Japan and its Navy had now been handed their first smashing defeat at the moment when Yamamoto made them overreach themselves. From now on they and the Army were to be on the strategic defensive. The Pacific Campaign had undergone a mighty sea-change heralding a shift in the balance of power across the ocean. But this was nothing like as clear-cut to either side at the time. The numerically much superior Japanese had lost four fleet carriers with 225 aircraft and their precious pilots plus one heavy cruiser, with another and a destroyer badly damaged. The Americans lost one carrier, one destroyer and 113 carrier planes, plus the thirty-three Marine aircraft that failed to return to Midway. One Navy and four Army land-based planes were also lost from there. The Americans had won by a mixture of superb intelligence work, resolute decisions by Nimitz and Spruance – and a mighty infusion of luck. Nagumo's change of plan at the key moment offset the uncoordinated nature and the ineffectiveness of most of the American air attacks. All the many errors by the Americans were not enough, however, to offset Yamamoto's strategic blunder and Nagumo's tactical one. The overall determination of King and Nimitz to counterattack the aggressor as hard as they could as soon as they could was magnificently and deservedly vindicated.

There remained for the Japanese, after the failure of MO and MI, only Operation AL. The Aleutians, with their wild climate and inhospitable, rough terrain covered in tundra, extend into the unfriendly North Pacific as forerunners (or afterthoughts) to Alaska. They were to be taken by the Northern Area Force, commanded by Vice-Admiral Hosogaya Boshiro of the Fifth Fleet. The Second Mobile Force, of Admiral Kakuta Kakuji, would cover occupation forces for Adak and Attu plus Kiska islands. Kakuta approached the area on the night of 2–3 June, and planes from his two light carriers, *Ryujo* and *Junyo*, bombed Dutch Harbor, Alaska, with impunity during the night. American reconnaissance was looking for the expected invaders but did not find them. Twenty-five people were killed on the ground, and some installations damaged, for the loss of two Japanese aircraft. A second wave was driven off by bad weather and US Army P-40 Tomahawk fighters, which shot down two. The next day Kakuta attacked Dutch Harbor again, in the afternoon, starting a fire at a tank farm. American Army and Navy efforts to find and bomb the carriers with the small forces available came to nothing, for the loss of eleven US planes.

On the afternoon of the 4th Kakuta was ordered by Yamamoto to head south to reinforce Nagumo's shattered First Mobile Force at Midway. By the time Kakuta's aircraft came back from Dutch Harbor, Yamamoto had changed his mind about canceling the occupation of the Aleutians: he signaled that it should go ahead. The Second Mobile Force eventually took up a deep covering position six hundred miles south of Kiska, where it was later reinforced by fleet carrier *Zuikaku*, light carrier *Zuiho*, and four destroyers – Yamamoto's last attempt at revenge on the Midway victors, who did not respond to the invitation. Hosogaya canceled the occupation of Adak because it was in range of a US air base on Umnak Island, of which the Japanese had not known (they discovered it when its planes attacked them). An Army detachment of twelve hundred men was landed on Attu on the night of 6–7 June. There was no resistance, and forty-one people, nearly all Aleuts, were taken prisoner. Kiska was taken on the early afternoon of the 7th by the 350 men of a Special Naval Landing Force. The small staff of a US weather station were taken prisoner. There was no resistance.

Rear Admiral Theobald, skeptical of Ultra and convinced that the Japanese would go for the Dutch Harbor area, wrong-footed himself by staying in the waters well south of the Alaskan mainland with his five cruisers. He thus lost any chance of disrupting them at their most vulnerable moment, which, as we have seen, was invariably at the time of landing. The toings and froings of the Japanese forces in response to the vacillations of Yamamoto did nothing to help Theobald. A completely false 'sighting' on 5 June by a US patrol aircraft of two carriers, two cruisers and three destroyers in the Bering Sea, between Soviet Siberia and Alaska, caused serious disruption in Nimitz's command in the immediate aftermath of Midway. The freshly restored *Saratoga*, racing west from San Diego to reinforce the Pacific Fleet, was diverted to join *Enterprise* and *Hornet* and go north to tackle the phantom threat. The frustrated Theobald learned on the 10th that the carriers were not coming, that the Bering Sea force did not exist – and that Attu and Kiska were firmly in the hands of the Japanese. This fact was discovered three days after the event, when the weather reports stopped coming in from the occupied islands.

Confused and inconclusive air skirmishes went on for nine days in poor weather. The lack of immediate clarity in both Japan and the United States about the strategic significance of what had happened in the first days of June obscured the irrelevance of Operation AL. Americans were most upset by the occupation of two of their islands (of which hardly anyone had previously heard), and the Japanese were correspondingly satisfied.

Very little was revealed about the Battle of Midway in Japan. On 18 June the high command announced the loss of one carrier and bad damage to another and a cruiser. In Tokyo, IGHQ (Navy) revealed the awful truth only to the highest levels in IGHQ (Army), where a Secret War Diary recorded 'a state of tension never experienced since the outbreak of war' on 6 June as IGHQ waited for news. 'The Navy Section is showing anxiety for the first time since the beginning of the war' (8 June). And on the next day: 'It appears that the Battle of Midway had ended in the defeat of our Navy . . . [putting it] in a difficult internal predicament.' The Army already scented profit in the Navy's discomfiture, as usual, as plans to attack New Caledonia, Fiji, and Samoa were abandoned. The Navy quietly dropped any idea of trying to take Port Moresby by sea, leaving the operation to an overland thrust across New Guinea by the Army. The Navy, having lost a hundred of its best attack pilots at a stroke, without advance provision for replacements because defeat was unthinkable, concentrated on reorganizing and re-equipping its carrier forces. Everything possible was done to prevent the news from spreading from the lower decks; sailors returning to Japan from the lost battle were kept isolated for months. But the truth would out; rumors spread instead of facts, and the public henceforward started to lose confidence in the bombastic announcements from IGHQ.

The Americans were unsure of the scale of their victory, having cautiously claimed two or three enemy carriers sunk and one or two damaged on 7 June. Fistfights broke out between Army and Navy men in Honolulu over the Army Air Force's completely unjustified claim to have played a major role in the sinkings, which was disproved only much later.

Colonel McCormick's Chicago *Tribune*, last heard from in these pages trumpeting the (fortunately superseded) US war plan, got a scoop out of the battle. On Sunday, 7 June 1942 – fast work by any standards – the paper carried this banner headline: 'NAVY HAD WORD OF JAP PLAN TO STRIKE AT SEA.' The dateline was Washington; there was no byline. The story also appeared, uncoincidentally, in the New York *Daily News*, owned by McCormick's cousin Joe Patterson, and the Washington *Times-Herald*, owned by Joe's sister 'Cissy' (Eleanor). The articles stated in alarming detail that the US Navy had obtained the Japanese order of battle and details of the enemy operational plan in advance of Midway. Admiral King was incandescent with rage; Nimitz once again worried about leaks and garrulity. The author was, on investigation, found to be Stanley Johnston, a reporter of professional repute and personal courage who had helped to rescue men trapped in the blazing *Lexington* at the Coral Sea. He got his story while covering that battle, from which he returned on a requisitioned merchantman which docked at San Diego on 3 June. Someone from the carrier appears to have shown him the Pacific command's pre-Midway intelligence appreciation, drawn from Ultra and intended for task-force commanders. The *Trib* pretended that Johnston got the story by straightforward journalistic digging, but Navy Secretary Frank Knox knew otherwise from internal evidence – even CINCPAC's errors had been faithfully reproduced – and set out to prosecute Johnston under the Espionage Act.

Not surprisingly, the Navy's intelligence staffers resigned themselves to a total change of codes and ciphers (there had been one anyway at the end of May, causing a temporary but worrying Ultra blackout). In August 1942, when this JN25C was replaced yet again, the assumption was that the Japanese had read their copy of the Chicago *Tribune*. But when the cryptanalysts found that the cipher was of the same old 'family' and therefore susceptible to the traditional methods of penetration, the case against Johnston was dropped like the hot potato it was, before it came to trial. The attendant publicity would have been a major threat to Ultra. American courts, admirably, did not sit *in camera* even in wartime, even when dealing with espionage, and there were justified fears of what might come out. The British chiefs-of-staff delegation in Washington, steeped in their country's love of secrecy for its own sake and unable to appreciate the American taste for openness, were worried witless. For once King agreed with them, and the incident was followed by a considerable tightening of procedures, distribution, and 'need-to-know' rules in the United States, very much along British lines but without the attendant paranoia. Perhaps the most remarkable aspect of this incident is the fact that the Japanese do not appear to have noticed it at all. At any rate, nothing they did was measurably influenced by it – not even when the conduct of the Chicago *Tribune*'s publisher was roundly and publicly belabored in Congress, to the collective wincing of the intelligence community!

6
Papua and Guadalcanal

After Coral Sea and Midway the Japanese settled on a policy of aggressive defense of what they now held. They had not been driven out of anywhere but only prevented from expanding much beyond the objectives laid down and swiftly gained in their original phase-I plan. They had taken Tulagi in the Solomon Islands on 4 May to guard the left flank of the seaborne invasion of Port Moresby, which was aborted. They were in control of strategically worthless Attu and Kiska. For their part, the Americans and their most important ally in the Pacific decided to roll back the Japanese by pushing from the southeast to the northwest, to prevent them from cutting off and threatening Australia – the ally in question.

Both King and MacArthur had their eyes on Rabaul, New Britain, in the Bismarck Archipelago, northeast of New Guinea. The Japanese had built this up as their strategic air and naval base in the area after seizing it on 23 January. On 18 February King proposed to Marshall jumping off from forward bases in the New Hebrides to the Solomons and northern New Guinea, and from there to the Bismarcks. After Midway, MacArthur confidently advanced a plan for snatching Rabaul in three weeks if he could just have a Marine division and a couple of carriers. Both plans were at least as much concerned with grabbing the greatest possible slice of the logistical cake at the expense of the rival service (and of the rival European theater) as with Pacific strategy. The Navy had only one amphibious division, not yet fully trained, and was not about to let its carriers get stuck in narrow waters in reach of so many Japanese airfields. The two services therefore agreed on the step-by-step advance on Rabaul advocated by King. The sticking point came with the question of command. An amphibious operation was obviously naval, but the objectives were almost wholly in SOWESPAC, MacArthur's command. The row went all the way up to the

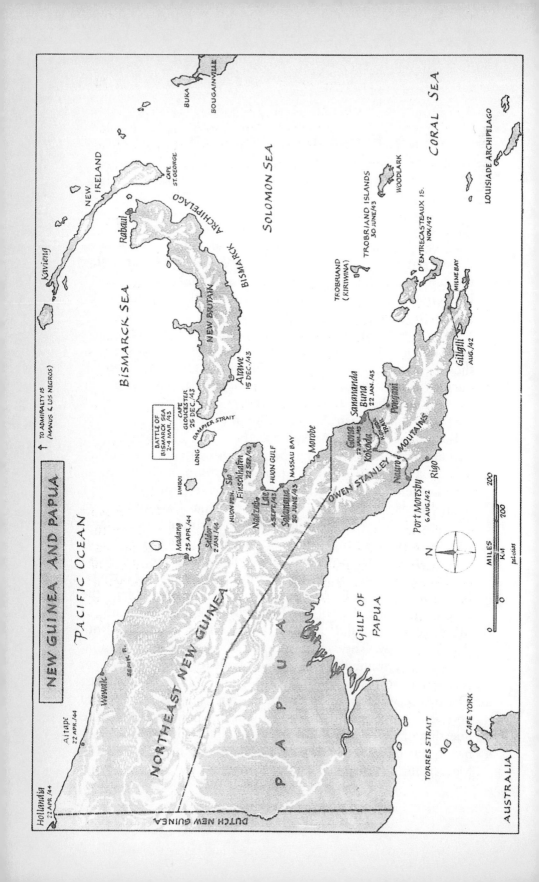

NEW GUINEA AND PAPUA

PACIFIC OCEAN

BISMARCK SEA

SOLOMON SEA

CORAL SEA

BUKA

BOUGAINVILLE

NEW IRELAND

CAPE ST. GEORGE

BISMARCK ARCHIPELAGO

Rabaul

Kavieng

NEW BRITAIN

Arawe
15 DEC./43

CAPE GLOUCESTER
25 DEC./43

DAMPIER STRAIT

LONG

BATTLE OF BISMARCK SEA
2-4 MAR./43

UMBOI

TROBRIAND ISLANDS
30 JUNE/43

TROBRIAND (KIRIWINA)

WOODLARK

D'ENTRECASTEAUX IS.
NOV./42

LOUISIADE ARCHIPELAGO

MILNE BAY

Gili Gili
AUG./42

Sanananda
Buna
22 JAN./43

Popondetta

Gona
9 DEC./43

Kokoda
2 NOV./42

KOKODA TRAIL

OWEN STANLEY MOUNTAINS

Nauro

Rigo

Port Moresby
GAUG./42

GULF OF PAPUA

Morobe

NASSAU BAY

HUON GULF

Salamaua
30 JUNE/43

Lae
4 SEPT./43

Nadzab
5 SEPT./43

Finschhafen
22 SEP./43

Sio

HUON PEN.

Saidor
2 JAN./44

Madang
25 APR./44

NORTHEAST NEW GUINEA

SEPIK R.

Wewak

Aitape
22 APR./44

Hollandia
22 APR./44

DUTCH NEW GUINEA

PAPUA

TORRES STRAIT

CAPE YORK

AUSTRALIA

TO ADMIRALTY IS
(MANUS & LOS NEGROS)

N

MILES   200

KM   200

0

pel-1005

Joint Chiefs in Washington before a compromise was struck. The drive on Rabaul was divided into three 'tasks.' The first required Nimitz to take Tulagi; to this end, the dividing line between his command and MacArthur's was moved one degree of longitude to the west! Task two was a double advance along the northern New Guinea coast and up the Solomons; three was the culminating attack on Rabaul itself; all this fell to MacArthur. Navy and Army contributions to the task forces would be settled by the Joint Chiefs of Staff. This deal, for all the world as if it were the Japanese reaching one of their many solemn Army-Navy agreements, was the immediate basis for the divided American advance on Japan itself: the two prongs would not come together for three years.

Task one, Operation Watchtower, was set to begin on 1 August 1942. A new sense of urgency was added to the planning when intelligence revealed that the Japanese in Tulagi had built up their garrison and laborers were slowly but surely constructing an airstrip on the larger island immediately to the south, called Guadalcanal. MacArthur's staff, meanwhile, were hard at work in Australia, planning task two at his headquarters in Melbourne, Victoria, which were about to be moved north to Brisbane, Queensland. The Japanese, it will be recalled, had occupied Lae and Salamaua, on the northern side of the Papuan peninsula of New Guinea, by 8 March and were well dug in. The Allied staff decided to seize Buna, well down the coast from the two enemy enclaves, and build a large air base to be used both against them and to cover Port Moresby, on the other side of the peninsula. The 10th of August seemed as good a day as any to make the move. Unfortunately, as so often in these cases, the enemy was thinking for himself and preferred to write his own script. When MacArthur arrived in Brisbane from Melbourne – in a borrowed royal train – on the morning of 22 July, a dispatch rider was waiting on the platform. The Japanese Army had landed in strength at Buna the day before, from a convoy that had been seen assembling at Rabaul. Indeed, Ultra had been yielding hints since 18 May of a possible Japanese overland advance on Moresby after the Coral Sea setback. Early in July the cryptanalysts had penetrated a long operational message foreshadowing a landing on 21 July; it is not known whether SOWESPAC ever got to hear of it.

Nimitz put Vice-Admiral Robert L. Ghormley, his commander, South Pacific, based in New Zealand, in charge of task one. The attack on the Solomons – Guadalcanal now as well as Tulagi because of the airfield – would be executed by the new First Marine Division under Major General Alexander Vandegrift. This formation was on its way to New Zealand by late June but not yet fully trained in the art of the opposed amphibious landing. The general had previously been told to expect no action until 1943. Ghormley flew to Melbourne early in July to consult with MacArthur; their meeting started well but subsided into a morass of cosmic pessimism. MacArthur, who had so recently offered to take Rabaul in three weeks with one assault-division, now urged the postponement of Watchtower, a much more limited operation, for lack of sufficient logistical support. King was scathing; the Joint Chiefs permitted a postponement of the Solomons invasion until 7 August. Since the Japanese got in first by pre-empting the Buna landing and the Americans were behind schedule in the Solomons, we can start with the Papua campaign. But before we do that we need to examine the untested alliance which uneasily took it on.

Australia played broadly the same role in the war in the Far East as Canada did in the struggle against Germany. Each was a geographically vast, demographically small, and strategically important dominion, an independent British Commonwealth nation whose head of state was King George VI of England. Each loyally declared war when Britain did, in September 1939, and again on Japan in December 1941; each had to see its armed forces serve, often far from home, under first British and later American commanders-in-chief. Like Canada's, therefore, Australia's steadfast and usually unglamorous contribution to ultimate victory has been consistently undervalued. Because there was no human agency capable of resisting the MacArthur promotional steamroller, least of all the weak Australian government, until Harry S. Truman came along, Australia has fared even worse than the Canadians at the hands of most

Anglo-American historians. And for Australians the war came much closer to home than it ever did for Canada; there was something very near to panic as the Japanese advanced toward them with such frightening speed, especially when northern towns were bombed in February 1942. What is more, with not untypical perversity, the electorate had chosen, in the middle of a war, a pacifist as prime minister in October 1941, in the shape of the inexperienced Labor Party leader, John Curtin. This was largely a vote of protest against the perceived selfishness of the British in 'stripping' Australia of troops to shore up their threatened interests in the Mediterranean as war loomed in the Far East – a poor return on prewar promises of Royal Navy protection and on Australia's show of loyalty against Germany (and the origin of the postwar severance of the apron strings).

When, in March 1942, MacArthur breezed in as Supreme Allied Commander, he also took control of the Australian Army, his main source of troops for some time, even after the first two US Army divisions arrived, as well as the small Royal Australian Navy and the rapidly expanding Air Force. Three Australian divisions had been in the Mediterranean, fighting well and gaining invaluable experience, on 7 December 1941. Two were recalled at once and came straight home instead of going to Singapore; the third returned in February 1943. The one division of the Australian Imperial Force that was to have been kept at home went to Singapore – and therefore into captivity when it fell. The AIF was made up of volunteers and always would be; conscription for Britain's faraway wars was political anathema, but was acceptable for the Militia, the Australian equivalent of the American National Guard or the British Territorial Army, primarily intended for home deployment. Perversely again, there were so many volunteers for the AIF that the Militia had chronic difficulty in finding men.

The man chosen to lead Australia's Army in what Curtin called its 'gravest hour' was the unlikely figure of General Sir Thomas Albert Blamey (1884–1951), a man only four years younger than MacArthur, his new and well-nigh impossible boss. Small, fat and red-faced, Blamey was an object lesson in not judging a book by its covers, a mistake many American senior officers made, sometimes to their cost. Major General

Robert Richardson, for example, sent on a fact-finding tour of Australia by Marshall in mid-1942, described MacArthur's commander, Allied Land Forces, as 'a non-professional Australian drunk.' As it happens, Blamey, who had to be given the job for political reasons, was not only bibulous but also a philanderer. He was not interested in winning popularity among inferiors or superiors and cultivated – insofar as he cared about the opinions of others – the caricature image of a philistine Australian. This left him almost, but not quite, defenseless against MacArthur's brand of manipulatory public relations. When people underestimated him, as they often did, it was not always or entirely their fault. All this made him no worse than, say, Ernest J. King, whom he closely resembled in manners – and also in intelligence and total devotion to the service. Even so, with the insensitivity of a superpower at its worst, Richardson, assiduously briefed by MacArthur's staff, snobbishly concluded that putting US troops under Blamey constituted an 'affront to national pride and the dignity of the American Army.' He appears not to have considered what it might be like for Australians to serve under a foreign egotist like MacArthur.

Blamey gave up teaching to join the Army in 1906. By May 1918 he was chief of staff to the brilliant Australian commander in France, General Monash, who gave him a 'rave' report. He 'retired' in 1925 to become chief of police in Victoria, retaining a major general's commission in the part-time Militia. Leaving the police under a cloud after attempting to cover up a scandal by lying, he was put in charge of Army recruitment in 1938. His wide experience, quicksilver mind and grasp of detail could not be spared after all. In 1939 he was made commander of the Sixth Division, the first to be raised for the AIF. When it was decided to raise a corps of two divisions, Blamey was promoted to command it, and in due course commanded the Australia-New Zealand Army Corps (ANZAC) in the Middle East under the British. He led a good fighting retreat from Greece – and brazenly put his son in the last spare seat on the last plane out! Opinions of his proficiency among professionals varied from the adulatory to the dismissive, but he was promoted to full general in September 1941. When Japan went to war he was made not only MacArthur's land-forces chief, responsible for the ground defense of Australia and offensive operations, but also C-in-C of all Australian forces in the areas of expansion and training – and chief military adviser to the uncertain Curtin government. Small wonder that this impossible pile of hats sometimes interfered with his vision. It is not the size of his responsibilities but the impossibility of reconciling them all which stands out. How fortunate that he had no fear of controversy, lawful authority or the enemy.

After negotiations between the Washington and Canberra governments, in which the US called nearly all the shots and a nervous Curtin did all the acquiescing, MacArthur formally took over supreme command of Australian forces on 18 April 1942 and, typically, demanded direct and exclusive access to Curtin – which made an American general, not Blamey, the government's principal military adviser through the prime minister's War Conference, usually made up of Curtin, MacArthur and Frederick Shedden, permanent secretary (administrative chief) of the Defense Department. Blamey naturally wanted to resign, until Curtin hurriedly offered him separate personal access on administrative, as distinct from operational, military matters. Americans – Vice-Admiral Herbert F. Leary and Lieutenant General George H. Brett – served as Allied naval and air commanders under MacArthur, whose chief of staff, Lieutenant General Sutherland, was also American. Only the air command and the joint intelligence effort among the major segments of SOWESPAC appear to have operated as genuine inter-Allied outfits. MacArthur surrounded himself with his 'Bataan Gang' and regarded Australia as a country and a nation created for his and America's convenience. In the circumstances it seems little short of miraculous that he and Blamey actually got along quite well for a surprising proportion of the time they were locked in their unequal and awkward relationship. MacArthur blamed Blamey no more often than he praised him. They fell out only when Blamey made the mistake of trying to exercise his right to command American troops.

The second Japanese invasion of Papua, on 21 July 1942, was based on a misapprehension. The Southern Army command

believed that there was a 'road' from Buna across the Owen
Stanley range and also seriously underestimated the treacherous
horrors of the terrain. The only overland route on offer was
the Kokoda Trail, itself an overgenerous misnomer for a
winding track usually two feet wide which passed through
the village of Kokoda at the halfway mark between Buna
in the northeast and Port Moresby to the southwest. In the
post-Midway reorganization of the Japanese Navy, the Eighth
Fleet was created under Vice-Admiral Mikawa Gunichi, based
at Rabaul. He had five heavy and two light cruisers, destroyers,
five submarines, garrison troops and Marines, a naval air group
and construction units to build airfields, all to support both the
New Guinea and the Solomons operations. But the Buna landing
was the responsibility of the Army, specifically Major General
Horii Tomitaro's elite South Seas Detachment, a reinforced
division of sixteen thousand men, nearly all combat troops.
There was no opposition as they occupied a beachhead at Buna
and also took the next settlement up the coast to the northwest,
Gona. By the 27th the Japanese had marched through steaming
jungle and steeply rising ground to Kokoda, which they also
occupied.

The original Port Moresby garrison of one Australian regular
brigade plus small Australian and American support units had
at least been reinforced after the Battle of the Coral Sea
with another brigade, albeit an inexperienced one from the
Australian Militia, on MacArthur's order. The Seventh Militia
Brigade was also sent to guard the Milne Bay area at the end
of the Papuan peninsula. In view of the importance of New
Guinea to Australia's defense, Blamey seems to have been
unduly parsimonious with his AIF troops; but the Australian
C-in-C initially preferred to keep his best available regular
division, the Seventh, intact in Australia for later operations.
Further, nobody on the Allied side thought the Japanese would
attempt the daunting overland route, even though by this
time there could be no possible excuse for underestimating
the enemy. MacArthur proposed sending the green US 32nd
Division, formed from the National Guard, to Papua under
his direct control, which would have made a nonsense of the
entire command structure. Instead, Blamey sent the entire

Seventh Division, under Major General Arthur Allen; one brigade reinforced Milne Bay, one Moresby and the third the defense of the Kokoda Trail. Lieutenant General Sydney Rowell of the Australian I Corps was sent to command in the field.

On 26 August, having built up his Kokoda force to eighty-five hundred men who were already suffering from shortages caused by the tenuous line of communication with the Japanese beach-head, Horii began his 'over-the-top' advance on Moresby. His attack was timed to coincide with a landing on the same day by fifteen hundred men from Mikawa's Special Naval Landing Force at Giligili on Milne Bay. Their task was to capture the airstrips there and establish local air superiority to cover a second line of attack on Moresby. Fortunately for the Allies, the simultaneous struggle for the southern Solomons decisively diverted Japanese air strength, enabling the American and Australian air forces to batter Horii's supply lines. The Australians, locally outnumbered, fell back from Kokoda, bitterly contesting every step of the way; but the two brigades in Milne Bay outnumbered and eventually repelled the Japanese Marines, most of whom were killed by early September in savage fighting. The Australians thus won the first significant Allied land victory of the war against Japan.

MacArthur's attention was, however, focused on the crisis threatening Port Moresby, toward which the Japanese were advancing relentlessly all through August and well into September. As one setback succeeded another on the main front in Papua, MacArthur and his claque began openly accusing the Australians of poor generalship and lack of fighting spirit. This shabby disparagement of their ally, whom almost no senior officer had visited at the front, did not prevent them from succumbing to panic when the main Japanese column, down to five thousand men, reached the Nauro area, just twenty-five miles from Port Moresby, on 16 September. Curtin put his faith in MacArthur and not Blamey, who on his return from a visit to Moresby the next day was ordered by both men to go back there and take personal command of 'New Guinea Force.' Rowell was understandably furious but lost an argument which split the entire Australian military establishment. He was replaced by Lieutenant General Edmund Herring on 1 October. By this time the aforementioned US 32nd Division, under Major General Forrest Harding, USA, was arriving from Australia, without both heavy artillery and armor. The next day, Supreme Commander MacArthur deigned to pay his first visit to New Guinea as the Allies began their counterattack up the Kokoda Trail. In fact, unbeknown to them, the Japanese had already begun to withdraw on 24 September because they were starving; some had reportedly been reduced to cannibalism. But they fought just as hard on their retreat as they had on their advance, even after Horii, a general who led from the front, was killed in action late in October.

Under constant pressure from MacArthur in Brisbane, and not personally familiar with the atrocious fighting conditions, Blamey in Port Moresby lost patience with General Allen and, still without going to the front to see for himself, replaced him with Major General George Vasey. It is hardly surprising that Blamey, for all his achievements in and between two world wars, has remained a subject of raging controversy in Australian military circles ever since. Whereas MacArthur can be blamed for using the Australians as scapegoats for his command's difficulties, and Curtin for falling all over himself to meet MacArthur's wishes, Blamey did use two of his own entirely competent subordinates as scapegoats in his turn. He was trying to save his skin in the impossible situation in which he found himself – partly as a result of his temporary docility in obeying orders that he could and should have challenged, but mainly because of the strains of a grossly unequal alliance.

He was doubtless relieved when MacArthur, fearful of dismissal in the event of a defeat, set up his own advanced headquarters in Port Moresby on 6 November, letting Blamey off the hook: each general was as bad as the other in his inability to leave an assigned task to the man on the spot. The Australian Seventh Division had already driven the Japanese back over the Owen Stanleys into their beachheads at Buna and Gona, and the American 32nd Division was also in place for the final assault on them, scheduled for 16 November. By sea would have been better, but there were no landing craft for a seaborne assault because of Guadalcanal; and the Navy shied away from the shoals and the risk of Japanese air attack in New Guinea waters.

But on the 19th the American troops were hurled back from Buna in their first taste of action. Not even a direct order from MacArthur could make them try again. At this stage of the campaign, Allied air forces had not yet absorbed the new skill of close support of ground troops, and tanks and flamethrowers (not available to the 32nd) were the best weapons against the formidable Japanese defensive positions. Amid reports of US troops dropping their weapons and fleeing the murderous network of Japanese bunkers, MacArthur followed Blamey's questionable example and replaced Harding with Lieutenant General Robert L. Eichelberger, USA, an officer who was to play a leading role in subsequent SOWESPAC campaigns. But the bitter Australians seized the shining hour: when MacArthur nervously proposed bringing up the American 41st Division from Australia, his land-forces commander coolly remarked that he would rather have Australian troops – 'as I know they will fight.' It must have been a sweet moment for the put-upon Blamey, especially when MacArthur gave Eichelberger the ridiculously overblown (and illegitimate) order to 'take Buna or not come back alive.' It was even more galling for MacArthur because his operation was faltering while his compatriots were doing so well on Guadalcanal. The American corps commander duly survived and captured Buna on 2 January 1943. The Australians had already taken Gona on 9 December, leaving the Japanese in their last redoubt, a beachhead at Sanananda, halfway between the other two places. After an almighty row between the rattled MacArthur and the newly reassertive Blamey about who was to get the reinforcing regiment from the 41st Division just landed, it was assigned not to Eichelberger but to Vasey, as Blamey wished. The Australians, thus strengthened, captured Sanananda on 16 January.

And so it was that more than 40,000 Allied troops with air superiority and much better supplies took six months to defeat fewer than 20,000 Japanese troops who spent most of the campaign starving: two-thirds of them were killed. The Australians lost 2,163 killed and 6,533 wounded or sick and the Americans 903 plus 4,232 respectively. The jungle had proved as formidable an enemy as the vaunted but vanquished South Seas Detachment; and the Allied effort had undoubtedly been sapped as much by the curse of internal wrangling among and between the Australians and the Americans at the highest levels as by disease. But Papua was saved, and with it access to northern Australia, leaving the Japanese in possession only of Dutch New Guinea, the western half of the enormous island. The recovery of that was left to a later date.

It is time to turn to the different but equally grim struggle which began and ended a little later than the campaign for Papua, the multiple battle for the southern Solomon Islands summed up in the name Guadalcanal, an overwhelmingly American affair involving Navy, Army, both air forces, and Marines: task one.

Vice-Admiral Ghormley, COMSOPAC since mid-April 1942, moved forward from Auckland, New Zealand, to Nouméa, capital of French New Caledonia, on 17 May to direct the assault on the Solomons, the first American strategic offensive of the war against a specific objective. Not the most forceful of men, the admiral allowed himself to be confined to his headquarters-ship rather than offend the sensibilities of the Free French administration by requisitioning offices ashore. In the first week of March, King had proposed using as the starting line Efate in the New Hebrides, now independent Vanuatu but then a jointly administered Anglo-French hybrid of a colony three hundred miles northeast of Nouméa. The first element of the new 'Americal' (American New Caledonia) or 25th Infantry Division (Major General Alexander M. Patch) was already garrisoning Efate before the end of March, alongside Marine ground and air units. A second base in the archipelago was built on Espíritu Santo Island, to the north. In the US the Navy was rapidly developing a remarkable organization for the swift construction of airfields on remote islands where little or no labor was to be had. The lion's share of this work fell, often in atrocious conditions and under fire, to the naval Construction Battalions (CBs – the famous 'Seabees,' who soon came to exemplify American 'can-do' at its formidable best).

At the end of May, MacArthur had vetoed Nimitz's suggestion of a commando raid by Marine Rangers on the already functioning Japanese seaplane base at Tulagi while the enemy was engrossed in his MO-MI-AL operations. Hindsight shows this to have been a proposal that might have shortened the war considerably, but MacArthur's objection, seconded by Ghormley, that the Americans did not yet have enough strength to hold on to such a position once taken, was entirely rational. The Japanese, meanwhile, having agreed on Operation F-S – Fiji, Samoa, and New Caledonia – in mid-May, first postponed and then abandoned it two months later. The new Eighth Fleet command, hived off from Admiral Inouye's Fourth Fleet area run from Truk in the Carolines, arrived in Rabaul by the end of June to cover the Army's assault on Papua and to guard the Solomons.

Nimitz made Spruance his chief of staff after Midway as the Pacific Fleet received its first major reinforcements. The new carrier *Wasp* arrived from the Atlantic, restoring the total available to four; the new fast battleship *North Carolina* and cruisers and destroyers also joined CINCPAC's flag very soon after Midway. For all that Washington was committed to 'Germany first' and the November landing of US troops in North Africa (Operation Torch) was in preparation, the Pacific Fleet could not complain of an unfavorable distribution of naval resources. It had six fleet carriers to the Atlantic Fleet's one, ten battleships to two, thirteen heavy cruisers to four, ninety-six submarines to twenty-eight. The two fleets were on a par in light cruisers and destroyers; the Atlantic boasted seven of the new smaller escort carriers to the Pacific's three, the better to help guard the vital transatlantic convoys.

But King's Pacific bias led him to starve the Atlantic theater, already overstretched for Torch, of Very Long Range Liberator bombers and merchant shipping just as the German submarine onslaught on the Anglo-American lifeline was at its height. And though American Army commitments to Europe were much greater in the longer term than to the Far East, especially in the air, nearly a quarter of a million troops were now in the Pacific, and the dispatch of two hundred thousand more was already envisaged. Clearly 'Germany first' was a strategic and not a tactical undertaking; there was never any question in Washington of a purely defensive, holding operation in the Pacific. The Americans had chosen an assertive policy which had already yielded significant dividends. But the British had a fair complaint about the diversion of vitally needed resources, especially when it came to VLR planes and above all shipping. Much of the latter was undoubtedly wasted on long periods of idleness in remote Pacific ports, as needlessly as it had been by King's refusal to use convoys in American Atlantic waters in the first half of 1942. His successfully fought campaign to get Allied sanction for 30 percent of combined resources for the war against Japan masked a significantly higher allocation of men and materiel to the Pacific from American resources – which of course constituted the bulk of what was available to the Allied cause. The US Navy never allocated less than two-thirds of its strength to the Pacific in the war.

King and Nimitz were in total agreement on the need for task one, especially after aerial reconnaissance showed an airstrip in an advanced stage of construction on Guadalcanal on 2 July, and earmarked the partly trained First Marine Division, already on its way to New Zealand, for the job. King adamantly refused to allow any delay and rode roughshod over the reluctance of MacArthur and Ghormley to see such huge resources committed so soon despite competing demands in New Guinea and the Atlantic: the threat to Australia and New Zealand had to be removed. There were by the end of July eight Allied island bases in the South Pacific, apart from Australia and New Zealand, from which to mount the attack on the southern Solomons.

Ghormley assigned the tactical command to Vice-Admiral Frank Fletcher at the head of the specially formed Task Force 61. His resources encompassed fifty-four American and three Australian warships, including three carriers, a battleship, and eleven heavy cruisers. There were also six US submarines, twenty-three transport and supply ships, plus five tankers. Under Fletcher's protection but virtually autonomous was Task Force 62, included in the above figures. This was the South Pacific Amphibious Force, led by Rear Admiral Richmond Kelly Turner, the omniscient martinet long since chosen for this key post by King himself (and offensively convinced for rather longer that he was the cleverest officer in the United States

Navy). Born in Portland, Oregon, in 1885, Turner, nicknamed 'Terrible' for his foul mouth and temper, was expert in both gunnery and aviation and was director of the War Plans Division and assistant chief of staff to King before his appointment as COMAMPHIFORSOPAC. Commanding the First Marine Division since only March 1942 was the softly spoken Virginian, Major General Alexander Archer Vandegrift (1887–1973), who seems to have inherited a full measure of stubbornness from his Dutch forebears – a modest model of a fighting general who was also no slouch at staffwork. He had the First Regiment of the Second Marine Division and a few small specialist units under his command as well. Operation Watchtower was supported by Rear Admiral J. R. McCain's South Pacific Air Forces (Navy, Marine, and Army) and MacArthur's 19th Bombing Group of two dozen Army B-17s. Everything in the Pacific Fleet apart from the older battleships and the *Hornet*'s Task Force 17 was committed to the Solomons, either under Fletcher or in escorting convoys. Nimitz was following the military precept of concentration of force, but if anything went wrong there could be no doubt that King's head would be on the block. He was the prime mover behind the first naval assault on the Japanese line in the Pacific. His plan determined the shape of the main campaign against Japan's southern front for two years.

The assault force assembled four hundred miles south of Fiji from half a dozen ports across the Pacific on 26 July. A conference on USS *Saratoga* the next day was marked by an almighty slanging match between Fletcher and Turner about tactics. The former did not conceal his doubts about the whole operation, while the latter was scathing about the plan to withdraw the carriers after only two days. Ghormley, their superior officer, who should have settled it, was not present: he had decided to command at a distance, from Nouméa, but neglected to tell Nimitz. As a result, he never set eyes on 'his' fleet and was absent when rows, almost invariably involving Turner, and other crises blew up. Pausing only to let the Marines rehearse their landing in Fiji for a couple of days ('a complete bust' – Vandegrift), the formation sailed northeast.

* * *

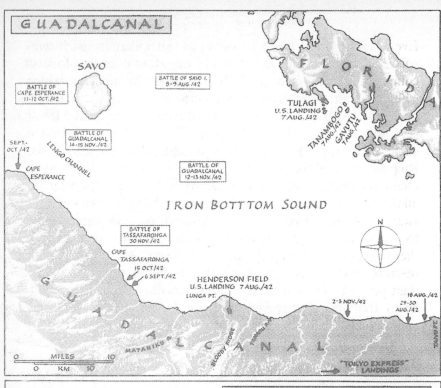

GUADALCANAL

SAVO

BATTLE OF
CAPE ESPERANCE
11-12 OCT./42

BATTLE OF SAVO I.
8-9 AUG./42

FLORIDA

TULAGI
U.S. LANDING
7 AUG./42

TANAMBOGO
7 AUG./42

GAVUTU
7 AUG./42

SEPT.-
OCT./42

LENGO CHANNEL

CAPE
ESPERANCE

BATTLE OF
GUADALCANAL
14-15 NOV./42

BATTLE OF
GUADALCANAL
12-13 NOV./42

IRON BOTTTOM SOUND

N

BATTLE OF
TASSAFARONGA
30 NOV./42

CAPE
TASSAFARONGA
15 OCT./42

6 SEPT./42

HENDERSON FIELD
U.S. LANDING 7 AUG./42

LUNGA PT.

2-3 NOV./42

18 AUG./42

29-30
AUG./42

TAIVU PT.

G U A D A L C A N A L

MATANIKU

BLOODY RIDGE

"TOKYO EXPRESS"
LANDINGS

0 MILES 10
0 KM 10

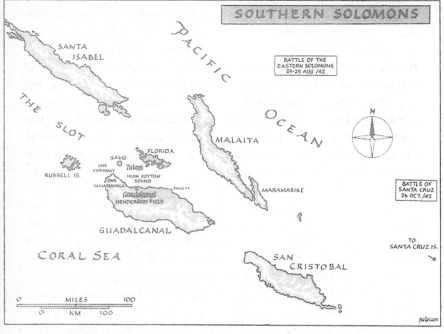

SOUTHERN SOLOMONS

PACIFIC

SANTA
ISABEL

BATTLE OF THE
EASTERN SOLOMONS
24-25 AUG./42

THE
SLOT

OCEAN

N

MALAITA

SAVO

FLORIDA

CAPE
ESPERANCE

Tulagi

RUSSELL IS.

IRON BOTTOM
SOUND

CAPE
TASSAFARONGA

TAIVU PT.

MARAMASIKE

BATTLE OF
SANTA CRUZ
26 OCT./42

Guadalcanal
HENDERSON FIELD

GUADALCANAL

TO
SANTA CRUZ IS.

CORAL SEA

SAN
CRISTOBAL

0 MILES 100
0 KM 100

palacios

Even before one drop of blood had been spilled on its fecund soil or a single corpse buried in it, Guadalcanal stank. Morison coined the word 'faecaloid' to describe the miasma which greeted the nervous Americans as they stole into the sound between Florida and Guadalcanal islands, a queasy mixture of superabundant vegetation, swift to rot, on a bed of primeval slime humming with malarial mosquitoes and nameless bacteria. Rich in mud and coconuts, the Solomons were wet from May to October, wetter from November to April and diabolically humid all the time. Today the jungle has gone the way of much greater swathes of virgin forest round the world, making it impossible to visualize what so many young men were about to endure there. Transport Group X was the larger of the two deployed, because much more resistance was expected on Guadalcanal than proved to be on hand (mostly laborers plus 550 troops). There were some nineteen thousand Americans. Brigadier General William Rupertus, USMC, took the equivalent of a brigade group ashore on Tulagi from Transport Group Y, fewer than five thousand men. The Japanese had fifteen hundred troops well dug in. On 7 August 1942, before dawn, the Guadalcanal campaign began with a coastal barrage against that island and its companions, Tulagi (off the much larger island of Florida), Gavutu, and Tanambogo. The Marines established themselves on all four within twenty-four hours, meeting little resistance on Guadalcanal and tough but brief opposition on the other three. Vandegrift's first consolidating move was to establish a firm grip on the nearly finished airstrip on the north coast of the largest target island, Guadalcanal. It was named Henderson Field in honor of the Marine bomber-commander lost at Midway, and it was to be the key to the whole campaign.

The Japanese reacted quickly from Rabaul, even though the base was bombed by SOWESPAC's B-17s on 7 August; forty-three bombers escorted by eighteen Zeros attacked various targets during the afternoon of the invasion day. The Americans were given fair warning by a clandestine force of 'Coastwatchers' which now came into its own. These Australian, New Zealand and British prewar residents of the Solomons stayed on in hiding, supported by the indigenous Melanesian inhabitants, after the Japanese occupied the islands, to report on enemy movements. Fighters from the US carriers were directed against the attackers by a team of officers specially installed for the purpose on cruiser *Chicago*, lying offshore. The Americans lost twelve and the Japanese sixteen planes. On the 8th the Japanese tried torpedo-bombing; twenty-six 'Bettys' came in, but only nine went out again. They and a subsequent dive-bomber attack were frustrated by well-prepared shipboard anti-aircraft fire and Turner's masterly ship-maneuvers. The Americans lost a transport. At 6:00 P.M. Admiral Fletcher went back even on his cautious undertaking to cover the landings for just forty-eight hours; alarmed by the swift airborne response of the enemy, and discounting the effectiveness of the American defense, he decided to pull out his carriers, which had lost twenty-one of their ninety-nine fighters, twelve hours early, falsely reporting to Ghormley that they urgently needed refueling and not bothering to wait for his commander's approval. Turner was furious; despite sigint and Coastwatcher warnings of a Japanese naval force approaching from Rabaul, however, he kept his supply ships at their task of unloading, because the Marines had very little food, heavy equipment or ammunition ashore. Even so, the first American amphibious operation of the twentieth century had gone off pretty well. But that was only the easy part.

The first act of Admiral Mikawa in Rabaul was to order air strikes. The second was to send six transports of troops to reinforce the southern Solomons. These were sighted by the American submarine *S38* (Lieutenant Commander H. G. Munson, USN) in the Solomon Sea. One ship was sunk and the rest were promptly recalled by Mikawa, who now decided to take personal command of his cruiser squadron and stage a massive attack on the enemy invasion force while it was in its most exposed position, in the act of unloading. He boarded the *Chokai* and collected four other heavy and two light cruisers plus a single destroyer. His plan was to attack at night on 8–9 August, because his men were well trained for this and the risk of American air attacks would be minimal. American and Australian reconnaissance planes, as well as the

Coastwatchers, saw him coming. Even so, he steamed boldly southeast down the middle of the double line of islands of the Solomons, along the seaway soon to be known as the Slot, to the waters between Florida and Guadalcanal, which were about to earn their sobriquet of Iron Bottom Sound (because of all the ships sunk there). Savo Island stands guard at the northwestern end of the sound.

Hard intelligence took its time reaching Turner, thanks to a combination of inefficiency, poor communication procedures between MacArthur's and his commands, and bad weather farther afield which kept most of McCain's planes down. Responsibility for protecting the transports under Turner lay with Rear Admiral V. A. C. Crutchley, RN, and his Australian and American cruisers and destroyers. He put one group of ships (Northern Force) between Florida and Savo – three US heavy cruisers and two destroyers – and another between Savo and Guadalcanal (Southern Force), comprising USS *Chicago* and HMAS *Canberra* plus two destroyers. Two more destroyers acted as advance guards for both groups; Eastern Force protected the Guadalcanal coast well to the southeast of Savo. Crutchley borrowed HMAS *Australia* from Southern Force to meet Turner and Vandegrift for a council of war off Lunga Point, on the north coast of Guadalcanal, west of Henderson Field, very late on the evening of the 8th. They agreed that Turner would have to pull out his transports the next morning to avoid a major Japanese attack, even though this would leave the Marines desperately short of many items. Turner had been misled by poor sighting reports into believing that the Japanese naval forces posed no immediate threat but would probably set up a seaplane base for an air attack the next day.

A Japanese aircraft, clearly identifiable by its floats as a scout from a cruiser, was sighted by the Americans over Savo Island. Turner was not told, even when it was joined by a second. Midnight came and went. So did a rain-squall which did nothing to relieve the sticky oppression of the equatorial night.

At 1:43 A.M. local time several things happened at once. The Japanese aircraft dropped dazzling parachute flares. Admiral Mikawa ordered his single line of the five heavy and two light cruisers and one destroyer, led by the *Chokai*, into battle.

The *Canberra*'s lookouts sighted her as a strange ship dead ahead. The destroyer USS *Patterson* sounded a general alarm by radio. Two torpedoes and twenty-four eight-inch shells hit the *Canberra*. The *Chicago* saw flashes and flares ahead. By 1:49 the *Canberra* was immobile and burning and the *Chicago* had her bow blown off by a torpedo. Mikawa was simultaneously giving serious attention to the Northern Force, now ahead of him as he swung northward to the east of Savo Island. At 1:48 the *Chokai*, still leading, fired a salvo of four torpedoes 'up the skirt' (at the stern) of the three American heavy cruisers as his other ships opened fire with their guns at about ten thousand yards. The last in line, USS *Astoria*, opened fire on the *Chokai*, which returned the compliment and set the American ablaze, helpfully illuminating her companions for the Japanese gunlayers. The *Chokai* took a few hits but fought on. The *Astoria* became her own funeral pyre. Next in line for the attackers was USS *Quincy*, which was pounded to pieces in crossfire from the Japanese cruisers, now in two lines on either side of her. Caught in this deadly sandwich, she was soon burning ferociously. Even so, she managed to get off rather more shells than any other Allied ship that night before she went down by the bow. Next and now last in line was USS *Vincennes*, which was soon dead in the water, on fire and doomed. She and the *Quincy* were gone before 3:00 A.M. The *Canberra* had to be put out of her misery the next morning, the 9th, after the crew was taken off; the *Astoria* sank at noon, despite a desperate fight to save her, just as Turner withdrew his transports and left the Marines to their own devices for the time being.

Such was the worst defeat at sea in the history of the United States Navy in 130 years: four cruisers lost with 1,023 sailors and one cruiser damaged (plus two destroyers badly damaged and 709 men wounded) at the Battle of Savo Island. There were no enemy losses to show for it. Another destroyer, the *Jarvis*, crippled by one air attack before the battle, was sunk by a second, late on the 9th, with the loss of all 247 aboard. The subsequent inquiry into the disaster apportioned no blame. But Mikawa, the outright victor, had failed in his primary mission, a specific order from Yamamoto to attack Turner's transports, which escaped unscathed. In the confusion of the night action, the Japanese ships, though only lightly damaged if at all, had lost their cohesion, and Mikawa decided to quit while ahead. The American screen had done its job, at staggering cost: it had protected the transports from harm. A consolation prize was claimed by the old submarine *S44* (Lieutenant Commander John R. Moore) on the 10th: it sank the heavy cruiser *Kako* in five minutes with four torpedoes as the victorious Japanese were sailing back to port.

Ten days after their counterinvasion of the Solomons, the Americans tried a diversion in one of their most elaborate special submarine missions of the war. Their largest, twenty-seven-hundred-ton fleet boats, *Nautilus* and *Argonaut*, took Lieutenant Colonel Evans F. Carlson, USMC, and 222 Marines of his Second Raider Battalion to attack Makin Island (now Butaritari); the president's son, Major James Roosevelt, was executive officer and second in command of the landing force. The task force was led by Commander John Haines, USN, on *Nautilus*. Makin is one of the Gilbert Islands, now the independent state of Kiribati; its Japanese garrison consisted of a platoon of forty-three men without officers. Nimitz knew the islands were weakly guarded and hoped to cause confusion by exploiting the fact. The two boats set off from Pearl on 8 August and reached the Gilberts, twenty-two hundred miles away, a week later. *Nautilus* arrived first to reconnoiter, which was fortunate because the attackers had been misinformed about the treacherous local tides round the atoll. As the Marines were to go ashore in rubber boats powered by outboard motors, the potential for disaster was already quite sufficient without hidden hazards. In the choppy waters off Makin, the right men did not always manage to get into the right rubber landing craft, and the surf drowned the officers' orders. Carlson therefore decided to make a single landing, rather than two at different spots. One boat out of seventeen did not hear the changed plan and became separated by a mile. However, all the men were ashore by 5:00 A.M. on 17 August.

An accidental shot having alerted the tiny garrison, Sergeant Major Kanemitsu organized a determined last-ditch stand round the radio station. *Nautilus*, with her two six-inch guns, fortuitously sank a pair of boats bringing sixty reinforcements from other islands in the atoll while she was bombarding shore positions. But the sergeant major's force was eventually doubled in strength, reducing the Marines' advantage to less than three to one. The submarines went under when Japanese bombers came over. The twelve Americans from the separated boat were able to attack from the rear; the raiders' difficulties only came to a head when they tried to get back to the waiting submarines that evening. The surf persistently retarded their weakly powered boats, and not even half the party got back to safety. By nightfall on the 18th seventy men still remained ashore, despite efforts, interrupted by constant air raids, to evacuate them all. The survivors ashore panicked, concluding they had been marooned, and decided to surrender – but could find nobody to accept. This was surely very fortunate for them, as the Japanese had shown scant respect for the concept of honorable surrender in defeat before and were unlikely to have changed their ways now. Most of the 'marooned' Marines managed to get back to the submarines on the third evening, and they left, well pleased with themselves, at midnight for a heroes' welcome in Hawaii.

But nine Marines had managed to get themselves genuinely marooned, bringing American losses in the raid up to thirty. When their Japanese opposite numbers arrived a few days later, they took their prisoners to Kwajalein, the local headquarters in the Marshall Islands, to the north. They were decently treated pending transfer to Tokyo – until the Marshalls commander, Vice-Admiral Abe Koso, had them beheaded, apparently because they had become an administrative nuisance. Abe was hanged for this at the end of the war.

The strategic consequence of the Makin Raid was to prompt the Japanese to build up their garrisons all over the Marshalls and Gilberts; the Navy persuaded the Army to help. This may have been useful to the Allies in the short term, because it weakened the enemy elsewhere and inconvenienced them generally. But in the longer term it meant that when the Americans came calling again they were met by much stronger defenses and resistance of a different order of magnitude altogether.

Pinprick raids may be good for morale and sometimes offer useful practice; but they can also anger and alert the enemy or backfire in other ways. The Makin operation was completely unnecessary and was not repeated elsewhere.

The 10,900 Marines now on Guadalcanal thought at first that they were about to relive the experiences of the US Army in the siege of Bataan and Corregidor. They found themselves in conditions worse than they could have imagined or been prepared for, cut off by an enemy who apparently controlled sea and air, with half-rations for five weeks and ammunition for just four days of combat. They had no radar, heavy guns, aircraft, or, as far as they could tell from their dugouts, prospects of relief. They had to put up with high-level bombing from beyond the range of their light anti-aircraft guns, and also with bombardment from submarines out of range of their own light artillery. The general feeling was one of having been left in the lurch by the US Navy – expressed rather more strongly at the time. In fact Admiral Fletcher, having retreated southeastward for twelve hours without permission from Ghormley, had reversed course on hearing of the massacre at Savo Island; two and a half hours later, when Ghormley's belated approval for the withdrawal arrived, Fletcher put about once more, despite pleas from Captain Forrest Sherman of *Wasp* for permission to race northward and help Turner and the Marines. Even the Japanese, unaware of just how many Americans had gone ashore, thought that the US Navy had given up the whole idea.

That this was not so was first shown on the night of 15 August, when four old destroyers converted into fast and maneuverable transports came up with aviation supplies – fuel and munitions for the planes to be based on Henderson Field as soon as the Marine engineers finished the work abandoned by the Japanese. An amphibious Catalina had already made a trial landing and takeoff on the 12th; the first Marine planes – nineteen fighters and twelve dive-bombers flown off the escort carrier *Long Island* – arrived on 20 August, on which day a second destroyer-delivery of supplies was made. By then 120 ground crew were already in place, as well as hard shelters for the planes, fuel dumps in the jungle, and air controllers in the 'Pagoda.' This was what the Marines called the airfield tower left behind by the Japanese.

Vandegrift, meanwhile, decided on a perimeter defense of the area round the airstrip as probing patrols tried to find out where the main Japanese forces on the large island might be. Colonel Frank Goettge led a party by boat to the Mataniku River, well to the west of the airfield, on 12 August; assuming that what appeared to be a white flag in the distance signified a Japanese surrender, they were massacred (there was no trick: lack of wind had concealed the Japanese sun on the white cloth). Over to the east of Henderson, on the Tenaru River, Colonel Ichiki Kiyono was preparing to mount the first Japanese attempt to recover the airstrip. He had landed with nine hundred men from destroyers on the night of the 18th and marched westward until they collided violently with two battalions of the First Marine Regiment, which was dug in with light tanks in support and on the alert, thanks to patrols. The outnumbered and outgunned Japanese were slaughtered almost to a man in a fierce fight at close quarters, for the loss of thirty-five Marines killed and seventy-five wounded. Colonel Ichiki committed suicide.

Major John Smith's Wildcats of Squadron VMF-223 and Major Richard Mangrum's Dauntless dive-bombers of VMSB-232 were in action on 21 August, the day after they arrived. These were the first of sixteen Marine, nine Army, ten Navy, and two Royal New Zealand Air Force squadrons that fought from 'Cactus' – the code name for Henderson Field – in the next five and a half months. They quickly established and resolutely sustained local air supremacy by day, forcing the Japanese to reinforce and supply after dark by what soon became known to the Americans as the 'Tokyo Express' – destroyer-convoys coming down the Slot to Iron Bottom Sound by night, unloading, and dashing away again. The struggle for Guadalcanal, a singularly obscure fleck of mud on one of the remotest parts of the earth's surface, was fought in four dimensions: in the air, on the ground, at sea and under it, as well as in logistics. The rapidly escalating and in the end wildly disproportionate effort invested by each side was really a battle of wills on which the United States as much as Japan staked its national prestige. Yet the American preparation for the ordeal in the air had been precariously sketchy. This is how Major Robert E. Galer – commander of VMF-224, which was formed on 10 July and joined the battle on 30 August – described his contribution just before his death in action at the end of the year:

First we had to draw airplanes and attempt to get spare parts. The fact that we were going into a combat area didn't seem to draw the weight that we thought it would in getting supplies. We ran into several difficulties. I believe we took a few short cuts in getting ready to go out that we hoped were legal. With fifteen days to outfit a squadron and qualify on the carrier with new pilots, the work got rather out of hand, and I believe was directly responsible for the death of a couple of pilots later on. Some of my people had exactly two weeks in which to get familiar with the airplane, carrier-qualify, do their gunnery and in general get rounded out from having just graduated from a training school . . .
    Our first combat flight was the next day [after arrival]; for eleven out of my nineteen pilots it was the second time they had ever been on oxygen . . . Because of either lack of experience or defective masks or equipment . . . we lost two pilots.

Galer must have been one of the first to make a complaint that would be heard times out of number in the air war over the Pacific – the absence of fixed tours of duty owing to the relentless demand for air crew. Galer's statement of the problem will be familiar to any reader of Joseph Heller's *Catch-22*:

As it is, you go in, you're in for an indefinite period. You begin to crack up a little bit, and there is no way of getting out unless you actually humble yourself, swallow your pride and say, 'I can't take it; take me out.' If you get your outfit out sooner, you'll have a hell of a lot better outfit! My squadron – I'm proud of them – did a darn good job out there, but there are only five people [out of nineteen] in that squadron I'd take back out to a combat area, for the simple reason that they were there too long. If they had gone out at the end of one month, got a month's rest and gone back in again, nine out of ten of them would still be good combat pilots.

Until short-range, very-high-frequency radio came into use, radio silence was often required from the carriers as insurance against Japanese radio direction-finders, probably the best in the world at the time. This could easily become another trap for the unwary airborne greenhorns, as Commander L. J. Dow, communications officer of USS *Enterprise* during the Guadalcanal campaign, recorded:

We have planes [that] get lost quite often. It's something to hear young pilots getting lost and running out of gasoline and not being able to give them any instructions because we have to keep radio silence. It's terrible. As soon as we get this *VHF* equipment, along with our radar organization, there is no reason why any planes should be lost [in this way].

The problems of supplying the ever-expanding American occupation force as it beat off successive Japanese counter-attacks, each bigger than the last, collided head-on with the laws of economics:

Robinson Crusoe should be required reading for anyone who is setting up an advanced base in the South Pacific islands [said Captain M. B. Gardner, chief of staff to Admiral McCain, South Pacific air commander]. There is nothing there. Everything they eat, everything they wear, every place that they live has to be brought in from the US. There is no such thing as living off the country in the South Pacific, unless you live on coconuts alone . . . There are no dock facilities, no cranes and . . . very poor beaches for landing craft. Heavy engineering equipment . . . must be brought ashore so that you can build your field and all the other facilities which must be built. Everything is manhandled and that requires a tremendous manpower . . . so that some balance has to be struck. It's quite conceivable that you could have so many men to do the work that the transporting of supplies for their actual living brings you to a point where the law of diminishing returns sets in.

McCain's staff worked out how much aviation fuel would be needed for the first two weeks of air operations from Henderson. Then they doubled it to be sure. It was put in drums and delivered by destroyer from Espíritu Santo – and it ran out in ten days. Meanwhile, backlogs built up: deliveries of twelve hundred tons of fuel a day were arriving at the base, but only four hundred were going on to Guadalcanal until the lessons were learned – sometimes only in time for later operations:

Bottoms were lying in Segend Channel for weeks before they could be touched. One ship arrived about 10 August and was still only 20 percent unloaded by 18 November . . .

I think that, viewed dispassionately, there were at least three occasions when the chance of our holding Guadalcanal was not worth five cents. But the Marines on Guadalcanal didn't figure that way.

After the 'white-flag' shock mentioned above, which went round the ranks in no time at all, the Marines were not inclined to take prisoners, even on those few occasions when the opportunity presented itself. A useful reminder of what they were fighting against, and of how the Japanese Army reduced atrocity to routine, is contained in the following extract from a diary found on the body of an unnamed officer, probably on the staff of Major General Kawaguchi Kiyotake, who served on Guadálcanal from the end of August to just before the Japanese withdrawal:

29 September: Discovered the captain and two prisoners who escaped last night in the jungle and let the guard company guard them. To prevent them escaping a second time, pistols were fired at their feet, but it was difficult to hit them . . .

The two prisoners were dissected while still alive by medical officer Yamaji and their livers were taken out, and for the first time I saw the internal organs of a human being. It was very informative.

Close by, the furious sound of cannon and rifle-firing could be heard, while the guns of naval vessels also shelled us. Tonight made preparations so that we would be able to start action at any time, and went to sleep.

Colonel Ichiki's doomed detachment was intended only as the advance guard of a rather larger force; there are grounds for suspecting that the colonel exceeded his orders in hope of glory by getting involved in combat so soon after landing on 18 August. Clearing the Americans out of the southern Solomons had been assigned to Lieutenant General Hyakutake Harukichi's 17th Army (corps), based at Rabaul – which was also responsible for the simultaneous operations in Papua. Hyakutake moved his underemployed 35th Infantry Brigade from garrison duty at Palau, five hundred miles east of Mindanao, Philippines, to Rabaul as the core of a force of six thousand men to be landed on Guadalcanal. The expedition, timed to start on 23 August, was to be covered by the bulk of the Combined Fleet, its post-Midway confidence largely restored by the Savo Island victory. But the preparatory movements for this Operation KA alerted American intelligence analysts. The circumspect Admiral Fletcher was therefore ordered back into the fray from the South Pacific with his three-carrier Task Force 61 to break up the Japanese attempt at reinforcement. *Saratoga*, *Enterprise* and *Wasp* each had a powerful escort of cruisers and destroyers. Nimitz sent to sea *Hornet*, his fourth carrier, out of Pearl Harbor, plus two battleships and supporting forces, as distant cover.

Anticipating such a move, Yamamoto assigned Vice-Admiral Nagumo Chuichi, now in somewhat reduced circumstances as commander of the Third Fleet, formed after the Midway disaster, to tackle Fletcher with the fleet carriers *Shokaku* and *Zuikaku*, supported by two battleships, three cruisers and destroyers. Admiral Kondo of the Second Fleet was in overall charge of this phase of the naval side of the Guadalcanal operations: he had five cruisers and five destroyers under his direct command, supported by a battleship, a seaplane carrier and destroyers, and the light carrier *Ryujo* as kernel of a 'diversionary group' to attack Henderson Field. Admiral Mikawa's Eighth Fleet contributed the light cruiser *Jintsu* and destroyers plus four cruisers and a Special Naval Landing Force. Various supporting ships, a dozen submarines, convoy escorts and land-based naval air forces to bomb the Americans ashore completed this overintricate organization of naval forces under the supreme direction of Yamamoto, aboard the battleship *Yamato* with escorting forces off distant Truk in the Carolines. All this vast effort was intended to put only fifteen hundred Japanese troops ashore, about half Army and half Navy.

The two navies clashed again on 24 August in the Battle of the Eastern Solomons – after Fletcher had detached *Wasp* and her escorts to the south to refuel, reducing his main forces by a third. The small carrier *Ryujo*, having launched an attack on Henderson which was beaten off by the Marine fighter defense, came under attack herself in her other assigned role as bait for the American fliers. Aircraft put up by *Saratoga* and *Enterprise* duly sank her. While these were thus engaged, their carriers were attacked by strikes from *Shokaku* and *Zuikaku*. The Americans had learned from experience. Each US carrier was in the center of a circle formed by its escorts, whose anti-aircraft guns were ready to put up an intense defensive fire while a strong fighter 'cap' flew high overhead to pounce on enemy raiders. These well-laid plans were adversely affected by the garrulity of the pilots, not only from the combat air patrol but also from the bombers that had gone after the Japanese carriers: the available radio channels simply could not cope with the traffic, especially when several separate dogfights developed. Heavy intermittent cloud gave the attackers on both sides a chance to hide – and to get lost – at the last minute.

But the *Enterprise* was soon hit fair and square by three bombs, which passed through several decks aft and exploded in the bowels of the carrier, causing serious fires. Well-drilled damage-control parties performed miracles of improvisation to have her ready for landing-on her own aircraft within an hour. But a steering engine broke down and her rudder jammed, forcing her to run clockwise in circles for more than half an hour before the failure could be corrected. Had Nagumo's second wave been able to find her, the lucky *Enterprise* could well have been lost. As it was, the Japanese pilots withdrew for lack of fuel. Some of her planes, having failed to find a target, were obliged to land at Henderson, providing a welcoming temporary reinforcement. Her chief escort, the battleship *North Carolina*, managed to evade the bombs while putting up a curtain of anti-aircraft fire. Others from the *Enterprise* crippled the seaplane carrier *Chitose*, and planes from the *Saratoga* bombed but missed Kondo's cruisers. The Americans lost a total of seventeen carrier planes in the confused, hesitant, and inconclusive clash, less a battle than a skirmish; the Japanese significantly more.

The loss of the *Ryujo*, whose role as a decoy was not conclusively exploited by the two large Japanese carriers, gave the Americans, who saved the damaged *Enterprise*, a tactical victory by a small margin. The *Chitose* also survived – as did

Rear Admiral Tanaka's reinforcement convoy of the light cruiser *Jintsu* with eight destroyers, four converted destroyer-transports and a large merchantman, none of which came under attack in the Slot on the 24th. Fletcher withdrew southward from eastern Solomons waters after dark to avoid a night gun action with superior enemy surface forces. Since this was precisely what Admiral Kondo intended, Fletcher's habitual caution was justified, if only from hindsight. The American admiral had shown no more lust for battle than his opponent.

On the 25th both sides groped for each other like tired wrestlers with smoke in their eyes. Two Japanese air raids caused minor damage to American positions on Guadalcanal. Marine dive-bombers from Henderson found Tanaka's convoy on the way back from failing to find the Japanese carriers, set the merchantman, a large troop-transport, on fire, and scored a damaging hit on the flagship *Jintsu*. Later in the morning eight B-17s came up from Espíritu Santo in the New Hebrides, flew over the Slot at their customary great height, scored rare direct hits on a destroyer, the *Mutsuki*, which sank, and damaged another. Tanaka was ordered to withdraw his convoy and reorganize it for a Tokyo Express night run on the 28th. USS *Wasp* moved up to a position east of Guadalcanal and searched for targets in all directions, finding none, as Fletcher's two other carrier groups refueled. The *Enterprise* then parted company to return under escort to Pearl for major repairs. Both sides went on the defensive at sea for the next six weeks.

On land, General Hyakutake decided to commit the 35th Brigade – thirty-five hundred men led by Major General Kawaguchi Kiyotake – to the recovery of Guadalcanal. The Tokyo Express was sighted before dark on its approach down the Slot from the north in unusually good visibility by Marine pilots. Henderson mounted a dive-bomber attack on the evening of the 28th. One escorting destroyer was sunk, and two out of four were damaged. The convoy withdrew. But on the 29th, 450 men were landed successfully east of the American perimeter, and on the 30th a Japanese air raid sank the American converted destroyer *Colhoun* in Iron Bottom Sound. After dark the Japanese landed the rest of their second troop-reinforcement at Taivu Point, well to the east of the US Marines. On the morning of 31 August the carrier *Saratoga*, on patrol 250 miles southeast of the Solomons, was sighted by Commander Yokota in submarine *I26* and hit amidships on the starboard side by one of his salvo of six Long Lance torpedoes. She was, if anything, less seriously damaged than in her first submarine attack, on 11 January, but could make only twelve knots by herself. A dozen men, including Fletcher, were wounded, none seriously. Towed by a cruiser, *Saratoga* got up enough speed to fly off her valuable aircraft to Espíritu Santo, where they refueled and flew to reinforce 'Cactus' (Henderson). Once again USS *Saratoga* withdrew from the fray for three months, saved from destruction by her battle-cruiser hull.

That night General Kawaguchi landed on Guadalcanal from a Tokyo Express with the last of his force, some twelve hundred men, well to the west of the American beachhead. Now and later, the American garrison had the greatest difficulty in latching on to these fast Japanese night runs. Radar was no real help with so much land in the background: the thick tropical nights and unpredictable weather made them difficult to spot by eye. Admiral King in particular waxed furious, especially in the early days of the long fight for Guadalcanal, demanding to know how it could be that the Expresses got away with it time after time when the Marines had established local air superiority. A pattern had developed whereby the Americans could do more or less what they wished during the day and the Japanese by night. To try to break the Solomons deadlock, Ghormley as South Pacific commander called a conference at Nouméa which was attended by Nimitz as well as General Arnold, the Army Air Force chief, Generals Sutherland (chief of staff) and Kenney (air commander) from MacArthur's command, plus Marine officers and Rear Admiral Turner, the amphibious commander. Tension was high as the Japanese threat to Port Moresby was rising to its peak; each of the two American Pacific supreme commands saw the other as a drain on resources it badly needed, just as Operation Torch, on the far side of the distant Atlantic – the planned US landing in North Africa – was demanding the lion's share of US material resources.

Nevertheless, it was decided to continue with both approaches to the intermediate Allied objective of Rabaul. Guadalcanal had ceased to be a sideshow for the Americans, who went there only because the Japanese had been building an airfield that could threaten Australia (but in fact was conceived to protect the flank of the Japanese drive on Port Moresby). It therefore took the Japanese longer to realize that their secondary operation had perforce become the primary one because of the very large resources being invested in it by the enemy. For many weeks the reaction at IGHQ in Tokyo to the American effort in the Solomons was one of paralyzed amazement that the Americans were already staking so much on so little. By the time the junta woke up to the need to stop wasting manpower by reinforcing in penny-packets, the Americans were too well dug in to shift.

During the night of 4–5 September, a Tokyo Express bringing more men and supplies to Kawaguchi's force caught and sank the American destroyer-transports *Little* and *Gregory* in a nasty little night action. It was triggered off by a patrolling US Navy Catalina, which saw the gun flashes of the three Japanese covering destroyers bombarding the shore to mask the landing and dropped flares. These revealed the presence of the two American ships to Japanese eyes.

Kawaguchi was planning an attack on Henderson Field from east, west, and south simultaneously on 12 September, to be followed by a seaborne assault on the beachhead from the north. Only when the constant threat posed by the airfield was eliminated would the Japanese Navy have the freedom of action it wanted for a 'decisive battle' at sea. Meanwhile, some six hundred American Marines under Colonel M. A. Edson landed on the northeast coast of Guadalcanal from Tulagi at dawn on the 8th and managed to destroy one of Kawaguchi's main supply dumps and capture a battery of artillery well to the east of Henderson. He stuck to his plans, however, and attacked on 12 September, his main line of advance being what the Americans soon named 'Bloody Ridge,' a small, paddle-shaped escarpment to the south of the airfield. Edson's Rangers and paratroops, using their vastly superior automatic firepower, successfully defended the ridge against yelling Japanese infantry brandishing long bayonets on the ends of bolt-action rifles almost as tall as themselves. Two nights of murderous closequarters fighting ensued, relieved only by skirmishing in daylight hours.

The eastern and western Japanese thrusts were beaten off with less difficulty, and Kawaguchi's forces retreated into the jungle, diminished by casualties of about one-third.

One night, at this or some other battle on Guadalcanal, an American sentry heard movement in the jungle and issued the customary challenge. He was rewarded with the following unusual reply: 'Hold your fire! We are American Marines and wish to report our evening's activities.' The sentry promptly blew his magazine. 'I fired right away because I knew no Marine would talk that way,' he explained. Usually the English of Japanese soldiers ran to little more than bloodcurdling threats like 'Marine, you die.' The first serious close combat between Japanese and American troops, in the Solomons and New Guinea, had in fact killed rather more of the former than the latter. The ground fighting in the Pacific Campaign had already become a confrontation between fanaticism and firepower – a pattern which would only intensify on both sides as the war continued.

7

At Sea in the Solomons

On 14 September, the day after Bloody Ridge, the Seventh Marine Regiment from the Second Division left Espíritu Santo to reinforce Vandegrift's force, in a convoy commanded by Turner and covered by the only two battle-worthy carriers in the Pacific Fleet, *Wasp* and *Hornet*, with their customary surface escort. On the afternoon of the 15th Lieutenant Commander Kinashi Takaichi, IJN, captain of submarine *I19*, got the *Wasp* (Captain Forrest P. Sherman, USN) in his sights and fired four torpedoes at her, about halfway between the New Hebrides and the Solomons – the 'Torpedo Alley' of the Coral Sea. Two struck home on the starboard side, and the new carrier went up in a welter of explosions and flame. The third torpedo ripped a gash thirty-two feet by eighteen in the port side of the battleship *North Carolina*, and the fourth tore the bow off the destroyer *O'Brien*. The most remarkable fact about this unsurpassed torpedo attack is that the two latter ships were escorting the *Hornet*: they were a good six miles farther away from the *I19* than the *Wasp*! The Americans, unaccustomed to such astonishing torpedo-performance, wrongly insisted on crediting the hits on *North Carolina* and *O'Brien* to *I15*, which was known to be in the vicinity at the time – but confirmed *I19*'s claim to have stung the *Wasp*. Japanese records attribute all four hits to *I19*. The *Wasp*, twenty-one thousand tons all up, was abandoned after an hour and had to be dispatched by US torpedoes after dark. The *North Carolina* patched herself up with remarkable aplomb and proceeded to Pearl for repair. The *O'Brien* got to Nouméa, received first aid, and sailed for the US West Coast for permanent repair, but the damage was even worse than it seemed: the destroyer broke in half and sank on the way across, though her crew was saved. The US Pacific Fleet was now down to one operable carrier and one new-generation battleship, the *Washington*.

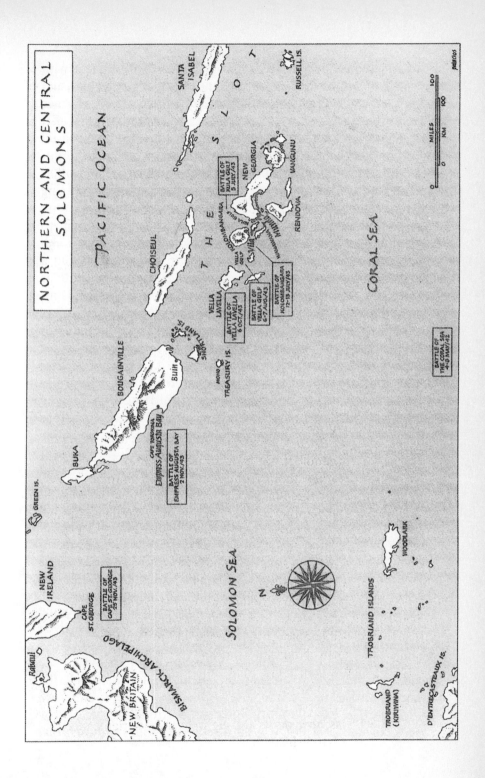

NORTHERN AND CENTRAL
SOLOMONS

PACIFIC OCEAN

SANTA ISABEL

T H E S L O T

RUSSELL IS.

NEW GEORGIA

VANGUNU

BATTLE OF
KULA GULF
5 JULY/43

KULA GULF

RENDOVA

CHOISEUL

KOLOMBANGARA

VELLA GULF

BATTLE OF
KOLOMBANGARA
12-13 JULY/43

CORAL SEA

VELLA LAVELLA

BATTLE OF
VELLA GULF
6-7 AUG/43

BATTLE OF
VELLA LAVELLA
6 OCT/43

MILES 100
0 KM 100

p8frc01

BOUGAINVILLE

BUKA

GREEN IS.

BUIN

CAPE TOROKINA
Empress Augusta Bay

BATTLE OF
EMPRESS AUGUSTA BAY
2 NOV./43

TREASURY IS.

MONO

SHORT

LAVELLA IS.

BATTLE OF
THE CORAL SEA
4-8 MAY/42

NEW
IRELAND

CAPE
ST.GEORGE

BATTLE OF
CAPE ST.GEORGE
25 NOV./43

BISMARCK ARCHIPELAGO

Rabaul

NEW BRITAIN

SOLOMON SEA

N

WOODLARK

TROBRIAND ISLANDS

TROBRIAND
(KIRIWINA)

D'ENTRECASTEAUX IS.

Turner coolly decided, having come so far, to press ahead with his delivery despite the debacle a hundred miles behind him, and deservedly got away with it. The four thousand fresh Marines went ashore at dawn on 18 September with all their gear, as if on maneuvers, while destroyers pounded the Japanese coastal positions to the east and west of the landing at Lunga Point, close by Henderson Field. From there the Marine and other pilots beat off one Japanese air attack after another, claiming six enemy aircraft for each US plane lost. The Japanese took to mounting raids by one or two planes at a time at night to keep the garrison awake. They would lob the occasional bomb or light up the area with a flare, to help the warships find a target to bombard from the Slot. The Marines drew on their cornucopia of instant nicknames to call one of these planes 'Washing-Machine Willie' because of its asthmatic engines and another 'Mayday Charlie' because its engines spluttered. But the generic term for the airborne nuisance-raider was 'Louie the Louse.'

Heartened by the arrival of their reinforcements, the Marines decided to go over to the attack, in a westward direction from the defended perimeter round Henderson Field. Three battalions tried to trap elements of the main Japanese force in impossible terrain on 26 September but were forced to withdraw the next day. The Americans doubled their stake to six battalions and launched a more successful though still not decisive attack on 7–9 October, pre-empting a Japanese attack from the west. On the latter date a small force of Marines shipped over from Tulagi destroyed the main surviving Japanese base east of Henderson, some thirty miles away, leaving Vandegrift free to concentrate on the main threat from the west. This was now considerable. The Tokyo Express had steadily built up General Hyakutake's forces on Guadalcanal to twenty-five thousand men, including the whole of the Second Division. The general himself arrived to take charge of the big push on the 9th.

On the same day, it fell to Admiral Turner to put to sea with the first US Army reinforcement for Guadalcanal, the three thousand men of the 164th Infantry Regiment, Americal Division. The two transports and eight destroyers of the convoy were covered at a distance by three task groups – the *Hornet* with escorts, the battleship *Washington* with escorts, and Rear Admiral Norman Scott's Task Force 64 of two heavy and two light cruisers plus five destroyers – largely the *Wasp*'s old group without the *Wasp*. Scott's secret weapon was his single-minded pursuit of excellence in what had hitherto been a Japanese preserve – the night action by surface ships. His task was to operate in the Slot, disrupting enemy movements. This was no different from Vice-Admiral Mikawa's order to Rear Admiral Goto Aritomo with his three heavy cruisers and eight destroyers, covering the latest Japanese landing on the night of 11 October. Scott heard from aerial reconnaissance during the day that the Japanese force was approaching down the Slot, and raced north to meet it. The desired encounter took place to the northwest of Savo Island and the northeast of Cape Esperance, a promontory of Guadalcanal, after which the ensuing battle was named.

The Americans were thrown into confusion – and could have been exposed to another cruiser defeat – by their own planes and radar. Two Japanese forces were in the Slot that night, one to unload troops and supplies, the other, much more powerful, to cover the first and to bombard. Scouting seaplanes from the US cruisers and radar picked up two separate forces all right, but it was not at first possible to identify them precisely. By placing his line between Savo and Esperance, Scott felt certain he would catch the Japanese force approaching Guadalcanal, or the other group as it left, but he did not know which was which: the cruisers might have passed ahead of their convoy into Iron Bottom Sound, or the convoy might have gone ahead, leaving the cruisers northwest of Savo. To retain his position between the two groups, Scott had eventually to reverse course, which he ordered just before midnight on the 11th. His flagship, the heavy cruiser *San Francisco*, was not using her old SC radar for fear of Japanese detectors; light cruiser *Helena*, the Pearl Harbor survivor, had the latest SG equipment and confidently reported a large target to the west – just as the light cruiser *Boise* seemed, in a garbled radio report, to be sighting another to the northeast. With the course reversal in progress, however, they might only be picking up their own destroyers, three of which were making a wider and faster turn round the rest of the column in order to resume their position at its head. The general radio-signals mix-up gave Captain Gilbert Hoover of the *Helena* the impression that he was free to open fire on his persistent contact, which he did at a quarter to midnight, from five thousand yards. The heavy cruiser *Salt Lake City* followed suit from ahead at only four thousand yards. Soon all the Americans were blazing away at whatever it was to their right, northeast of their line.

Scott, however, believed they were shooting at their own destroyers and ordered a cease-fire. Fortunately for him, Admiral Goto on the other side believed he had come under fire from his own Tokyo Express convoy and ordered his cruisers, led by the *Aoba*, into a starboard turn which set them up one by one like ducks in a shooting gallery. Goto was killed on the *Aoba*'s bridge. Scott, meanwhile, established by direct radio communication the exact whereabouts of his three 'missing' destroyers and ordered the resumption of fire before most of his command had obeyed the order to stop. But the American destroyer *Duncan* had been caught between the two lines and was sunk by the shells of both sides; her sister ship, *Farenbolt*, was damaged by American shells also. *San Francisco* blasted the Japanese destroyer *Fubuki* to destruction before Scott, on the bridge, ordered a cease-fire at midnight to sort out his line. The Japanese heavy cruiser had fallen out, mortally stricken, and later sank on the far side of Savo Island. *Aoba* burned fiercely, while the US cruisers were only lightly damaged if at all. But when firing resumed in the opening minutes of the 12th, USS *Boise* caught fire and turned aside to save herself.

By half past midnight it was all over. The Japanese had lost one heavy cruiser out of three and one destroyer out of two engaged, with one cruiser badly damaged, to the American loss of a single destroyer with another crippled. Of the four US cruisers, only the light *Boise* was seriously damaged, with a smashed bow section, while the heavy *Salt Lake City*, which had interposed herself to save her, was moderately damaged: a clear victory for the US Navy – when it was able to sort fact from fiction in the claims of damage to the enemy – but only a tactical one. The Japanese too thought, or at least reported, that they had done far more damage than was in fact the case, but their Tokyo Express completed its delivery, including the first Japanese heavy artillery, despite the punishment administered to its covering force. On 12 October two destroyers from the transport convoy were caught and sunk by Cactus aircraft. The airfield was heavily if inaccurately bombed in its turn by Japanese aircraft on the same day, but Turner too managed to deliver his reinforcements without interruption.

On the 13th the bombers came in again and did rather more damage, and later the same day a heavy gun, immediately nicknamed 'Pistol Pete,' started a leisurely, carefully aimed shelling of the American airstrip. In the early hours of the 14th Vice-Admiral Kurita Takeo's two battleships, *Kongo* and *Haruna*, hurled fourteen-inch shells at Henderson, destroying half the Americans' ninety aircraft. For those who experienced it, this remained '*the* bombardment,' no matter how long they served on Guadalcanal. After daylight, Japanese bombers appeared, adding to the damage and knocking out the original strip, now replaced by a new reserve strip which immediately drew the attention of Pistol Pete. The naval bombardments continued: two cruisers expended 750 eight-inch shells on the night of 14 October, a prelude to the daylight landing of the last elements – forty-five hundred men – of the new Japanese attacking force and their supplies only ten miles west of Henderson, where there was a desperate shortage of aviation fuel as well as planes. That night, the 15th, another fifteen hundred eight-inch shells came in. The Japanese still had control of the sea; the Americans had not established a monopoly of airpower, even by day, at this delicately poised stage of the stupendous struggle for steaming, stinking Guadalcanal. American ships, barges and aircraft stole in whenever they could to deliver ammunition, supplies and fuel. Both sides sent in carriers to raid each other's supply lines, with considerable effect.

If the best efforts of Marines, soldiers, airmen, naval sailors and merchant seamen could not break the deadlock, perhaps a new admiral could. Nimitz decided after a tour of the South Pacific that the exhausted and demoralized Vice-Admiral Ghormley was not aggressive enough, and on 18 October replaced him with Vice-Admiral 'Bull' Halsey. If aggression was the missing ingredient in the American campaign for Guadalcanal, the deficiency was now more than made good. The fleet and the Marines were jubilant. Nimitz also borrowed the Army's 25th Division from Hawaii, assigned it to reinforce the Marines, and sent in more submarines, Army fighters, and B-17s from central to South Pacific command. Halsey's first order – and his slogan – was 'Kill Japs, kill Japs, kill more Japs!' Roosevelt's 'hold Guadalcanal' message to the Joint Chiefs of 24 October underlined how high the stakes had become politically: this huge and still-growing American effort to hold an island hardly anyone could find on a map while the U-boats were still roaming the Atlantic and the US Army was preparing to land in North Africa could only be justified by positive results. But exhortations, whether from Halsey's new office, brusquely commandeered from the difficult French in Nouméa, or from the Oval Office at the White House, were not enough.

The Japanese were busy organizing their third and largest attempt to dislodge the Marines from the Solomon Islands under General Hyakutake, a struggle that now far surpassed the stalled Papuan operations in importance for them as for their enemy. Their plan was essentially a repeat of Kawaguchi's – coordinated attacks on Henderson from west and south – but with four times the number of men. The Combined Fleet had agreed to send large forces to deal with any American naval presence. The dreadful terrain held up the advance of Lieutenant General Maruyama Masai's Second Division from the south, but Major General Sumiyoshi Tadashi's smaller detachment, unaware of this, came in from the west on schedule on 23 October behind nine light tanks, only to be savagely mauled and driven off by the Marines, who were fully prepared. Late that night, however, the thinned-out defense of the southern perimeter came under attack from Maruyama's men. They formed up in two columns, led by Major Generals Kawaguchi and Nasu Yumio, and threw themselves at the 164th Infantry Regiment, USA, and the single Marine battalion guarding a four-thousand-yard front. Japanese losses, at over a thousand killed, were five to one. The Japanese tended to leave their dead on the battlefield – because there were usually not enough left alive to bury them. Their second and last assault in this third attempt to take the airfield was no less ferocious – and no more successful. It came twenty-four hours later and probably killed twice as many.

It was during this second round of the third Japanese counterattack that Sergeant John Basilone, USMC, won the Congressional Medal of Honor in one of the most extraordinary displays of sustained personal valor on record. He led a section in a weapons company of the First Battalion of the Seventh Marine Regiment, First Marine Division. His command consisted of four teams, each manning a Browning heavy machine gun, part of the force guarding the western, landward approach to Henderson in the vicinity of Lunga Point. He had hardly put down the field telephone after a call from battalion headquarters warning him of a Japanese attack when the enemy arrived, in the middle of a tropical cloudburst. He picked up one of the guns, which weighed a hundred pounds, and ran through the mud to strengthen the threatened right flank of his position thirty yards away. The two machine guns there had gone silent, and their crews were defending themselves with rifles and pistols. Basilone cleared the area with long bursts of fire, to gain the sixty seconds he needed to unjam one of the other weapons.

The Japs kept charging, the rain had stopped and the moon was just coming out and we could see them when they got up to the wire in front of our position, about thirty feet away from us. We could hear them too. Every time they started a charge they'd scream. That night they hollered, 'Marines, you die!' but they died . . . I fired both machine guns. I'd fire one and then roll over and fire the other. Every now and then someone would yell, 'Look out!' so I'd grab my automatic pistol which I kept on the deck beside me and used it. Then we ran out of ammunition for the machine guns. [Having crawled 150 yards to get more,] I got back [and] we kept on firing all night.

Basilone, nicknamed 'Manila John' as he had served three years with the Army in the Philippines until 1939, joined the Marines out of boredom with civilian life in July 1940. Each gun had started the battle with 126 belts of ammunition, each of 250 rounds of .50-caliber. After this frantic night there was one belt left at his position. Promoted to platoon sergeant, Basilone was decorated by General Vandegrift in Australia in May 1943. He was killed in 1944; it was widely believed in the Marine Corps that he died trying to win a second Medal of Honor.

The Japanese did better at sea. Yamamoto ordered Vice-Admiral Kondo Nobutake's Second Fleet to support the Army on Guadalcanal. At his disposal were two fleet carriers, two light carriers, four battleships, four heavy cruisers, more than two dozen destroyers and supporting vessels, including Nagumo's striking force of one light and two fleet carriers with the customary escort. The Americans had two carrier task-forces at sea built round the repaired *Enterprise* and *Hornet* respectively and a third led by battleship *Washington*. Halsey stayed in Nouméa like his predecessor, but issued a simple, all-embracing order: 'Attack – repeat, Attack!' After preliminary clashes between light forces in Iron Bottom Sound cost the Japanese a light cruiser and the Americans a tug and a patrol boat on the night of the 25th, the two fleets searched for each other all day, the Japanese ships waiting for the news that never came – the capture of the airfield – and the Americans trying to find them under the tactical command of Rear Admiral Thomas Kinkaid on the *Enterprise*. Both forces were in the waters north of the Santa Cruz Islands, well to the east of the southern Solomons, when their scouts found each other's carriers some two hundred miles apart, the Americans to the southeast, the Japanese to the northwest. Each launched a carrier strike, the Japanese with sixty-five, the Americans with seventy-three planes twenty minutes later – they passed and saw each other. Six out of nineteen *Enterprise* planes in this wave were lost, and two were forced to withdraw, while the Japanese lost two Zeros. When they arrived over the American force, the *Enterprise* was lucky once again, passing into a concealing squall.

Of the two carriers only the *Hornet* (Captain Charles Mason, USN) was visible. Kondo's scouts had therefore reported only one such enemy ship. All the incoming planes therefore concentrated on the *Hornet*, which was struck within minutes by a bomb and the blast from two near-misses;

a crashing dive-bomber with two bombs still aboard, which promptly exploded below; and two torpedoes which hit the engine rooms. *Hornet* went dead in the water and started to heel to starboard. Three more bombs and another crippled aircraft scored direct hits shortly afterwards – all this despite the destruction of twenty-five Japanese planes by fighters and gunfire from two cruisers, two anti-aircraft cruisers and seven destroyers. *Enterprise* planes blew a large hole in the deck of light carrier Zuiho; *Hornet*'s badly damaged the fleet carrier *Shokaku*, out of commission for nine months after this Battle of Santa Cruz, and caused major damage to the *Chikuma*, a heavy cruiser.

The Japanese, now fully aware of the presence of a second US carrier, tried but failed to get her as well, as the struggle to save the blazing *Hornet* faltered. The *Enterprise* was hit three times in two major attacks but stood up to the pounding until she could withdraw in her own time in the afternoon. The American destroyer *Porter* was ruined by a torpedo-bomber and had to be sent to the bottom by her own side after the crew were taken off. The destroyer *Smith* was badly damaged when a burning Japanese plane hit her bow section. In mid-afternoon the Japanese planes came at the *Hornet* again as cruiser *Northampton* tried to tow her to safety, and once more in the early evening. By that time it had become impossible to explain how she could still be afloat. Indeed, when the time came for her own destroyers to sink her, sixteen American torpedoes were fired, nine hit – and she still held up, even after more than four hundred rounds of gunfire. In the end the Japanese came up, drove off the US destroyers, and dispatched the American carrier with four surefire twenty-four-inch torpedoes at 1:30 A.M. on 27 October. The pain of losing *Hornet* was compounded by the embarrassment of being unable to put her out of her misery because American torpedoes were so poor. Not displeased, Admiral Kondo withdrew his forces to Truk, Japan's 'Gibraltar' in the Carolines, after winning another tactical victory in the fourth carrier-battle in six months. Once again the Americans were left with only one at sea, the indestructible *Enterprise* – and she was damaged. So was the *South Dakota*, in a collision with a destroyer after the battle as they dodged torpedo tracks. The

Americans appealed to the British for the loan of a fleet carrier, a request which could not be met for some months.

At the end of October both sides were as determined as ever to get the upper hand on Guadalcanal and in the surrounding waters in the coming month. The Americans delivered long-range, 155mm artillery pieces to the Marines so they could outshoot Pistol Pete and his colleagues; the Tokyo Express came in and went out at night almost with impunity, bringing twenty thousand fresh troops and Marines. Each side sent cruisers and destroyers to bombard the shore positions of the other as the dogfights continued in the air and skirmishes erupted on the ground. Admiral Turner's Task Force 67, as the South Pacific Amphibious Force was for the moment styled, brought in another three thousand men of the Americal Division and some Marine reinforcements on 12 November, covered by Rear Admiral Daniel J. Callaghan's five cruisers and eight destroyers in Task Group 67.4. More distant cover was provided by TF 16 – Kinkaid with the patched-up *Enterprise* and escort – and TF 64, made up of two battleships and four destroyers under Rear Admiral Willis A. Lee. Admiral Kondo led an attack group of two heavy and one light cruisers and eight destroyers toward the southern Solomons from the northwest; under him Vice-Admiral Abe Hiroaki had a raid group of two more battleships, a light cruiser and fourteen destroyers. Vice-Admiral Kurita led the air support aboard the two light carriers *Junyo* and *Hiyo*. All this was to cover another Tokyo Express due to unload on the 14th: the Americans found out by air reconnaissance that they were coming.

Callaghan's two heavy and three light cruisers were the only US force near enough to Turner to help him. The former's flagship, *San Francisco*, had already been moderately damaged in an air raid. Turner pulled out his transports after dark on the 12th and left the field to Callaghan, reinforced at the last minute by Rear Admiral Norman Scott, this time in the light anti-aircraft cruiser *Atlanta* with two more destroyers. Inspired by Scott's success with the 'line of battle' at Cape Esperance, Callaghan followed suit: four destroyers led five cruisers and four more destroyers. Admiral Abe's raiding force came down in three lines, led by the light cruiser *Nagara* followed by battleships *Hiei* and *Kirishima* and flanked by three destroyers to starboard with six to port and two as advance guards. His purpose was to bombard the shore; he therefore passed between Savo Island to the north and Guadalcanal to the south of Iron Bottom Sound on a moonless night. Three of his destroyers to port became separated during maneuvering.

The American radar found Abe's squadron at 1:24 A.M., some fourteen miles to the northwest. Callaghan's line went straight at them, but the Americans were still surprised when they made visual contact less than twenty minutes later. So were the radarless Japanese, but they were trained to react in split seconds during a night action. Their searchlights found *Atlanta* first, and immediate cruiser gunfire wrecked her bridge, killing Admiral Scott. Japanese destroyers followed up with torpedoes and crippled the ship terminally. A confused battle followed in the acrid dark. The destroyer USS *Cushing* fired on the battleship *Hiei* and was sunk; *Laffey*, the next in line, met the same fate. *Hiei* was hit and moderately damaged by eight-inch cruiser shells and showed flames but carried on fighting. Callaghan ordered a general cease-fire on realizing that *San Francisco* was firing at the doomed *Atlanta*; the pause assisted battleship *Kirishima* and two other Japanese ships to get *San Francisco* in their searchlights and bring her up short. A heavy shell on her bridge killed Callaghan, the second American admiral to die in the battle, and several other officers. Heavy cruiser *Portland* took a torpedo hit on her stern but sailed on in circles, her steering jammed. The habitual survivor, light cruiser *Helena*, was lightly hit and survived again. The last cruiser in line was the light anti-aircraft ship *Juneau*.

We were in a column and we went in between these Japanese ships and we got word down from the bridge to stand by [said Seaman Allen C. Heyn, whose battle station was a heavy machine gun on the 'fantail,' or stern]. And it wasn't but a few minutes when everything just broke loose, flames and shots and gunfire all over . . . You could see the Jap ships so close that you'd think you could almost throw something and hit them . . . I don't know whether they knew there was a fish coming or what, but all at once a fish hit. It stunned us, like, and knocked us down and the propellers didn't seem to turn.

The paralyzed *Juneau* drifted toward the open sea, slowly going down by the bow. Fire Controlman First Class Douglas J. Huggard was in the last ship of all in the American line, the destroyer *Fletcher*, the only US ship not hit in the battle.

Later on in the day [Friday the 13th], as we were steaming southward for our rendezvous, the . . . *Juneau* was struck by an enemy torpedo and blew up. Within the space of less than a minute, the smoke had completely cleared away, the turmoil in the water had subsided, and all that was left of the *Juneau* and the crew was a little debris and fuel oil floating on the surface of the water. That was the most horrible thing that I have [ever seen] in this war.

I gave up [Seaman Heyn, dragged down with his ship by a trapped foot, went on]. I just thought that there wasn't a chance at all; everything just ran through my head. And you could see all objects in the water, all the fellows and everything, and after we were under the surface . . . the sheet of iron or whatever it was, was released and my foot came loose.

The luckless ship had been caught limping along with the other floating wounded by *I26* and Commander Yokota, who had knocked out the *Saratoga* in August. As if that were not enough, the 140 men who had got off the *Juneau* spent nine days adrift on the life rafts and flotsam or held up by life jackets, subject to sun, exposure, rough seas, delirium – and sharks. Only ten out of a crew of nearly seven hundred survived. Heyn spent four months in South Pacific hospitals, made a complete recovery – and volunteered for the submarine service in order to get his revenge.

Destroyers *Barton* and *Monssen*, ahead of *Fletcher*, were both hit hard by Japanese heavy guns; the former sank almost at once, and the latter was set adrift burning, abandoned by her crew. The final score for this ferocious and disorderly action was two light cruisers and four destroyers lost by the US Navy and every surviving ship damaged except one, against two Japanese destroyers sunk and battleship *Hiei* left battered by fifty hits, helplessly vulnerable to an air attack.

On the morning of 13 November Kondo sent his cruisers into the Slot to the rescue of the stricken, thirty-two-thousand-ton battleship and to bombard Henderson again. Rear Admiral Tanaka prepared to take his reinforcement convoy of a dozen transports escorted by eleven destroyers, held up by the night-action phase of the naval Battle of Guadalcanal just described, into Iron Bottom Sound. These operations were to be covered by Kondo's two battleships and the two light carriers on hand. Meanwhile, Admiral Kinkaid was bringing *Enterprise*, two battleships, and assorted escorts from New Caledonia in the south. As *Enterprise* still had trouble with her hydraulics after the eastern Solomons battle, fifteen of her planes – nine Avenger torpedo-bombers and six Wildcats – were sent to reinforce Henderson. On the way they sighted *Hiei* and attacked, jamming her steering and making her steam in circles inside her screen of destroyers. After collecting ten more planes from Henderson when they refueled, the *Enterprise* pilots, backed by Marine planes, made another attack, bringing *Hiei* to a standstill. By the time a squadron of B-17s arrived from the New Hebrides to make the usual high-level misses, the old Japanese battleship, completed in 1914 and modernized in 1932 and 1939, was going down by the stern five miles northwest of Savo Island.

The American main body was too late to prevent the Japanese from coming in to bombard Henderson again on the evening of the 13th. The Americans wondered if Callaghan's last fight had been in vain; in fact, the Japanese would have used battleships rather than cruisers to shell the field but for his ships. The two bombarding cruisers made a lot of noise but did little substantial damage with five hundred rounds of eight-inch. They then withdrew up the Slot and toward the Bismarck Archipelago in the early hours of the 14th. After daybreak Tanaka's transports began their approach, while American planes from Guadalcanal flew northwest in pursuit of the Japanese cruisers, two of which they damaged and set on fire. A second raid shortly afterward completed the destruction of heavy cruiser *Kinugasa*. All this temporarily drew American attention away from the transports, which were found by scouts from *Enterprise*, now less than two hundred miles off the Solomons. Aircraft from Henderson were sent in at noon to attack them, causing widespread damage but no sinkings. Heavily reinforced by the *Enterprise* air group, they returned within an hour to cause havoc among the transports, despite interference from Zeros sent in from Kondo's carriers. Tanaka, having lost seven troopships, transferred as many soldiers from the stricken merchantmen as he could to his undamaged destroyers and defiantly pressed on toward western Guadalcanal.

Rear Admiral Willis Lee had detached his two battleships, with four destroyers as escorts, from Kinkaid's *Enterprise* force on the evening of the 13th to race north to the Solomons and plug the gap left by Callaghan's stricken group. Kondo sent battleship *Kirishima*, two heavy and two light cruisers plus nine destroyers into the Slot on the morning of the 14th to help Tanaka and his transports with another bombardment. By evening the two main opposed surface-forces were approaching Iron Bottom Sound from opposite directions. Before midnight their destroyer screens were involved in a fierce fight in the narrow waters between Savo Island and Guadalcanal. All four US escorts were soon out of action (two sunk; one fatally and one moderately damaged), against only one out of nine Japanese damaged. Lee now had two new sixteen-inch battleships, *Washington* (flag) and *South Dakota* (nine big guns each), against the eight fourteen-inch guns of *Kirishima* (slightly younger sister of *Hiei*) and the two heavy cruisers *Atago* and *Takao*, mustering a total of twenty eight-inch. *South Dakota*, lit up by Japanese searchlights, drew all the enemy's fire, leaving *Washington* to blast *Kirishima* out of the battle from the darkness in little more than five minutes. *South Dakota*'s superstructure was badly damaged, knocking out most of her control systems and communications, but Captain Thomas Gatch was able to withdraw her at full speed for an eventual refit in California. *Washington* chased off the rest of the Japanese group and herself withdrew unharmed; Kondo called off his bombardment. *Kirishima* went down northwest of Savo at 3:00 A.M. on 15 November, and one Japanese destroyer had to be abandoned and sunk by her own side.

But the indefatigable Rear Admiral Tanaka managed to land some two thousand troops near Cape Esperance, the northwestern extremity of Guadalcanal, by dawn on the 15th, and then to get away with his eleven destroyers to the Japanese base at Shortlands in the northern Solomons. They grimly shrugged off constant harassment by destroyer USS *Meade* from Tulagi and planes from Henderson which had forced the grounding of all four remaining transports. This may have been a modest fraction of what he intended to put ashore, but Tanaka surely rates an award of some kind for sheer persistence. The naval battle of Guadalcanal came to an end at last on the afternoon of 15 November, and its result was a strategic victory for the Americans. They had got all their reinforcements ashore while preventing the Japanese from delivering more than a few troops and a small quantity of ammunition and food. At the same time, the US Navy, at the cost of two light cruisers, seven destroyers, many hundreds of men, two admirals and much damage and injury, had seen off two battleships, one heavy cruiser, three destroyers, and a dozen transports – seventy thousand tons of precious, top-quality shipping. The Americans had also handed out rather more punishment in the air than they received from the Japanese air forces, which were now beginning to feel the pinch.

The battle for Guadalcanal at sea, on land and in the air had become one of attrition, and in such a contest there could be only one winner. Japan's resources in capital ships, aircraft, pilots and above all merchant shipping were finite, because she did not have the capacity to provide replacements for any of these commodities at anything like the pace at which they were now being used up. The Japanese, who would have liked an extra seven hundred thousand tons of shipping for the Guadalcanal operation alone, asked their German allies for merchantmen but were told on 26 September 1942, that there were none to spare. In the constant dispute among Army, Navy and government over the already dwindling national tonnage stock, the Army threatened to close down its successful Burma front unless it got two hundred thousand tons more than the 1.04 million it had already been allocated. General Tojo was forced to step in with a promise to scrape the barrel to meet Army requirements.

The Navy was most worried about the shortage of fuel tankers and would have liked up to three hundred thousand tons more at this stage of the war.

All in all the first half of November 1942 was a period of almost undiluted good news for the Allies. Not only had the Americans got the upper hand in the Solomons, but Rommel was on the retreat from the British in North Africa after his defeat at El Alamein; the British captured Madagascar; the Americans successfully made their landing in North Africa and the British recaptured Tobruk; the Russians held the Germans at Stalingrad before launching a mighty counterattack; and the Australians and Americans were making gains in Papua. But the Japanese had not given up on Guadalcanal – far from it. Their troops in Papua were more or less left to their own devices from October, so that all efforts could be focused on winning back Guadalcanal, which was now designated 'the decisive battle' at IGHQ (many more were labeled thus in advance by the Japanese junta; when one was lost, the next looming clash was awarded the dubious title, making it clear that only a victory truly merited this description: there was no such thing as a 'decisive defeat' – unless, of course, it was suffered by the enemy).

As the ground forces continued to hammer each other in jungle actions and skirmishes round the American perimeter, held by two divisions, or thirty-five thousand men, against barely twenty thousand starving Japanese, Admiral Tanaka was bracing himself for another reinforcement-run by Tokyo Express. Only destroyers, whether converted as transports or not, could now be used to deliver troops and supplies, supplemented by barges over short distances. The Japanese were reduced to desperate measures, such as floating rice and other supplies ashore in drums from ships, an exercise which commonly resulted in wastage of 80 percent. There was no question of living off the land on Guadalcanal; as we have seen, the only diet on offer was coconuts. Meanwhile, a shipload of turkeys reached the American troops in time for Thanksgiving.

*   *   *

20. Up from under:
an alert gunner on
a US submarine.

21. Machine gunner on a PT boat in New Guinea waters.

22. Up periscope: aboard a US submarine in 1942.

23. The man who lost Singapore: the British Lieutanant General Arthur Percival after his release from a Japanese prison camp.

24. Taking the salute "Down Under": General Sir Thomas Blamey, Australian C-in-C and MacArthur's land commander.

25. The Bataan Death March.

26. Help for a hurt torpedo-bomber pilot on USS *Saratoga* after a raid on Rabaul late in 1943.

27. Here come the marines: a typical beach landing in the Pacific Campaign.

28. General James H. Doolittle checking out a B-29 Superfortress.

29. One-way ticket: one of Doolittle's B-25 bombers taking off from USS *Hornet*.

30. Tokyo, as seen by Doolittle's raiders in April 1942.

31. US Army Air Force P-40 Lightnings "somewhere in China."

32. American dive-bombers go in at the Battle of Midway in June 1942.

33. One of the *Mogami*-class heavy cruisers hit by the Americans at Midway.

34. Last farewell to the *Yorktown* after Midway.

35. Piling ashore in the Pacific: a typical post-invasion scene with "amphtracks," men, and munitions.

36. The most famous plane of the Pacific Campaign: the Japanese Zero. This Zero, shown with US markings, was pieced together by the Americans with parts from five captured and damaged Japanese planes.

37. Talisman of the US Navy: the ever-lucky carrier *Enterprise* "landing-on" a scout/dive-bomber in 1942.

38. Master Marine: Lieutenant General Alexander A. Vandegrift, as Commandant of the Corps in 1944.

39. Mighty Marine: Platoon Sergeant John Basilone after being awarded his Congressional Medal of Honor.

40. The US Army moves up on Guadalcanal, 1942.

On 23 November 1942, Major Donald Dickson, USMC, adjutant of the Fifth Regiment in the first Marine Division and a useful amateur artist, completed his tour of duty on Guadalcanal and was flown out, personally ordered by Vandegrift to report to Washington on conditions on the ground. Dickson left an unusually lengthy, vivid, and articulate report of his experiences in the archives very soon after he had lived through them. Dickson was struck by the lack of imagination of the Japanese enemy: troops would make a suicidal frontal assault on an American position because an officer had ordered it, even though they could have walked through a neighboring gap in the American line. 'They don't apparently use a great deal of imagination and I am satisfied that all during the fighting in Guadalcanal they underestimated us. All the maps and plans we ever saw, any of the Japs that we talked to [showed that] they didn't realize how many people we had on the island.'

The Marines felt bitter about not being relieved or reinforced by the Army for so long, and also being 'let down' by their own Navy, which was not stopping Japanese deliveries of troops and supplies (such as they were). 'The thing I think hurt everyone more than anything else was . . . the fact that they had to stay there so long, and undergo this strain for such an extended period. We heard of a number of men who eventually cracked up under that. The boys called them "psychos" among themselves. There wasn't any question about these people being yellow at all – everyone knew how they were because everyone felt the same way . . . It was generally . . . the older men, as far as I could see, who cracked up first' – such as the National Guardsmen from the Dakotas in the Americal Division.

Dickson felt it was not the fighting but the waiting, in appalling conditions, for the enemy's next move that was worst. Malaria was a very serious problem which affected the majority, debilitated many, and was even more prevalent than dysentery. Shell shock was commonplace. There were so many insects that they flew into men's mouths when they spoke. Turkey dinners notwithstanding, there was constant hunger, especially in the early days, when captured Japanese rations were eaten with something short of relish, cows were shot for their meat, and TNT was thrown into rivers to kill the unpalatable fish. There was no means of satisfying the men's appetite for candy or any sweet item (probably reflecting high usage of glucose in energetic combat).

The Americans were struck by how often the Japanese enemy used English to deceive or frighten them. One infantryman jumped up and shouted 'Blood for the emperor' as he charged, and an educated Californian Marine answered 'Blood for UCLA' and shot him, Dickson recalled. When the Americans called on their hidden enemy to surrender, they would be told to 'F*** off.' Mindful of the notorious difficulty all but the most fluent English-speaking Japanese have in differentiating between the two liquid consonants 'L' and 'R' (there being only one in Japanese), the Americans used passwords with at least one 'L' in them. The enemy would pick them up on hearing Americans challenging and answering one another. Dickson remembered one night when the password was 'hallelujah.' A Marine challenged a shadowy figure, got no audible answer in the noise of battle, and fired. A voice cried, 'Hallelujah, goddamnit, hallelujah!' again and again, and receded rapidly into the distance like a demented revivalist.

The First Marine Division was relieved by the Second, and the Army's American Division was up to full strength (three regiments), all under General Vandegrift. The 'Cactus Air Force' on Henderson was increased by almost half, from eighty-five to 124 aircraft of several types, operating now from one bomber and two fighter strips. Halsey was promoted to full admiral in November as the carrier *Saratoga*, completely repaired, came to the aid of the *Enterprise*. Five battleships also assembled in the South Pacific command, and Callaghan's cruiser group was replaced by a stronger one, Task Force 67, under Rear Admiral Carleton Wright, who was new to the campaign and the area. He took over one light and four heavy cruisers plus four destroyers on 29 November – and led them into battle the next day, for the last of the six major naval actions involved in the struggle for Guadalcanal – the Battle of Tassafaronga.

'Tenacious Tanaka,' as Morison called him, came into Iron Bottom Sound with six destroyer-transports escorted by two destroyers on the evening of 30 November, to offload his supply drums for collection by the small boats of the Japanese Army on Guadalcanal. Admiral Wright's cruiser squadron was expecting a Tokyo Express and was, on Halsey's order, heading east into Lengo Channel, between Savo Island and Guadalcanal, when it met an incoming American convoy of three transports and five destroyers. Two of the latter were temporarily transferred to Wright's command, at the rear of his line. In the van were his original four destroyers, followed by three heavy cruisers in one division, with light cruiser *Honolulu* and heavy *Northampton* in a second division under the flag of Rear Admiral Mahlon Tisdale. American reconnaissance seaplanes were unable to take off from the glassy-calm waters of the Slot that night (they needed to bounce into the air off waves), so Wright had no time to prepare when USS *Fletcher*, his leading destroyer, detected the enemy by radar, ahead and to port at seventy-five hundred yards. The American admiral hesitated for four minutes before sanctioning a general launch of torpedoes from the destroyers ahead of him. By this time the Japanese force was already passing on a divergent course with the leading vessel nearly ten thousand yards away, a daunting range indeed for feeble American torpedoes. Twenty were launched at the unsuspecting Japanese, who as usual had no radar. They got their first intimation of an enemy presence when the American cruisers opened up with their guns from up to five miles and lit up the sky with star shells. It was not quite 11:30 P.M. local time.

Japanese training in night fighting still counted for a great deal. Tanaka's standing orders in the event of meeting enemy warships were to sow the sea with torpedoes, refrain from telltale gunfire, and retire at speed. The lead destroyer, *Takanami*, however, showed up best on the American radars and drew the brunt of the gunfire, to which she replied in self-defense – which only brought down more shells upon her, turning her into a floating inferno. Astern, the transports were already committed to releasing their drums. The Japanese line crumpled as individual ships took evasive action, but managed to launch a total of two dozen Long Lances, somewhat more torpedoes than the Americans had fired. Two of them struck Wright's flagship, *Minneapolis*, on the port side, almost severing the bow. Next in line, *New Orleans* was hit by only one torpedo but actually had her bow blown off by an ammunition explosion. Both heavy cruisers burned brightly. Third in line was *Pensacola*, hit next square amidships and crippled in her turn. The relatively nimble light cruiser *Honolulu* was able to jink her way out of trouble and took no hits. Fifth and last in the American cruiser line was USS *Northampton* (Captain Willard Kitts). She was struck twice in more or less the same spot, aft and on the port side; her fuel tanks ruptured and exploded, turning the wretched 'tail-end Charlie' into a giant torch. Fortunately, the two 'guest' destroyers at the rear had become separated from the rest of Wright's force and got away unscathed.

It all amounted to a brilliant victory for Tanaka Raizo, who did not even have his light-cruiser flagship on hand, against a vastly superior force which took him by surprise yet succumbed to Japanese torpedoes fired by reflex action in desperate circumstances. Seven of his eight destroyers were undamaged; *Takanami* sank, as did *Northampton*. The other heavy cruisers reached harbor at Tulagi, opposite Guadalcanal. All three lived to fight another day, though not for a long time. The Americans still had a lot to learn about how to make the best use of destroyers, about gunnery at night – and about accurate enemy-damage reports. On this occasion each side was as imaginative as the other, with the Japanese claiming a battleship and a cruiser sunk and three destroyers hit, and the Americans boasting of four destroyers sunk and three damaged. The scale of this American defeat emerged only much later; for the Japanese the victory was merely tactical and would have been no more significant even had they sunk every cruiser they met. The Americans had won strategic control of the seas round the southern Solomons just as they had in the air over the fetid islands. It remained for them to dispose of the Japanese troops in the jungle, whose naval lifeline had now all but withered away. The Japanese were reduced to using submarines as supply vessels, just like the Americans in the Philippines. But the boats could not begin to alleviate the sufferings of the troops on the island. Disease and hunger had killed as many as had the Americans in battle.

Yet Lieutenant General Imamura Hitoshi, the very able

conqueror of Java, was assembling no fewer than fifty thousand men of his Eighth Area Army in the Rabaul area for the real 'decisive battle' for Guadalcanal, early in the coming year. On 9 December General Vandegrift was relieved in command of what was now the American XIV Corps by the US Army's Major General Alexander Patch: the new corps had one Marine division (the Second) but two from the Army (Patch's Americal and the freshly delivered 25th Infantry), and protocol plus the usual considerations of interservice rivalry decreed that an Army officer should be in charge. The Americans lost nothing in the exchange, as Patch proved an excellent front-line commander, every bit as tough as his predecessor.

Rather late in the day, the Japanese built a new airfield in the Solomons at Munda, on New Georgia Island, some 175 miles northwest of Henderson. Had they done this when they took Tulagi seven months earlier, they would have been able to swamp Henderson with fighters in its shaky early days and make it unusable, as well as being able to offer useful air support to their troops in New Guinea. As it was, until the end of November the nearest land-based air cover had to come from Buin at the southern end of the large northern Solomons island of Bougainville. By now the shoe was on the other foot, and Munda was an attractive extra target for the unshakably strong American presence at Cactus and heavy bombers from farther afield. The Americans, including MacArthur's B-17s, exploited this new opportunity on a daily basis. Shattering air and naval bombardments prevented the Japanese from turning New Georgia into a mirror image of Guadalcanal.

As the bidding in the deadly auction for ridiculously overpriced Guadalcanal reached its climax, the Americans accumulated invaluable experience in an area of minimal interest to the Japanese ruling class: the husbanding of human resources. A wounded Japanese soldier was as good as dead, whereas a wounded American had a remarkably good chance of survival, despite the shocking conditions. The expertise gained in the treatment of wounds and disease on Guadalcanal (and in Papua, to a lesser degree) was of enormous importance for the rest of the Pacific Campaign. The First Marine Division had the First Medical Battalion attached to it, made up of six companies, one with each regiment (three infantry, one artillery), and two companies in reserve. Each company had five doctors, one dentist, a warrant officer, corpsmen (we would now call them 'paramedics'), and Marines.

When we found ourselves in the jungles of the islands of the South Pacific [said Commander Don S. Knowles, M.D., USNR] the first thing everybody did was to throw the book away. We had to learn all over again. There was no front as such, there was no rear . . . We had about as much privacy as you would expect had we been in a large goldfish bowl.

There were no hospital ships in the highly dangerous Slot, so the wounded had to be lifted out each day at dawn by Douglas DC-3 aircraft. Ambulances formed up to wait for them to land at Henderson, even under Japanese bombardment: the medical companies had their own transport and could set up a field hospital of a hundred or even 250 beds in short order as required. Men in the care of such units were moved to underground dugouts during air raids or shelling, as in the First World War. But unlike that earlier conflict, in which 9 percent of the wounded died of their injuries, the Pacific Campaign reduced such losses to less than 1 percent, thanks to blood plasma, sulfa drugs, modern sanitation and transport, and highly trained medical personnel. The last phase of the Guadalcanal campaign saw the introduction of the 'Dumbo,' the amphibious naval Catalina with doctor and paramedic aboard which rescued sailors and airmen from the sea, often in highly dangerous conditions.

Of the thirty thousand Japanese who fought in the ground campaign on Guadalcanal, roughly one-third were killed in battle or died of wounds, one-third died of starvation or disease, and one-third were evacuated. Those who got to know of the American way with the sick and wounded were astounded. One Japanese officer, commenting on the primitive to nonexistent medical facilities in the Army (the Navy was rather better), and the high American hygiene standard, remarked: 'It became evident that our military and government leaders never really understood the meaning of total war.'

Another Japanese officer, anonymous but obviously senior and experienced, had by December 1942 fought in both Papua and Guadalcanal and wrote down his impressions of the enemy.

Characteristics of American Troops
  1) They possess strong national unity.
  2) They like novelty and adventure.
  3) They excel in the technical field.
  4) They are boastful but are inclined to carry out their boasts.
  5) They are optimistic and lack patience.
  6) American troops are very good fighters when possessing superior firepower.

The same commentator found American marksmanship excellent but hand-to-hand fighting poor; communications excellent but scouting and security poor; ground-to-air communications, artillery, and handling of vehicles all good or better. He thought Americans relied on firepower and positional warfare with their troops disposed in depth, but 'despised hand-to-hand combat,' were vulnerable to attacks from the rear, and would not attack without strong artillery support including heavy preliminary bombardment. He found senior officers inexperienced in the control of large units and junior commanders lacking in initiative but thought American staffwork very good. Americans were good at planned attacks but not so good at defense, easily distracted, and slow in attack. 'It is best to cut the enemy's lines of communication because they live in luxury,' he noted enviously (a hungry officer, in New Guinea, wrote in a diary: 'There is a saying that the samurai displays a toothpick even when he has not eaten!'). 'American troops are simple-minded and easy to deceive . . . They will attack regardless of the length to which their lines are extended,' the analyst concluded. One more unnamed veteran wrote this judgment on Japan's Pacific enemies in the back of a message log found in the Solomons: 'American soldiers are generally weaker than Chinese, but Australians are the strongest.'

*   *   *

In December 1942 the Japanese junta, after bitter feuding between the admirals who wanted to pull out and the generals who wanted to dig in, gave up hope of recapturing Guadalcanal. Their staffs tried to find a 'solution' to the problem of dislodging the Americans by war games but failed to construct even a hypothetical scenario leading to such a conclusion. The fundamental reason was the Americans' airpower, which also gave them ultimate control of the sea and prevented the Japanese from significantly reinforcing or properly supplying their troops from about the middle of October 1942. The huge tonnage of irreplaceable shipping lost in frustrated attempts to strengthen their position on land was a decisive factor. An imperial conference on the last day of the year formally confirmed the IGHQ decision to abandon the southern Solomons, but the order had gone out to the Eighth Area Army three days before 'to direct the Seventh Army to reorganize the existing front-line positions and occupy a strategic rear line suitable for future operations, in a preparatory step for the evacuation of Guadalcanal.'

Once again Army and Navy concluded a 'general agreement' for Operation KE, the evacuation of Guadalcanal. The plan was to fight to the bitter end, to step up air attacks from the northern Solomons, and to evacuate the remaining troops. The northern Solomons, Rabaul and northwestern New Guinea were to be held, the 20th Division, originally earmarked at Rabaul for the relief of Guadalcanal, was to go to New Guinea, as was the 41st Division. The withdrawal was recognized by staff officers as 'an unprecedented event in the annals of the Japanese Army.' The galling retreat from the southern Solomons was superbly done, complete with a nasty sting in the tail for the US Navy, which lost the heavy cruiser *Chicago* to an airborne night torpedo attack by a dozen 'Bettys' on January 29, just south of Guadalcanal. She was part of a task force bringing supplies to the American garrison; ten of the Japanese planes were shot down. As General Patch's westward drive, begun on 10 January 1943, moved forward along the northern coast of Guadalcanal and inland, some ten thousand men were spirited away from the island on three nights in the first week of February. The US Navy, whose sources of information now included the 'Black Cat,' yet another use of the versatile Catalina as a radar-equipped night-patrol-cum-nuisance-raider, was aware of heightened enemy activity offshore in the Slot at night. But Halsey concentrated on defending Henderson against what he assumed was one more impending Japanese onslaught, just as Kondo concentrated on covering the withdrawal. In fact, the Tokyo Express had gone into reverse, collecting rather than delivering passengers. When Patch and his men got to the western tip of the island on 8 February, they found nothing except the detritus of war, and abandoned barges floating offshore. They had expected to take eight weeks longer to clear the island, but the enemy had decided to cut his losses at last and retreat.

The fight for 2,500 square miles of hostile territory in the tropical Pacific cost the US Navy and its allies twenty-four destroyers and larger warships totaling 126,000 tons, including two fleet carriers; the Japanese lost two battleships and six submarines among a similar total of twenty-four warships sunk, with a combined tonnage of 135,000. Of the 36,000 Japanese who fought on or over Guadalcanal (excluding shipboard personnel), 15,000 were dead or missing, 10,000 died of sickness, and 1,000 were captured. The US Marines and Army committed a total of 60,000 men of whom 1,600 were killed. Neither side ever computed its human losses at sea. The Americans' most decisive victory in the southern Solomons was in the air. The Zero was still the best fighter in the theater but was already outnumbered by planes with pilots who had learned from experience or from other pilots how to contend with it.

In Papua 13,000 Japanese were killed out of about 20,000 committed. The Australians and Americans had 3,000 fatal casualties and 5,500 wounded – plus 37,000 cases of tropical disease, including 27,000 malaria victims, out of about 60,000 troops engaged.

*   *   *

The Japanese war machine, checked and forced to change its plans for the first time at the Coral Sea and decisively repelled at Midway, was now forced to go into reverse by its failure on Guadalcanal and simultaneously in Papua. From now on it was the United States and its allies who were strategically on the offensive, in the air, on the ground, and at sea.

But, as we have noted, there was one arm of the American forces that had been on the attack from the moment of first response to Pearl Harbor: the submarine service of the US Navy. Before we turn to the massive double counterattack across the Pacific Ocean, we need to review the gradual growth of this war-winning weapon – and the great American torpedo-crisis which hampered its efforts.

After the fall of the Philippines, the American submarine campaign continued to be conducted in two separate but mutually supportive parts: COMSUBPAC under Nimitz and COMSUBSOWESPAC under MacArthur, the latter subdivided into western and eastern Australian contingents. From June 1942 Nimitz's boats used Midway as a small forward base eleven hundred miles closer to Japan than Pearl, and Johnston Island served a similar purpose for boats bound for Japan's mandated islands such as the Carolines. Pearl serviced the boats of both forces, and South-West Pacific submarines reconnoitered the Japanese Pacific islands for Nimitz en route between Pearl and their bases; boats were also rotated between the two supreme commands. Their main function was to blockade the Japanese Home Islands by sinking merchant shipping, but competing demands from operations around the Solomons and Aleutians cut the coverage of enemy home waters severely. The old S-class boats of the South-West Pacific were falling apart in 1942 and were withdrawn in September as more fleet boats became available; Admiral King took the opportunity to reduce the numbers operating from Australia by nine to twenty, but this still left many holes in the blockade. During the year, American submarines, now acquiring the new SJ radar specially developed for them, went closer and closer to such prime areas as the waters north of Tokyo and round Formosa more and more often.

They executed the Makin Raid, spied on the large Japanese Combined Fleet base at Truk in the Carolines, and sowed mines there and elsewhere. By the end of the year COMSUBPAC had fifty-one fleet boats and eight S-boats under his flag, with many more on the stocks from the beginning of 1943, when the Guadalcanal campaign ceased to divert existing submarines.

'Torpedoes' was probably the only branch of naval warfare in which the United States Navy was markedly inferior to the Japanese. The reason for this inferiority lies deep in the organization of the US Government and the psychology of its people. It certainly cannot be shifted off as solely the responsibility of a small group of technicians . . . Undoubtedly torpedo inferiority added months to the war and thus cost the US thousands of lives and billions of dollars of treasure.

So says the unpublished Submarine Operational History (SOH), on page 699 of volume two, a source that can hardly be gainsaid. Since 1869 America's torpedoes had come from the Torpedo Station in Newport, Rhode Island, a key contributor to the economy of the smallest US state, and, from 1940, a new station at Alexandria, Virginia. Both were Navy facilities, and between them they enjoyed a monopoly. Officers did a two- or three-year stint at Newport just as they did in other posts ashore between tours at sea, so there were few or no real technical experts in the service and none at all outside it. In practice firings, the expensive, complicated and slowly manufactured missile ($10,000 each) was deliberately set to pass under the target to preserve both for reuse! As long as the torpedo ran in a straight line, everyone was satisfied; no check was ever made on the depth at which it ran, difficult though this is to believe. The arrival of the magnetic pistol allegedly made accuracy of depth less important, but this exploder was never tested in advance in simulated combat conditions, and a really inaccurate depth vitiated the effect of a torpedo altogether; the depth problem was compounded, as we have seen, when heavier, live warheads replaced the practice dummies. 'The war began with an entire generation of submarine personnel none of whom had ever seen or heard the detonation of a submarine torpedo,' says the SOH.

Not only were the torpedoes feeble; they were also scarce.

The Navy started the war with a few hundred submarine torpedoes (Type X for the S-boats, Type XIV for the rest), of which nearly half – 233 – were lost through abandonment in the Philippines. Disgracefully, submarines had to be sent to sea with only two-thirds of their complement of torpedoes. Production capacity at the beginning of 1942 was all of sixty per month; orders to conserve torpedoes reduced or destroyed the effect of many attacks even when inherent inaccuracy was offset by skilled operators. COMSUBSOWESPAC had a standing order to keep torpedoes for large targets only; single shots instead of salvos, and small salvos instead of large ones, were so ineffectual that they actually increased the waste which the conservation measures were supposed to reduce; skippers were praised, in a war, for not expending their ammunition! Yet, over the year, 2,010 missiles were fired and 2,382 were manufactured.

But the resistance within the Navy's Bureau of Ordnance to the monopoly was coming to a head after eight years of bureaucratic guerrilla warfare; private manufacturers had already become involved in making torpedoes for planes and surface ships soon after America entered the war. Westinghouse Corporation was called in to make the new Mark XVIII electric torpedo. The Mark XXIII was introduced to simplify production – essentially the twenty-one-inch Mark XIV deprived of its near-useless slow-speed option, but the consequent loss of range on a missile already seriously underendowed with that commodity only showed up the overwhelming superiority of the Japanese Long Lance. In the South-West Pacific, submariners concluded from carefully collected evidence that the main problem with American torpedoes was that they ran deep. Lieutenant J. W. Coe of USS *Skipjack* fumed after a patrol in June 1942: 'To make round trips of 8,500 miles into enemy waters, to gain attack position undetected within 800 yards of enemy ships, only to find that the torpedoes run deep and over half the time will fail to explode, seems to me to be an undesirable manner of gaining information which might be determined any morning within a few miles of a torpedo station in the presence of comparatively few hazards.' Nevertheless, he sank four ships on his patrol.

In the same month, Rear Admiral Charles A. Lockwood,

commander of Task Force 51 (MacArthur's submarines) and
soon to become chief of Nimitz's Pacific boats, ordered *Skipjack*
to conduct such a test off Western Australia. A Mark XIV
torpedo set to hit at ten feet was fired just 850 yards into
a net – which it struck twenty-five feet below the surface. A
second missile set for ten feet hit at eighteen, a third set at zero
feet hit at eleven – and showed signs that it had hit the seabed on
the way to the target. Lockwood reported this to the Bureau of
Ordnance on 22 June. Two days later, COMSUBPAC reported
that a Mark XIV set in error at zero had struck at eight feet on
an exercise. The bureau at first replied that the test conditions
had been 'improper' and no conclusion was possible; conceding
shortly afterward that there were indications of deep running, it
invited submarine crews on active service to pay special attention
to 'variables introduced by operational conditions' – of which it
knew nothing. This shifted the burden of experimentation to
the front line, the most inappropriate address. On 21 July a
furious COMINCH intervened, abruptly ordering the bureau
to test the submarines' Mark XIV and the destroyers' Mark
XV torpedoes. Even so, they used tubes placed on a barge
rather than a real, live submarine or destroyer for the tests,
which, unsurprisingly, were inconclusive. Only on 1 August did
the bureau report that the torpedoes ran ten feet deeper than
set in tests at Newport; on the 24th proper tests by the bureau
itself confirmed this treacherous foible. A new depth-control
accurate to within three feet was hurriedly introduced from the
autumn of 1942. One of the many curiosities in this sorry tale is
that the Royal Navy had a similar problem for similar reasons
in the First World War and corrected it, a fact that should have
been well known in American submariner-circles.

Yet the difficulties continued, because one problem solved
tended to reveal the existence of another, previously masked by
the former. The detonators were even more unreliable than the
torpedoes. When the depth problem was solved, for example,
the number of premature detonations rose, because the missiles
were upset by having to run through the more turbulent water
close to the surface of the sea. When that was dealt with, the
Mark VI detonator's magnetic exploder was revealed as hyper-
sensitive: instead of going off a few feet under the enemy ship's

bottom, as intended, it might detonate as much as fifty feet short of her side, creating a lovely splash but no damage. However, the effectiveness of American torpedoes when they did hit was considerably enhanced in February 1943 on the introduction of the new and more powerful Torpex torpedo explosive. When an American interrogator asked a prisoner from a Japanese submarine whether his Navy had any problem with premature torpedoes, the sailor smiled and replied: 'We don't, but you do!' When desperate captains deactivated the magnetic exploder and relied on the simpler contact pistol in the Mark VI, the high proportion of dud torpedoes, which would not go off at all, was revealed in all its shame. As late as July 1943 USS *Tinosa* conducted a remarkably cool and dispiriting experiment near Truk. Having paralyzed a tanker with two torpedoes that struck home out of six fired, Commander Daspit fired nine more from right angles at 875 yards. All hit; none exploded. He took his last five home for tests while the Japanese came out of Truk and saved their tanker. Tests in Hawaii revealed that the firing pin of the contact pistol did not hit the primer with enough force to set it off. This fault was corrected by October 1943, at which time the torpedo crisis is usually regarded as having come to an end. Subsequent torpedo-performance does not bear that out: indeed, the US Navy can fairly be said not to have solved the problem at all during the war.

Meanwhile, captains were carpeted for poor attack records, fell out with their technicians on missions, were betrayed to the enemy by 'prematures,' lost their nerve or were ignominiously transferred because authority was so slow to acknowledge that American submarines had been sent to sea with an unreliable main armament. How the morale of the service survived the protracted ordeal is hard to imagine, but somehow it did, and the destructive harvest reaped by the American boats grew steadily with their numbers until it became the principal strategic factor in the decline of Japan's ability to wage total war. The American experience of 1941–43 closely parallels – where it does not surpass – the German torpedo problem of 1939–40. The Germans too gave their Navy a torpedo monopoly, had trouble with magnetic pistols and stole foreign (British) ideas as a stopgap until they came up with something better. The American crisis lasted rather longer, but early in 1942 they captured a complete German G7e electric torpedo and commissioned Westinghouse to copy it: easier said than done because of the very different technical standards and measurements of the two opposed belligerents. Further, the dead-handed Bureau of Ordnance was developing one of its own, the Mark II electric, and refused to cooperate with the company until King intervened. This was the origin of the Mark XVIII, but submariners were leery of it even after all its many teething troubles were resolved.

For a long time electric torpedoes, even though they left no telltale wake, were only carried voluntarily, and in the submarines' finest hour, 1944, less than one in three torpedoes fired was electric. Only in 1945, after much combat experience had been gained with them, were two electrics fired for every Mark XIV 'steam' torpedo. Remaining briefly ahead of ourselves, we may note here that, in the war as a whole, the Mark XXIII was the most effective American torpedo in terms of percentages of hits; that the Pacific Fleet consistently outshot the South-West Pacific command; that the Mark XVIII had the fewest hits; and that a Pacific Fleet submarine equipped with Mark XIVs was more likely to score hits than any other US boat. The overall hit-rate claimed was 34 percent, or one in three; one enemy ship was recorded sunk for every eleven fired.

PART THREE
America Resurgent

8
Intelligence Applied

The German submarine offensive threw a long and men-
acing shadow over the city of Casablanca in Morocco, where
Roosevelt, Churchill and their chiefs of staff opened another
summit conference on 14 January 1943. The venue was chosen
to underscore (a little prematurely as yet) the success of
Operation Torch, in which the Americans had landed their
troops in North Africa in November. But the dilution of the
Allied anti-submarine effort caused by this politically motivated
invasion (done as much as anything to justify the American
Army's massive mobilization program) was apparently threatening to sever the very lifeline between America and Britain
which had made it possible in the first place. The U-boat was
therefore uppermost in many British minds even as they debated
how to convert the growing successes on other fronts into victory
over the Axis – and how to distribute the ships and planes and
troops among the various theaters. The conference had a crucial
influence on the course of the war against Japan and proved to
be as much of a turning point for the United States diplomatically
as Guadalcanal was militarily at the same time. Casablanca is
therefore worth more than a fleeting visit.

The conference was aptly code-named 'Symbol.' What it came
to symbolize was the definitive passing of the torch as leading
world power from the United Kingdom to the United States.
This permanent shift in the balance of power began with the
intervention of the US in the First World War, grew uncertainly
with the American economy between the wars, and was clear
for all to see with the landing in North Africa (even though the
British Navy had covered that operation). Torch was executed
by a nation that was projecting its might across the Atlantic
against Germany even as it reached out across the Pacific to
dismantle Japan's conquests. Casablanca was the last occasion
in the war when Winston S. Churchill was able to exert decisive

influence on Franklin D. Roosevelt – and then only on some issues – and the last time a British prime minister swayed an American president over anything much more serious than the menu at a state dinner. After Casablanca, the US was the acknowledged senior partner in the Anglo-American alliance, the banker of the United Nations, the real 'arsenal of democracy' of which Roosevelt had spoken in 1940.

The British wanted to expand operations in the Mediterranean, exploiting their success against Rommel in the desert by crushing the remaining German presence in North Africa between themselves and the American Army advancing from the west. They wanted to move on from there to attack Churchill's 'soft underbelly' of the Axis in Europe – Sicily, then Italy, the classic British flank attack on the enemy's weak point. They already knew, because they had heard it before, that Marshall and King had said they wanted to cross the English Channel and attack the Germans in 1943 – the main strength of the main enemy, as recommended by Clausewitz. The British wanted to get onto the European continent to mollify the ever-demanding Stalin, but (not without logic) took the view that the Allies were not yet strong enough to take on the German Army in France – where they had lost a generation of young men trying that very recourse less than thirty years before. Marshall at least had some sympathy for their viewpoint. This was the main strategic issue; the Americans, however, also wanted to shift even more resources into the war against Japan, not only to boost the campaigns of Nimitz and MacArthur but also to help Chiang Kai-shek to tie down more Japanese troops in China – which also entailed a greater effort by the British in Burma.

The Allies were formally committed to 'Germany first,' but the Americans were in the embarrassing position of having eighty thousand more troops in the field against Japan than against the Germans and the Italians at the turn of the year – nearly half a million compared with under four hundred thousand, after more than a year of belligerence. Their Navy, of course, was overwhelmingly 'Pacific'; the British had larger fleets in both the Indian Ocean and the Mediterranean than the Americans had in the Atlantic and the rest of the world combined. Even more embarrassing was the fact that the lack of handy ports in the Pacific had led to an enormous backlog of undelivered supplies in idle ships desperately needed elsewhere. Meanwhile, the British were forced to limit themselves to fifteen million tons of nonmilitary imports per year – compared with fifty-five million before the war. A clear majority of American Lend-Lease supplies was certainly going into the war against Germany; but the promised preponderance of manpower for Europe was just that: a promise, not yet made good. From mid-1942 more and more future US divisions were being earmarked for the Second Front in Europe. Yet the American position was not irrational; as ever, they believed that where there's a will there's a way, and they had the utmost confidence in their ability to fight two transoceanic wars at the same time as they supplied all three others of the UN Big Four – Britain, Russia and China. But their initial setbacks against the Germans in North Africa should have led them to the conclusion that they were not yet ready to take on the enemy A-team or first eleven – the Wehrmacht, which would be even more formidable close to home. The only way to be sure of beating the Germans, as indeed the Japanese, was by the application of superior force.

King had an almighty row with the British at Casablanca about the deployment of very-long-range B-24 Liberator bombers. These were the only available Allied aircraft with the 'legs' to close the air gap in the middle of the North Atlantic, which was being so savagely exploited by Dönitz's U-boats. The air forces would not spare as much as a squadron from the build-up of bombers against Germany; King wanted to reserve his six dozen or so for the Pacific, and Eisenhower was not about to give up his two squadrons based in Morocco. Roosevelt, however, overruled COMINCH, and the British eventually got a couple of squadrons – in the nick of time to retain the initiative until American-built small escort carriers could be brought in to sail with the convoys. By contrast, the 'Bald Eagle' surprised his British opposite numbers by offering them all the landing crafts and escorts they could use for operations in Burma and even for invading Sicily. This was only because logistical problems in the Pacific prevented him from using them there, and because he wanted to embarrass the British over their lack of activity against Japan!

The British got their Sicily and Italy; the Americans got their Burma and more scope for the Pacific. The invasion of France was agreed, as was a joint staff to plan it, but the date was left open. More significantly perhaps, though only hindsight makes it so, Roosevelt – without the specific prior agreement of Churchill – announced at a press conference that the war aim of the alliance was the 'unconditional surrender' of both Germany and Japan. In fact, premier and president had been broadly agreed on this for at least six months, if only as a means of proving to Stalin that they were serious about defeating Hitler and would never do a separate deal with him, despite their equivocation on the Second Front. Occasional explosions from the Combined Chiefs of Staff notwithstanding, as a political summit Casablanca passed off in a cordial atmosphere, laced with Churchill's orotund compliments and Roosevelt's different but no less elaborate rhetoric.

The goodwill was diluted as time passed. The Americans on reflection felt they had been diverted from France into an irrelevant sideshow in the Mediterranean, while the British thought their ally was bent on transferring the brunt of his effort to the Pacific – insofar as he had not done so already. The agreement demanded from the Combined Chiefs of Staff by King that 30 rather than 15 percent of total combined resources should go to the Pacific looked like a cynical joke in London, where the strictly rationed British contemplated the vast American seaborne presence in the Far East – and its virtual absence in the Atlantic. The 15-percent figure seems to have come from those rarefied upper regions which only eagles can reach; what King was baldly requesting was a doubling of the effort in the Pacific, and that is what he got. King accused Churchill of wanting America to bail out Britain in the war against the European Axis, whereupon it would be only too pleased to leave the defeat of Japan to America as well. Churchill angrily offered a treaty committing Britain to throwing everything it had left into the Far East as soon as Hitler was beaten; Roosevelt restored calm. This issue was to come up again between the Englishman and the Anglophobe.

Casablanca produced no firm program for defeating either Germany or Japan. But it prudently agreed that the US should keep the initiative it was just winning at such huge cost in the Pacific and should also prepare for a full-blown counterattack on Japan as soon as Germany was defeated. The conference thus managed to give all parties what they wanted, including King. In other words, that ancient unwritten principle of Anglo-Saxon law came into play: whatever is not specifically prohibited is allowed. Provided they were not too blatant about it and kept their promises on beating Hitler, the Americans had carte blanche in the Pacific. They could go ahead and knock out Rabaul.

Back in Washington, the question now was, who would take charge of this Operation Cartwheel, as it was to be called. Not content with bearding the British prime minister, King next took on the United States Army, proposing that Nimitz should be made super-supreme commander of the entire Pacific.

King was trying to get out of the interservice agreement of July 1942, which required MacArthur to take the bulk of the responsibility for the execution of 'tasks two and three' as most of the targeted territories and the ultimate goal of Rabaul lay within his command. Casablanca had given a broad blessing to advances up the Solomons and across New Guinea, to the first assaults on Japan's mandated islands, the recovery of the Aleutians, and also of Burma. To hear a report on Casablanca and decide how to proceed, the Pacific Military Conference – a new liaison-and-coordination mechanism for the administration, the services and the Pacific commands – met for the first time in Washington. For MacArthur his chief of staff, General Sutherland, tried to impose a grandiose five-point plan for a two-pronged advance on Rabaul by MacArthur and Halsey, for which five extra divisions and nearly two thousand planes and pilots would be needed. Halsey's representative seconded. The meeting adjourned in disorder. After much horse-trading behind the scenes, a compromise was reached by the end of March 1943 on a modified task two, involving rather more modest reinforcement. Halsey's South Pacific command was to advance up the Solomons up to but not including the largest and northernmost island, Bougainville; MacArthur's SOWESPAC command was to complete the conquest of New Guinea and establish a beachhead at Cape Gloucester, at the opposite end of the long island of New Britain from Rabaul, which was left for later. Despite four months of often heated interchanges, the problem of unified command, which all and sundry professed to support (so long as it did not go to the other service), was adjourned *sine die*.

MacArthur was in charge of task two, but Nimitz kept control of the Pacific Fleet except for task forces assigned to specific operations in SOPAC and SOWESPAC. Surprisingly for two such assertive characters, MacArthur and Halsey got on famously when they met for their first tactical conference in Brisbane after the Joint Chiefs gave them their strategic orders. A month later, at the end of April, the intricate plan for Operation Cartwheel was ready, almost Japanese in its complexity, with thirteen distinct objectives.

Planning should have been much easier for the Japanese than for their enemies. Those running the war also happened to run the government, and there were no allies to consult, the Axis being a political alliance with few military dimensions. On 18 January 1942, a military agreement had been concluded by Japan, Germany and Italy, giving Japan complete freedom of action in all areas east of seventy degrees east longitude, including India. But we have already seen how the autonomy of the two Japanese armed forces, which went all the way to the top, made paralyzing disagreement inevitable even when the dominant institution, the Army, was united behind a policy objective. The supremacy of the divine emperor was an idea rather than a fact, and it was honored more in the breach than in the observance. Like other 'constitutional monarchs,' Hirohito had to choose very carefully the moments when he put his foot down, expressed disapproval or even concern; the effect could be devastating, but the 'February Incident' of 1936 had shown that a coup was not beyond the imagination of the military. Despite the emperor's 'concern,' the Army decided to hold out too long in Guadalcanal after the Navy had written it off. But after the double withdrawal from Guadalcanal and Papua, the two services were fully in agreement that the Navy should defend the rest of the Solomons and the Army New Guinea, broadly reflecting the American approach; each service had its headquarters in Rabaul, the midway point and strategic center of gravity between these two flanks.

The seriousness of the crisis was underlined in February 1943 when Tojo, already premier and Army (or war) minister, became minister for home affairs and chief of the Army General Staff, a sure sign of unease as well as inability or unwillingness to delegate. The Army reacted to its unprecedented twin defeats, and the new experience of losing the initiative to the enemy, by creating a new 18th Army (a corps of two divisions) to reinforce New Guinea, where Wewak, Lae, Salamaua, and Madang were still held, and neighboring areas including the Solomons and Bismarcks, against a predictable Allied advance. Air units were brought in from Burma and Manchuria. The new corps was formed out of units withdrawn from China and Korea and put under the command of Lieutenant General Adachi Hatazo. The next trick was to get the troops to where they were needed. Adachi decided to move 6,900 men of the 51st Division, a force of 400 marines, 2,500 tons of supplies, and his own headquarters from Rabaul to Lae in a convoy of eight Army transports with a combined capacity of 33,700 tons, escorted by a similar number of destroyers and covered by 100 planes. Rear Admiral Kimura Masatomi was in charge at sea, and he hoped to slip along the north coast of New Britain in poor weather, down through the Dampier Strait at its western end, round the Huon Peninsula on New Guinea, and into Lae, a distance of less than 500 miles. H-hour was midnight on the night of 28 February–1 March.

Dead on time, at dead of night, the convoy slipped out of harbor under threatening and extremely welcome cloud into the Bismarck Sea. The poor weather lasted into the afternoon of the 1st, but even so a long-range B-24 Liberator reconnaissance aircraft from V US Army Air Force, headed since August 1942 by Lieutenant General George C. Kenney, came through the cloud and saw the Japanese formation but lost contact with it soon afterward. The bad weather meanwhile capriciously shut

New Britain in, so that the Japanese were unable to put up most of their air umbrella, while Papua, where MacArthur and Kenney had more than three hundred planes, stayed clear enough for operations. Kenney, unsurpassed among American operational air commanders, had not failed to notice that high-level bombing at sea was a waste of bombs, fuel and flying time, rather like trying to hit a postage stamp with a shotgun. He had therefore ordered that B-25 Mitchell bombers, the versatile type which Doolittle had used to bomb Tokyo, be fitted with an orchestra of eight forward-firing .50-caliber machine guns and five-hundred-pound bombs with delayed-action fuses, and that these be used to practice attacking ships at masthead height – not dive-bombing but 'skip-bombing.'

Another Liberator rediscovered the convoy at breakfast-time on the 2nd, as it was about to 'turn the corner' round Cape Gloucester into the Dampier Strait. Two hours later, twenty-nine B-17s and Liberators attacked in two waves, from high level and then from five thousand feet, sinking one and damaging two other transports. The thin Japanese 'cap' of two dozen fighters could make little impression on the well-armed and armored attackers. Kimura, pointlessly adhering to radio silence, left two destroyers to pick up 811 survivors and then rush them ahead to Lae while he stoically, stupidly, steamed on. An Australian Catalina kept contact through the afternoon and night.

The two destroyers came back just in time for Kenney's third strike. This time about forty Zeros were on hand, circling high to pounce on high-flying heavy bombers. But the B-17s attacked from medium height; and this time the adapted B-25s, accompanied by similarly flown A-20s, came in just above sea level, escorted by a squadron of the new P-38 Lightnings – high-powered, twin-engined Army fighters – and another of Australian Beaufighters. The Japanese reacted as if to a torpedo attack, aligning their ships with the approach of the attackers, thus positioning themselves perfectly for combing from end to end by machine-gun bullets and hits by bombs dropped 'down the funnel' at the same time. Of the thirty-seven bombs dropped in the first wave, twenty-eight were claimed as hits. One B-17 and three Lightnings were lost, compared with five Zeros; but the convoy itself was reduced to a smoking shambles as attack after attack came in. Every single transport was sunk or sinking, together with half the destroyers. The four undamaged escorts picked up as many survivors as they could and fled the scene. The Americans were now masters of the new technique of land-based maritime air attack, of the kind that had destroyed the British capital ships off Kuantan in December 1941. In the Battle of the Bismarck Sea their aircraft had eliminated an entire military convoy and half its escort without assistance from any warship.

As darkness fell over a sea strewn with smoking hulks, wreckage and hundreds if not thousands of Japanese (in the water, in boats and dinghies, or clinging to life rafts and flotsam), eight American PT boats commanded by Lieutenant Commander Barry K. Atkins, USN, came in from Papuan bases. They dispatched with two torpedoes the last crippled transport still afloat. On the morning of the 4th, the Army airmen came back and sent the two damaged destroyers still afloat to the bottom. One US bomber was shot down, bringing the total American air losses to five out of 335 participating. The Japanese lost at least two dozen planes – together with Adachi and his divisional commander, their staffs and many seasoned officers, as well as 3,664 troops. Only 2,427 were saved over and above the 811 mentioned earlier. Here is how an unseemly proportion of them died, as described by Morison, the Navy's own historian:

Meanwhile planes and PTs went about the sickening business of killing survivors in boats, rafts or wreckage. Fighters mercilessly strafed anything on the surface. On 5 March the two PTs which had sunk *Oigawa Maru* put out to rescue a downed pilot and came on an enemy submarine receiving survivors from three large landing craft. Torpedoes missed as the I-boat crash-dived. The PTs turned their guns on and hurled depth charges at the three boats – which, with over a hundred men on board, sank. It was a grisly task, but a military necessity since Japanese soldiers do not surrender and, within swimming distance of shore, they could not be allowed to land and join the Lae garrison . . .

Several hundred swam ashore, and for a month there was open season on Nips in Papua; the natives had the time of their lives tracking them down as in the old head-hunting days. Many boatloads, spotted by planes, were destroyed by the PTs.

This was an isolated if also a very large incident, aggravated by its duration and by the number of 'Nip-hunters' involved. The Japanese may well have made something of a habit of shooting at survivors, but such barbarism was one of the main justifications for fighting them, and also Nazi Germany, whose troops (especially the SS) were guilty of atrocities on a routine basis, like Japan's. On 13 March 1944, Lieutenant Commander Heinz-Wilhelm Eck, commanding *U852*, sank the Greek steamer *Peleus* off West Africa. He surfaced and machine-gunned the survivors to prevent them from giving him away at a time when the Allies had an overwhelming advantage in the Atlantic. *U852* duly got away, but three Greek seamen lived to tell the tale. Eck and his officers, the only German sailors known to have killed survivors on purpose, were arrested, tried and shot by the British in 1945. A British submarine was involved in a similar massacre in the Mediterranean in 1943, but that was hushed up until 1988. Atkins, who was obeying orders, was not punished. Nor was his chief, Commander Morton C. Mumma, Jr., USN, the failed submarine-skipper mentioned earlier. Nor were the air crews or their commanders – one of the many ways in which it pays to be on the winning side in a war. The reluctance of the Japanese to surrender was certainly well known, and we shall soon be returning to that subject. But this has no bearing on the fact that the victims of the Bismarck Sea massacre were not offered the opportunity to give up before being killed in a defenseless condition in the water.

Their disaster in the Dampier Strait caused the Japanese forces to conclude a new 'Central Agreement' on 23 March, aimed at regaining air superiority over the New Guinea-Solomons region. The 18th Army was to be built up again for the defense of Japanese positions in New Guinea, while the 17th held the northern Solomons; the Sixth Army Air Division was to work with the Naval Air Service's 11th Air Fleet to break enemy airpower in eastern New Guinea. More than three hundred aircraft were assigned to the task. Admiral Yamamoto made a rare and, as it turned out, fateful decision to supervise personally the Navy's greater contribution to the joint effort in the air, and flew to Rabaul. His operational plan, code-named 'I-Go,' was in two phases and involved using carrier as well as land-based aircraft. Phase I, or Operation X, was to attack the Solomons between 5 and 10 April; phase II, Operation Y, was against New Guinea between the 11th and 20th. But all this, tactically sound though it might have been, was already too little too late. The American and Australian air forces already had numerical superiority, and their pilots were rapidly accumulating combat experience. Operation X involved well over two hundred planes against Guadalcanal and Tulagi; Operation Y flung nearly three hundred at a time into similar raids on ports and airfields in Papua. These were the largest Japanese air operations since Pearl Harbor. But the attrition of the pilots in the carrier battles, especially Midway, and over Guadalcanal and New Guinea before I-Go, and the willfully shortsighted failure to provide in advance for their replacement, had undermined this major new effort in advance. The total haul was two small Allied warships, two merchantmen, a tanker and some twenty-five aircraft, in exchange for losses of up to 10 percent on the Japanese side.

But the claims were something else, especially from the new and inexperienced pilots flung into the battle, who tended to overstate enemy losses by a factor of five. Yamamoto called a halt to operations and decided on 14 April, against the advice of his anxious staff, to make a morale-boosting tour of inspection of the triumphant air units, starting with those in the Solomons. For the protocol-conscious Japanese, an unprecedented flying visit to the front line by no less a personage than the C-in-C of the Combined Fleet entailed intricate preparations at minimum notice. These were broadcast by radio from Rabaul on the 14th to save time, in code JN25D, already in use for more than seven months. Among those who received them was Major Alva B. Lasswell, USMC, duty officer in the early hours of the 14th, Hawaiian time, at the Pearl Harbor cryptanalysis center. Commander Layton, Pacific Fleet intelligence officer, took the intercept to Nimitz that morning, and the two men discussed the advisability of shooting the admiral down in an air attack when he came within range of American fighters on his approach to Buin, at the southern end of Bougainville in the

Solomons. They concluded that Yamamoto was irreplaceable, but Nimitz, even as he alerted Halsey by top-secret message, felt it necessary to seek authorization from his own C-in-C to kill his opposite number on the enemy side.

The go-ahead from the White House came back via Frank Knox, the Navy secretary. Halsey was worried in case the trap now to be set for Yamamoto might reveal American penetration of Japanese ciphers; he was told to rely on disinformation and let it be reported to those involved in the operation that the tip had come from Allied Coastwatchers. Other factors in the American decision included strong personal animosity against Yamamoto, based on a distortion of something he had said before the war, to the effect that the kind of peace deal the Japanese Army required from America would have to be dictated in the White House. This was misinterpreted – by both sides – as a threat to invade the United States and subjugate it. Further, American commanders were intensely angered by the fact, not publicly announced until 20 April, that the three unreprieved Doolittle raiders had just been beheaded. The first anniversary of their exploit happened to be 18 April, the day the Americans could catch Yamamoto. The newly promoted Rear Admiral Marc A. Mitscher, naval air commander in the Solomons since 1 April, also wanted to pre-empt a feared Japanese special effort to mark the emperor's birthday on 29 April.

All these things came together to ensure that Major John W. Mitchell took his Army 339 Squadron of sixteen P-38 Lightnings from Henderson, flying low up the western side of the Solomons chain, to be over Buin at 11:35 A.M. – the moment when two 'Bettys' from Naval Air Squadron 705, escorted by nine Zeros, which had left Rabaul only half an hour earlier, began their landing approach. Captain Thomas Lanphier shot down the first 'Betty,' piloted by Chief Petty Officer Kotani and carrying Yamamoto with a handful of staff officers; Lieutenant Rex Barber got the second, flown by Petty Officer First Class Tanimoto with Vice-Admiral Ugaki Matome, Yamamoto's chief of staff, and other officers aboard. In the accompanying dogfight, three Zeros and one Lightning were also shot down.

Yamamoto's plane crashed in the jungle north of Buin;

Ugaki's into the sea. The chief of staff, though badly injured, managed to swim ashore with two other survivors. Yamamoto was killed outright. Second Lieutenant Hamasuna, an Army engineer building a road nearby, led a search party to the scene of the crash. He was shocked to find the instantly recognizable, diminutive and bareheaded figure of Yamamoto, attired in black flying boots and a simple green-drab uniform, with epaulettes showing the three cherry-blossom symbols of a full admiral. He was apparently sitting under a tree, his head on his chest as if sunk in thought and his body still firmly strapped into his aircraft seat, which had been flung clear. His hands in their white gloves, the index and middle fingers of the left one empty and tied up with thread, still clutched his ceremonial samurai sword. Or so Japanese sources reported; it is more than likely that the august corpse was tidied up by the search party out of respect, as none of the other bodies was so outwardly unscathed amid the wreckage. Yamamoto had two large-caliber bullet wounds, one in the head and one in the chest, which are rather difficult to reconcile with the suspicious neatness of his probably mythical final posture.

At any rate, his remains were reverently collected by Captain Watanabe of his staff and taken to Buin and cremated; the ashes were flown to the giant battleship *Musashi*, his last flagship, which took them from Rabaul to Tokyo Bay. The death of Fleet Admiral Yamamoto Isoroku, posthumously promoted and decorated with the Order of the Chrysanthemum, First Class, was announced only on 21 May, provoking an outburst of national mourning in Japan. A special train took his ashes into the capital; the track was lined with mourners. The state funeral ceremony was delayed until 9 June, the ninth anniversary of the death of Admiral Togo, victor over imperial Russia at Tsushima. Tens of thousands came to pay their last respects. Half the ashes were buried beside Togo's, half at Yamamoto's hometown of Nagaoka in a small memorial park.

Nimitz, Halsey, Turner, Mitscher and other American admirals in on the secret were overjoyed at the success of Major Mitchell's mission. Morison describes it as 'equivalent to a major victory.'

America Resurgent

Mature consideration suggests it was nothing of the kind. It was a perfectly legitimate operation in every respect, and a prime example of the intelligent use of intelligence, but the man thus surgically removed from the Pacific board, revered in Japan, first hated and subsequently admired in America, was by now a strategic irrelevance. The significance of the *coup de main* at Buin was surely limited to the psychological for both sides. Yamamoto, after all, had planned and ordered, even insisted upon, the Pearl Harbor operation which absolutely guaranteed American belligerence as nothing else could, not even the attack on the Philippines. Sound though it may have been strategically to stun the most powerful enemy by a pre-emptive strike, the move begged the underlying political question: it was no service to his country to force American entry into the war, which might otherwise have been delayed. Yamamoto also pressed for and led the disastrous Midway foray. He had just presided over the piecemeal operations in the Solomons, which frittered away Japanese maritime strength instead of concentrating it, with the huge initial advantages of superior strength and interior lines, against a growing enemy before he could establish himself. And with all his knowledge of naval aviation, for whose development in Japan he was principally responsible, to say nothing of his realistic prediction of a long war with America, Yamamoto must bear much of the blame for failing to procure a constant and much stronger flow of new planes and properly trained pilots for fleet and land-based naval air groups. A few more mistakes of this order and Yamamoto might have come to be seen, even by the Americans, as an asset to their cause.

Yamamoto was at his best before the Pacific theater opened, when he resisted the reckless rush into world war by the junta and warned against joining the Axis or getting involved in conflict with the Soviet Union, Britain and above all the United States. He clearly understood American potential strength and correctly predicted the consequences of such a war – which, however, as a pessimistic fatalist, he did nothing to allay. The personal hostility and stupidity of many of his senior colleagues at IGHQ had much to do with Yamamoto's ultimate ineffectuality. But his personal prestige as C-in-C, Combined Fleet, was second only to the emperor's, and, like Hirohito, he is open to criticism for failing to make more positive use of it. Yet it is not the truth but what you believe that counts, and Admiral Koga Mineichi, who was appointed Yamamoto's successor on 21 April and arrived at Truk four days later, said that 'nobody could take his place.' Certainly Koga, an uninspired and uninspiring conservative, was unable to, but there is nothing in Yamamoto's wartime record to indicate that he would have done any better against the rising tide of American air- and seapower, now well beyond the control of any oriental Canute.

The American strike over Buin, in which so much schoolboyish delight was taken by those in the know, was the second of its kind. Lieutenant General Imamura Hitoshi, commander of the Eighth Area Army in the Solomons, was flying in an unescorted naval plane which was jumped by Lightnings in the same area on 10 February 1943. The pilot just managed to escape into cloud. This knowledge led Captain Watanabe, Vice-Admiral Kusaka of the Southeast Area naval command at Rabaul, and Combined Fleet headquarters alike to the immediate conclusion, after Yamamoto's death, that the Americans must have broken into the ciphers. This suspicion was conveyed to the Fourth Department (Communications) of the Naval General Staff, responsible for codes and ciphers, which smugly insisted that such penetration was impossible and put the two enemy actions down to coincidence. Thus the Mitchell strike, even though many Lightning missions were flown in the area immediately afterward to allay suspicion of an ambush, came close to being extremely counterproductive by threatening Ultra.

That would have been doubly unfortunate, because the US Navy's intelligence apparatus was only just recovering from a major upheaval whose principal casualty was the amazing Joseph Rochefort. This was the result of a struggle for power in Washington pursued with the kind of single-minded ruthlessness that would have been better deployed against the enemy in the real world. At stake was control of sigint and cryptanalysis, the subject of bitter rivalry between the Office of Naval Intelligence (ONI) and the Office of Naval Communications (ONC). We saw in chapter two how the seeds of discord were sown by the division of responsibility which had the ONC gathering the raw material by radio interception and the ONI processing and assessing it. The objective of the struggle was Op-20-G (Pacific section), the unit of Admiral King's staff which did the decipherment work at Stations NEGAT, originally at the Navy Department in Washington, HYPO in Hawaii, and CAST, originally at Cavite in the Philippines and subsequently in Australia (the peculiar names came from the initial letters of each location in the bizarre phonetic alphabet then in US-military use; today they would be called 'November, Hotel and Charlie'). Two brothers, Captain Joseph and Commander John Redman, were in command of ONC and Op-20-G (Pacific) respectively at the time of the Midway sigint coup, which, as we have learned, was the work of Station HYPO under Rochefort. The Redmans not only falsely claimed credit for the breakthrough, which they disingenuously said was a joint effort by all three in which NEGAT had done the lion's share; they also used the unfair advantage of their presence at court to persuade Admiral King and his senior advisers to believe them.

So, when Nimitz recommended Rochefort for the Distinguished Service Medal for setting up the Midway victory, King refused (when Nimitz tried again after his own retirement, he was told by the Pentagon that it was too late). An insidious campaign against Rochefort, who was totally opposed to the ONC takeover bid, managed to smear him with some of the responsibility for the Chicago *Tribune* leak of the Coral Sea and Midway sigint secrets, and for the subsequent, almost certainly coincidental, Japanese changeover from JN25C to D, which caused Ultra virtually to dry up for several weeks from August 1942. The Japanese were not so complacent about their codes and ciphers that they neglected to change them from time to time – as indeed they did just before Midway, for no discernible reason other than routine prudence.

HYPO was also blamed by the whisperers for failing to predict Pearl Harbor, although, as explained in chapter three, the fault lay in Washington if anywhere. The upshot of the intrigue was the appointment, without reference to Nimitz, of Captain William Goggins, USN, a communications man and not a cryptanalyst, as head of HYPO in October 1942. HYPO was constitutionally an outstation of Op-20-G and therefore directly controlled by Washington. Rochefort was called to the capital for 'temporary' duty. He never came back, despite a very strong protest from Nimitz to King. In mid-November a furious CINCPAC was told – by surface mail – that Rochefort's temporary posting had become permanent. But he refused to cooperate and demanded a post at sea – playing into the character assassins' hands by 'proving' what an awkward customer he was! He got his wish eventually, in a sense, when he was given a command – of a floating dock, USS *ASBD Number 2*, on the California coast. Because he knew so many secrets, he was forbidden to take it to sea, an ambition he probably never entertained. Only in April 1944 was he recalled to Washington to work in two areas where his special talents were of value: strategic intelligence, and planning the (aborted) invasion of Japan. Such was the contemptuous and contemptible treatment of the man who opened the way to the Midway triumph.

But his influence survived at CINCPAC headquarters, to the extent that a reorganization of intelligence operations in the Pacific, as suggested by Rochefort, was put into effect in September 1943, with the creation of the Joint Intelligence Center, Pacific Ocean Areas (JICPOA). This was the combined-services version of ICPOA, formed exactly one year earlier to coordinate all the various intelligence efforts of the Pacific Fleet. Now Army and Navy pooled their resources in the interests of efficiency, and HYPO became FRUPAC – Fleet Radio Unit, Pacific Fleet – under JICPOA, whose first chief was Brigadier General Joseph J. Twitty, USA. A particular beneficiary of this joint effort was the submarine service. Boats could be guided to targets whose movements were revealed by sigint, the method used with even greater success by the Germans. A running plot was kept at Pearl of all known enemy naval and mercantile shipping and the movements of US forces, an idea first developed by the British. In Washington the sigint industry grew so rapidly that both Op-20-G (which also had a huge Atlantic section) and the Army's Special Intelligence Service burst out of their original quarters and developed a taste for commandeering girls' schools. The Navy sequestered Mount Vernon Seminary, near George Washington's family home, and the Army requisitioned Arlington Hall, just over the District of Columbia boundary in Virginia. In the South-West Pacific command, the old Station CAST became FRUMEL, Fleet Radio Unit, Melbourne, working with the Australian Navy's cryptanalysts, and MacArthur's Sixth Army set up its 'Central Bureau,' a joint but not overly successful operation with the Australian Army and Air Force. (Its information was not always reliable, and the standard of security was probably the lowest of any major Allied intelligence center worldwide. Only Japanese arrogance and rigidity may have prevented it from compromising Ultra and Magic.)

And what of the Japanese in this crucial field? Like that of every other major belligerent, Japan's intelligence-gathering system was complex and multifarious. The Army General Staff and the Navy's each had an intelligence service; the Foreign Ministry gathered diplomatic information from the whole world except the Far East, which was covered by the Greater East Asia Ministry's General Affairs Bureau. The Foreign Ministry's Investigation Bureau was divided in three: Section I covered the world except for East Asia and for the fief of Section II (Soviet Union and West Asia); Section III collated intelligence material from all sources. All this seems rational enough – until it is remembered that the foreign minister and the Greater East Asia minister were one and the same! Military intelligence came under Department 1 of the Army General Staff, and naval intelligence under the Navy General Staff's Department 3, with a separate Special Service Section for cryptanalysis.

The Army's main source of tactical information was enemy sigint. The Japanese did not penetrate the American or British high-level ciphers but did break deeply into lower-grade ciphers used for routine transmissions such as weather reports. Not surprisingly, the greatest cryptanalytic successes were achieved against the Chinese, with their related language. Much information was garnered from careful traffic analysis, which yielded a lot of information even if the messages themselves were unintelligible. Air-to-ground, air-to-air, and air-to-ship radio were also sometimes very fruitful tactically, until short-range VHF was in general use by the Allies for such purposes. Department 2 passed on its gleanings to Department 1 – war planning – but took no part in the latter's work. The Japanese Army called sigint 'special intelligence' and confined its distribution to divisional and higher commands. Only armies (corps in the West) and 'area armies' (armies) had separate intelligence staffs; lesser units had to make do with a few part-timers. Department 2 boasted five officers for the first year of the Pacific Campaign; its Section 6 (the Americas) expanded to a less-than-generous twenty-nine officers and half a dozen NCOs by the end of the war, an infinitesimal total compared with the tens of thousands working on sigint alone in the American and British forces. Other sources of information included old-fashioned spies, employed by the battalion, it seems; prisoners of war; people in occupied areas; shot-down enemy airmen and shipwrecked enemy sailors; aerial reconnaissance; and captured documents, which were particularly prized.

The Navy relied on similar sources but paid more attention to foreign press and broadcasting. It also located enemy ships by RDF – radio direction-finding – which the Japanese practiced with more skill than any other warring nation. But then they needed to, in order to offset their weakness in cryptanalysis compared with the Allies. Even so, naval intelligence was usually able to predict American naval moves three to seven days ahead and picked up very accurate information on enemy losses. The role of intelligence officer on a ship fell to the captain, except on a flagship, where there was an intelligence officer on the admiral's staff: his duties were strictly tactical, and he was also in charge of cryptography – encoding and enciphering messages to be sent. Formations led by officers below flag-rank made their communications officer double as intelligence officer. Fleet headquarters, of which there were fewer than a dozen, were also permitted the luxury of specialized intelligence officers – who nevertheless received no special training. Department 3, Section 5 (USA), consisted of a rear admiral and five other officers; Section 8 (Britain, Australia, and Indochina) had one captain and four other officers; the radio-interception operation engaged one captain and two other regular officers, with forty reserve officers chosen for their educational standard. A special section of Department 4 – communications – of the Navy General Staff handled sigint. Its Section 1 had three officers for general duties; the second handled cryptanalysis with four officers and thirty typists; the third was in charge of analysis and distribution, employing one captain, one commander, ten reserve officers and 120 enlisted men. Small change indeed by Allied standards.

Neither Japanese service established a separate section for air intelligence. The only liaison mechanism between the two intelligence departments was the weekly meeting between the rear admiral and the junior general in charge of each, so that they could prepare a joint and noncontradictory briefing for the War Cabinet. Because of their different traditions, historically and strategically, the Navy specialized in general intelligence on America and the British Empire, while the Army focused its main efforts on China and the Soviet Union. Each service would distribute urgent information by radio, which was a boon to Allied intelligence; less pressing material was distributed by aircraft and was therefore totally secure, unless captured by the enemy. Neither service produced regular intelligence bulletins.

In all, the Japanese intelligence effort was small, of indifferent quality and overreliant on espionage. Its sigint work was of a high standard but limited in scale and scope; the maximum intelligence was extracted from open sources, such as enemy publications and broadcasting. Totalitarian regimes are always unduly excited by democracy's open books and waste an enormous amount of effort on them for a relatively small yield, sifted from the indiscriminate consumption of a voracious vacuum cleaner. The absence of trained specialists meant that Japanese intelligence officers were unusually gullible and shaky on evaluation, with a tendency to overestimate the worth of information. The underlying weakness was the ignorant indifference of senior commanders to the value of intelligence: they prized their autonomy too much to have any more to do with it than they needed to. Their attitude was not so different from Henry Stimson's distaste for reading other people's mail, or from MacArthur's inability to appreciate the sensitivity of Ultra information, which he leaked more than any other senior Allied commander. Fortunately for him, the Japanese also undervalued eavesdropping and seriously underestimated their enemies' interest and ability in this fruitful field. The Allies found them easy to deceive and after the war discovered from their files how meager their information about America, Australia and other enemies had been.

If the Imperial Navy was not quite so bad as the Army in its neglect of intelligence, it was far worse than any other important wartime navy in its failure to recognize the potential of the submarine, whether in enemy hands or its own. For a state run by a military-naval junta which professed to have absorbed the lessons of the First World War this was astonishing enough; for a Navy built up on British lines and drawn, like the Royal Navy, from an island-nation, the simultaneous neglect of antisubmarine warfare is stupefying. If anyone on the Naval General Staff had read anything more demanding than a comic about the First World War at sea, he could hardly have failed to notice how close Germany came to starving out Britain, reduced to six weeks' supply of essentials by the spring of 1917, when the London Admiralty awoke in the nick of time to the necessity for mercantile convoys. Its Japanese opposite number, psychologically committed to early victory or death, had no plan for mercantile convoys – or for attacking American shipping – when it went to war against the US Navy. It sent its submarines to sea in 1941 without radar, sonar or even hydrophones (the passive listening device invented by the British in the First World War). These deficiencies were not fully remedied until well into the Pacific Campaign. As usual in their Navy, the Japanese boats were exceptionally well equipped with night optics and their crews well trained in night attacks. But in general sailors were trained like pre-1914 submariners to work with the surface fleet, to scout, to hunt and attack warships, not convoys or merchant ships. Captains were obsessed with sinking aircraft carriers, the best-protected enemy ships, and then other warships in descending order of size. They must have passed up many opportunities to sink lesser but strategically significant targets like merchant ships.

It took a German naval officer to appreciate in full the

opportunity his Japanese allies were willfully missing, given what the U-boats, sometimes only operating in handfuls, had been able to achieve in the Atlantic from 1939 to 1943. Vice-Admiral Paul Wenneker, Hitler's frustrated naval attaché in Tokyo, noted:

I suggested the desirability of attacking the route between Honolulu and the West Coast because that would force the use of convoys and would force the withdrawal of many escorts from the western Pacific . . .
   We arranged for one full Japanese submarine crew to be sent to Germany [via the west coast of France] for training. They had, I think, very good training in German boats and in German attack methods, but unfortunately they got caught in the North Atlantic . . . while returning [to Japan].

German urgings and appeals for attacks on American merchant shipping with the outstanding Japanese torpedoes (originally developed for surface vessels) persistently fell on deaf ears. Wenneker took the view that Japanese boats were too large for the contemporary technology. He regarded the highly successful German Type IXd submarine as best suited for the Japanese in the Pacific and had one sent over. It was moved to the Navy Yard at Kure, on the Inland Sea, methodically taken to pieces and studied. The Japanese concluded it was too complicated for them to copy. As it was, their own boats were too large and clumsy. Some carried recoverable float planes kept in a deck hangar and launched by catapult, and tried to operate as underwater cruisers; others sported a pair of remarkably large guns of 5.5-inch caliber for long-range shore bombardment.
   They had sixty-four boats in 1941 – seven more than the British and the Germans in 1939, who had gone to war with fifty-seven each. The Submarine Command operated three main types: the cruiser and fleet submarines (I-boats) of exceptionally large displacement, a medium type and a smaller, older coastal type whose numbers carried the prefix RO. In all, 126 boats were added to the fleet during the war, the last twenty-six as transports, reflecting the new needs created by the adverse course of the war. They were all ultimately derived from German and British First World War types.

There were also four inefficient mine-laying submarines in the fleet, several hundred underused and usually ineffectual two-man, two-torpedo midget submarines – and three fifty-two-hundred-ton underwater aircraft-carriers equipped with three small bomber-seaplanes each. These were the largest boats in the world at the time and were meant to attack the US West Coast and an obvious target mentioned earlier, the Panama Canal. But they were never deployed. Nor were two experimental types, one medium and one small and both extremely fast underwater, which only reached the prototype stage. The last addition to Japan's array of underwater weapons was the miniature suicide-attack 'boat' – essentially a torpedo with a saddle.

The Submarine Command told the Naval General Staff after the opening operations in the Pacific: 'We have discovered that it is very hard for submarines to attack warships and to blockade a well-guarded harbor. We believe that the principal targets of submarines should be merchant ships, not warships.' Fortunately for the Allies, the admirals, including Yamamoto, took no notice. The total haul of the submarine fleet was 184 merchantmen totaling 907,000 tons, plus two carriers, two cruisers, ten destroyer-escorts, and some smaller naval vessels, including submarines. The Japanese lost 129 boats out of 190 – seventy to surface warships, eighteen to aircraft, nineteen to enemy submarines and twenty-two otherwise, including collisions, explosions, accidents and unexplained causes. Considering the plethora of targets on offer, far more than the Americans had to choose from in their submarine campaign, the performance of the Japanese submarine arm was indifferent: under five thousand tons of merchant shipping per boat deployed, just over seven thousand tons per boat lost. The enormous length of Allied supply lines in the Pacific, and the near-infinite expanse of ocean and multiplicity of islands to hide in, could have opened the way to devastating campaigns against the Americans, and also the British in the Indian Ocean. For fear of the Soviet Union, Japanese boats took no action at all against the US trans-Pacific supply convoys which complemented the British ones to the Russian North Atlantic ports. All in all, the Japanese Submarine Command – through no fault of its own, because it clearly understood the possibilities – missed one of the outstanding strategic opportunities of the war.

In many respects, 1943 was the pivotal year of the war, when the United Nations threw the Axis onto the defensive on almost all fronts: against Hitler and Mussolini in the Soviet Union, North Africa, the Atlantic and Italy; and in the war against Japan, the North and South Pacific and New Guinea. Only in the ill-coordinated China-Burma-India theater was the worldwide struggle against the totalitarian powers bogged down.

Events in the Pacific involved limited offensives by the two Allied supreme commands, sometimes interlinked, sometimes independent but simultaneous. For the sake of clarity, chronological order has been set aside for the next chapter in favor of a region-by-region approach: starting with the North Pacific, followed by New Guinea, then the Solomons and lastly the Bismarck Islands.

9
Birth of the Leapfrog

The Japanese capture of the Aleutian Islands was rendered pointless by the pivotal American victory at Midway. Instead of forming the last link in a new, forward-defense perimeter, the dismal islands were an isolated outpost of no strategic value to Japan but difficult to maintain – and a standing offense to the Americans until they could be recovered. These were not colonies like Guam but United States soil, part and parcel of the territory of Alaska, as critics never tired of reminding the government. Their recapture took an unconscionably long time, because Nimitz felt he could not spare the disproportionately large force he thought he needed to dislodge the Japanese garrisons. They came under the command of Vice-Admiral Hosogaya Boshiro, C-in-C, Northern Area Force, based at Paramushiro, in the Kurile Islands, six hundred miles to the west of Attu and a thousand miles west of Kiska. The Japanese were in no mood, and no position, to use the far-northern toehold as a springboard for aggressive operations. For the Americans, Rear Admiral Robert Theobald had the task of holding them in check from his headquarters ashore on Kodiak. An enervating sideshow followed, involving carriers, surface ships, aircraft, submarines, transports, and supply ships on both sides for remarkably small returns in a series of frustrating operations. The Americans tried naval and aerial bombardments; the Japanese endured them stoically for nine months, until March 1943. Then at last there was a naval clash, but even that did not affect the stalemate.

Rear Admiral Charles McMorris was at sea toward the end of March in his elderly little flagship, the light cruiser *Richmond*, accompanied by the heavy cruiser *Salt Lake City*, fully recovered from the Battle of Cape Esperance, and four destroyers. He was looking for one of the Japanese arctic convoys which had been sailing to and fro feeding the garrisons and exchanging or augmenting their troops. They were about a hundred miles due south of the Komandorski Islands, a group about two hundred miles west of Attu, when they sighted a powerfully escorted convoy just before dawn. Admiral Hosogaya himself was at sea with his entire squadron of two heavy cruisers, a light cruiser leading four destroyers, and two armed merchant cruiser-transports. Heavily outgunned, McMorris nevertheless tried at first to sink the armed transports while himself under fire from the Japanese cruisers; Hosogaya blocked the move by sending the merchantmen to the rear and turned the fight into a gunnery match between his two heavies and the Americans' one. *Salt Lake City* (Captain Bertram Rodgers, USN) took on *Nachi* and *Maya* for three and a half hours. The American veteran was hit hard four times, developed steering trouble and a five-degree list when a plunging eight-inch shell passed right through her hull, went dead in the water, recovered full power and lived on behind her smokescreen to tell the tale. She hit *Nachi* twice, briefly knocking out her electric circuits; the destroyer USS *Bailey* also hit her, causing an ammunition explosion and much smoke. Hosogaya, correctly anticipating American air intervention from Alaska, withdrew before it arrived. The inferior American force had conducted a skilled fight to extricate itself in one of the very few 'fleet actions' on the open sea involving the US Navy in the twentieth century. Only one aircraft, a spotter from *Nachi*, took part. All US ships got to Dutch Harbor safely. Strictly medicinal alcohol was issued to the exhausted men of *Salt Lake City*, 'dry' Navy or no, after the action. Hosogaya was sacked.

By the time it came to invading Attu, Vice-Admiral Thomas Kinkaid had replaced Theobald as commander, North Pacific; Rear Admiral Francis Rockwell was his Amphibious Force commander, his task to land the US Army's Seventh Division, covered by three battleships and an escort carrier, three heavy and three light cruisers, nineteen destroyers, transports and auxiliaries. The XI Army Air Force performed 'softening-up' duty. The invasion fleet assembled in Cold Bay, at the western end of Alaska, and set sail on 4 May. This was timely because the Japanese had divined the American plan to go for Attu, bypassing Kiska which was closer to their Alaskan bases, and planned to strengthen it at the end of the month, when the spring fogs returned. On 11 May the first waves of two thousand US troops went ashore on the south side of Attu in Massacre Bay – happily a misnomer, as there were no Japanese yet to be seen, and near Holtz Bay to the north. The Japanese garrison of twenty-six hundred men was concentrated between the two bays at the eastern end of the island and soon made its presence felt. The Americans deployed eleven thousand men against them and took until 29 May to eradicate the resistance, which ended with a suicide charge by a thousand screaming Japanese. Those not mown down by massive American firepower pulled the pins from their grenades and blew themselves up. Fewer than three hundred Japanese survived. It was America's third amphibious operation of the war, after Guadalcanal and North Africa, and the first to meet opposition that was both early and determined. The US casualties, at six hundred dead and twice as many wounded, amounted to 16 percent – not an impressive achievement given their local superiority of more than four to one and their huge material advantage on the day. The Japanese Combined Fleet considered sending a huge force of carriers and capital ships to the relief of Colonel Yamazaki but wisely thought better of it. The forces gathered for the aborted move usefully took some of the pressure off the Americans in the Solomons. As a result the Imperial Navy took no part in the affair: Hosogaya's successor, Vice-Admiral Kawasa, was in no position to interfere unaided against the powerful American task force and came no nearer than four hundred miles.

The Americans spent the next two and a half months pounding Kiska preparatory to invading it. They assembled nearly a hundred ships under Rockwell at Adak, in the Andreanof Islands, to the east of Kiska, and set sail with twenty-nine thousand troops on 13 August. They had even invited their Canadian allies to make a unique appearance in the Pacific theater with an additional fifty-three hundred soldiers. They began landing on the 15th, after an almighty bombardment, and were all ashore in two days. They searched for the Japanese garrison for a week. Unfortunately they had ignored a suggestion from Major General Holland M. Smith, USMC, that a scouting party be sent ashore first by boat to assess the likely opposition. Had they done so they might have found out sooner that Admiral Kawasa had sent in Rear Admiral Kimura Shofuku on 28 July to lift the entire garrison of fifty-two hundred and take it home. Even though they waited for fog, this took courage under the constant pressure from massive American air raids, which had made the garrison's last weeks unbearable. This was a small but spectacular addition to the list of brilliant military evacuations of the Second World War; it will be remembered that the Japanese had already achieved something similar in their final withdrawal from Guadalcanal in February 1943. If the American commanders at Kiska felt foolish when they discovered that the nut had been whisked away before the sledgehammer struck, they thoroughly deserved to. The useless occupation of the Aleutians, ended by an invasion of empty space, cost the Japanese twenty-four hundred troops, three destroyers, six submarines and nine merchantmen. But North America had been divested of its minuscule Japanese irritant, whatever the cost in wasted effort and lost pride. At the opposite corner of the enormous Pacific theater, in New Guinea, the situation was rather more serious.

Despite the total failure of their reinforcement attempt at the beginning of March, the Japanese were still entrenched in northern New Guinea, the right flank of their 'Bismarck Barrier' (the Solomons being their left flank). One of their strongholds was Salamaua, on Huon Gulf. For the soldiery there was not much to do except dream of a decent meal and shoot at Allied air patrols. The daily ration at this time of minimal ground activity and acute shortage of supplies was down to four hundred grams (fourteen ounces) of polished rice, thirty grams of canned meat (one ounce), twenty grams of powdered bean mash and a similar quantity of powdered soya sauce, plus ten grams each of sugar and salt. Front-line soldiers got an extra two hundred grams of rice; sometimes there was an emergency ration pack, type B, consisting of dried biscuit and compressed food. Every now and again they shot down one of their airborne tormentors. The following account of what happened thereafter to the unnamed pilot of a Royal Australian

Air Force Douglas Dauntless dive-bomber, a flight lieutenant aged twenty-three based at Port Moresby, was recorded in an unknown Japanese soldier's diary found at Salamaua in October 1943. The entry covers 29 March 1943, and is headed 'Blood Carnival.' It is reproduced here, after due reflection, because in all the research for this book nothing illustrated the cultural and moral gulf between the two sides in the war in the Far East quite so graphically, not even the description above of the casual human vivisection in the jungle of Guadalcanal.

Tai- [detachment-] commander Komai, when he came to the observation-station today, told us personally that, in accordance with the compassionate sentiments of Japanese Bushido, he was going to kill the prisoner himself with his favourite sword. So we gathered to observe . . .

The prisoner . . . is given his last drink of water etc . . . The time has come, so the prisoner, with his arms bound and his long hair now cropped very close, totters forward . . . He is more composed than I thought he would be. Without more ado he is put on the truck and we set out . . . As I picture the scene we are about to witness my heart beats faster . . . The prisoner has probably resigned himself to his fate . . . [and] seems deep in thought. I feel a surge of pity and turn my eyes away . . .

Tai-commander Komai stands up and says to the prisoner: 'We are now going to kill you' . . . The Flight Lieutenant says a few words in a low voice. Apparently he wants to be killed with one blow of the sword. The [commander] replies, 'Yes' (in English).

Now the time has come and the prisoner is made to kneel on the bank of a bomb crater filled with water . . . He remains calm. He even stretches out his neck and is very brave . . . Ordinary human feelings make me pity him.

The tai-commander has drawn his favourite sword . . . It glitters in the light and sends a cold shiver down my spine. He taps the prisoner's neck lightly with the back of the blade, then raises it above his head with both arms and brings it down with a sweep . . . I had been standing with my muscles tensed but in that moment I closed my eyes.

Sssh . . . It must be the sound of blood spurting from the arteries. With a sound as though something watery had been cut, the body falls forward. It is amazing – he had been killed with one stroke . . . The head, detached from the trunk, rolls in front of it . . . The dark blood gushes out.

All is over. The head is dead white, like a doll. The savageness which I felt only a little while ago is gone, and now I feel nothing but the true compassion of Japanese Bushido . . . This will be something to remember all my life. If ever I get

back alive it will make a good story to tell, so I have written it down.
– Salamaua Observation Post, 30 March 1943, 0110 hours.

The first objective of the avenging Americans and Australians in Operation Cartwheel was two islands in the Trobriand group, north of the eastern end of Papua, to be used as air bases against the Japanese positions in the Bismarcks and the northern Solomons. 'MacArthur's Navy' was now styled the Seventh Fleet in King's naval reorganization of 15 March, when Halsey's force, the bulk of Nimitz's Pacific Fleet, also became the Third Fleet. Vice-Admiral 'Chips' Carpender commanded the heterogeneous collection of American, Australian, Dutch and New Zealand ships, which now had its own amphibious force under Rear Admiral Daniel Barbey, even if many of its vessels had to be borrowed from Turner's Third Fleet amphibians.

MacArthur, disparaging as ever about his laconic Australian allies, decided to bypass his land-forces commander, the Australian General Blamey, by creating a body code-named 'Alamo Force' – alias Lieutenant General Walter Krueger's United States Sixth Army, which, because it was styled a task force, could come under MacArthur's direct command. The two islands, Woodlark and Kiriwina, were taken without opposition; the only difficulties were logistical as the new amphibious force learned its complicated trade. Otherwise, the first of more than fifty landings by 'VII 'Phib' passed slowly but without interference from the enemy. He after all had to contend with a larger landing at Nassau Bay, to say nothing of the Third Fleet's assault on New Georgia in the Solomons, all on 30 June. The Nassau Bay landing was to reinforce and supply the Australian Seventh Division in its efforts to dislodge the Japanese in Salamaua. Commander Atkins, with four PT boats, led a flock of landing craft from eastern Papua through dreadful weather to a chaotic landfall forty miles westward. The three hundred Japanese Marines of the local outpost, having lost their commander to a bomb, staged a rare withdrawal without resistance.

Blamey, meanwhile, was planning to crush Lae between two Australian divisions, one to come by air, the other by sea, as the Americans and his Seventh Division attacked Salamaua overland. Once these places fell, the Allies could move east against the third important Japanese stronghold, at Finschhafen. All three, plus Japanese positions at Wewak and Madang to the west, were subjected to heavy and persistent aerial bombardment, to which they had no effective reply. The Japanese no less persistently sent penny-packets of reinforcements by barge trains from New Britain in the Bismarcks, keeping the American PT boats well supplied with easy targets; destroyers and submarines were able to get through on several occasions, but only a few thousand men could be added to the garrisons as they awaited the Allied assault they knew was coming. A Japanese attempt to wrest local control of the air from the Allies by assembling hundreds of planes at Wewak was broken up in mid-August by a strong American raid. Four US destroyers set a precedent at the same time by making the Seventh Fleet's first shore bombardment from the highly dangerous, sketchily charted waters of Huon Gulf. Barbey delivered the Ninth Australian Division for the attack on Lae on 4 September, against fierce but fragmented Japanese air attacks from New Britain and western New Guinea; the other Australian division arrived in a superbly organized airlift to an old airfield at Nadzab, west of Lae, the next day, and the Seventh Fleet cut off any hope of escape or relief by sea. This nut cracked even before the sledgehammer struck. The garrison of only two thousand chose to retreat on foot to Sio, fifty miles away, on the north coast of the Huon Peninsula. The majority died on the way. Lae fell into Australian hands on 16 September – one day after Salamaua succumbed.

MacArthur now had the bit between his teeth and ordered Barbey to be ready to land troops to capture Finschhafen within a week. Despite massive confusion caused by the unaccustomed night landing (for fear of enemy air attack), and despite one medium-sized air raid from Rabaul, the Australians and Americans got ashore. An Australian reinforced brigade took the town on 2 October. Memories of the Bismarck Sea disaster deterred the Japanese from reinforcing their small garrison, most of which retreated inland. A counterattack, by a Japanese division from Madang and other troops on 17 October, was beaten off. The sick and half-starved Japanese were unable to make an impression on well-fed, healthy Australians with tanks. Five days earlier, 350 American planes of MacArthur's V Army Air Force had staged the largest raid of the Pacific Campaign so far, against battered Rabaul. From now on the mostly Australian troops on New Guinea were engaged in mopping-up operations, protracted by the tortuous terrain and the cruel conditions prevailing on the great island. The Japanese had bigger fish to fry in the Solomons, which they were able to defend in depth, from southeast to northwest, with scope for naval action aplenty. We can therefore switch flanks and examine the progress of Admiral Halscy 'up the Solomons ladder.'

The first objective in the Solomons, after the straightforward American landing on the Russell Islands, immediately northwest of Guadalcanal, on 21 February 1943, was New Georgia and its associated islands, northwest of the Russells, to be attacked in Operation Toenails on 30 June. Before that the naval air forces of Admiral Koga, the new C-in-C, Combined Fleet, engaged the Americans flying from Henderson in a long series of bombing raids and dogfights in an attempt to disrupt the build-up on Guadalcanal. When the American naval pilots knocked out the New Georgia airfields of Munda and Vila by repeated bombing, the Japanese flew from Rabaul and Bougainville. In the dogfights the Americans usually defeated less experienced and less numerous opponents. The command at Rabaul was lulled into a false sense of security by the exaggerated claims of its pilots, and by the lull in American movements in the Solomons in the last week of June. When submarine *RO103* reported Admiral Turner's huge amphibious force approaching from the south on the 30th, intent on landing at five points in the New Georgia group, the Japanese were taken by surprise, with their main attacking air strength withdrawn from the northern Solomons to Rabaul. They were thrown into confusion by the first landing on New Georgia, because they had not had time to digest the alert from New Guinea, where Allied troops had begun their new landings.

But other Japanese submarines sank two freighters off New Georgia.

The invasion of New Georgia was all but complete by the time a Japanese air raid was mounted. Its only success was a torpedo hit on Turner's flagship, the transport *McCawley*, which was eventually abandoned, and then sunk in error by American PT boats. They had not been told of the landing in which she had taken part, at Rendova Island. Despite hazards, expected and unexpected, from reefs, mud, Japanese troops and bombardment by artillery, planes and destroyers, two battalions of Marines under Colonel Liversedge and two regiments of Major General John Hester's 43rd Army Division took on five thousand Japanese under Major General Sasaki Noboru and soon had footholds on the various islands. But their troubles were only beginning. The main objectives were Munda (especially) and Vila airfields, which the Americans soon expected to have back in working order. The Japanese had decided to reinforce the circular island of Kolombangara, where Vila lay, with four thousand troops and sent in a Tokyo Express from the Shortland Islands, just south of Bougainville, on the night of 4–5 July: seven destroyer-transports and three destroyers commanded by Rear Admiral Akiyama Teruo. But Rear Admiral Walden L. Ainsworth's Task Group 36.1 of three light cruisers and four destroyers got there first and was bombarding the shore. The Japanese retreated but sank the destroyer *Strong* by torpedo, at such long range that the Americans assumed it came from a submarine. The next night the Japanese came back and Ainsworth's group came out to meet them in Kula Gulf, between Kolombangara and New Georgia.

In the Battle of Kula Gulf, on the night of 5–6 July, the first of six major sea-actions in this second phase of the Solomons campaign, the Japanese once again proved they were not nice people to do business with at night. The Americans saw the enemy on their radar and gave themselves away by opening fire with their flashing guns at about six miles. The Japanese retaliated with their latest twenty-four-inch torpedoes. Three of them hit the hitherto fortunate USS *Helena*, tearing off her bow, folding her in two and sending her to the bottom in three minutes. The Japanese lost one destroyer (and Admiral Akiyama) to gunfire, and another, which ran aground landing troops at Vila, to bombing the next day. Three more were damaged. Some two thousand soldiers were landed. Nearly two hundred men of the *Helena* were left to their fate, presumed lost but in fact clinging to the still-floating bow section of the cruiser or to rubber boats nearby. The 165 survivors eventually got ashore on the island of Vella Lavella, north of Kolombangara, and hid in the jungle from the Japanese under the protection of Australian Coastwatchers and friendly local inhabitants. Ten American destroyers came to their rescue from Tulagi on 16 July, in a daring dash up the Slot to within sixty miles of major Japanese air and naval positions, and got away undamaged.

Elsewhere ashore, the Americans, especially the 43rd Division, were faring badly. Major General Oscar W. Griswold, Hester's chief as commander of XIV Corps, paid a visit to the front and concluded that the green division was effectively broken, despite modest casualties. Halsey decided to put two more divisions into the fight for New Georgia, told Griswold to take command on the spot, and sent his ground-troops commander, Lieutenant General Millard Harmon, to Guadalcanal to superintend what had suddenly become a desperate undertaking.

On the water, the two navies clashed again in the Battle of Kolombangara, when a Tokyo Express led by Rear Admiral Izaki Shunji, in light cruiser *Jintsu* with five destroyers and four destroyer-transports, met Ainsworth's restyled and heavily reinforced Task Force 18 on the night of 12–13 July. The New Zealand light cruiser HMNZS *Leander* had replaced *Helena*, and Ainsworth had two destroyer-squadrons, a total of eleven, as well as two other light cruisers. The Japanese still had no radar on their ships, but they had just acquired a radar detector, which found the Americans by their emissions before the emissions themselves located the Japanese. It was an extension of the art of direction-finding from radio to radar. The Japanese therefore had already fired torpedoes at long range before the Americans opened up with their own torpedoes and guns. The *Jintsu*, veteran of so many battles in the Slot, was demolished by twenty-six hundred rounds of six-inch cruiser shells, and Admiral Izaki went down with her; the *Leander* was knocked out but not sunk by a torpedo. A second torpedo attack hit the light cruisers *Honolulu* (flagship) and *St Louis* and sank the destroyer USS *Gwin*. All three Allied cruisers needed extensive repair. The Japanese landed twelve hundred troops on the west coast of Kolombangara in the early hours of the 13th, as planned, and withdrew to Buin in Bougainville.

The Americans misled themselves about both these short, sharp naval engagements: having claimed large enemy losses, they thought they had won tactical victories. In fact they had come off worse in both fights through overestimating the value of radar and gunnery and dangerously underestimating the mighty Long Lance torpedo. Not only could it run for twenty-one miles; the Japanese could also reload their tubes twice as quickly as the Americans. But at this stage of an increasingly unequal war, the Japanese needed rather more than tactical victories to save themselves.

It took American troops ten days short of three months to complete the conquest of obscure New Georgia and adjacent islands. The main objective, the Munda field, fell only on 5 August, after an inch-by-inch struggle over appalling terrain too forbidding even for tanks. Artillery, naval bombardments and air attacks expended mountains of ammunition pulverizing the well-dug-in Japanese, who managed to recover in time to launch another fearsome night attack with bayonets. Another six weeks of mopping up was necessary before General Sasaki's force was eliminated to a man by the numerically much superior American troops. At sea, more than fifty American PT boats did all they could to interfere with Japanese reinforcement runs by destroyer and barge-train. One of them, *PT-109*, was cut in half south of Kolombangara by the destroyer *Amagiri*, which thus came close to depriving the United States of a future president. Lieutenant John F. Kennedy, USNR, the skipper, had to swim for it with ten other survivors from the crew of thirteen. After landing with them on an islet, Kennedy swam for help. He was rescued by Solomon Islanders in a war canoe and smuggled to an American PT base on the island of Wanawana, southwest of Kolombangara, on 5 August, whence he led a PT boat to the rescue of his men.

The next naval action in the Slot came the next night, when Commander Frederick Moosbrugger, USN, led six destroyers (there being no cruisers on hand) from Tulagi against a Tokyo Express of four destroyers, three of them transports bound for Kolombangara, in the Battle of Vella Gulf. This time it was a clear victory for the well-drilled Americans, with all three transports (and nine hundred troops) sunk by torpedoes at no loss to themselves. Only the escorting destroyer got away. Japanese sailors and soldiers persistently refused rescue from the water, and some fifteen hundred died in the action. About four hundred US sailors, and three times as many troops, were killed in the long, messy and rather ineffectual struggle for Munda and adjacent territories. At this rate, the schedule for the reconquest of the Japanese-held islands of the Pacific could be torn up and thrown away. Given the chance of defense in depth, a few thousand Japanese ready to die could tie down for months huge forces led by men concerned to minimize casualties, offsetting the enormous numerical and material odds against the Japanese. At this rate, indeed, the strategy of sapping the American will to fight still had a fighting chance. On the Allied side there was an urgent need for lateral thinking to break the strategic logjam.

The answer had been lying around since 1940, when a staff study on war with Japan at the Naval War College proposed a thrust from Hawaii straight at the major Japanese naval base of Truk in the Carolines – bypassing the Marshall Islands which would otherwise have been attacked first in a step-by-step progress across the Pacific by the US Fleet. This stratagem of the bypass or leapfrog was under serious consideration by the staffs of both Nimitz and Halsey in January 1943, with the former wanting to bypass New Georgia in favor of Bougainville, and the latter Munda for Kolombangara. But the airfield at Munda was considered too important to forgo. Nimitz's staff also contemplated at this time a really daring proposal for a giant leap to the Bonin Islands, only a few hundred miles from Tokyo, but concluded they needed more carriers for the inevitable clash with the bulk of the Japanese Navy which such a move would provoke. By attacking Attu in the Aleutians before Kiska in May 1943, Nimitz's North Pacific command, because it did not have enough strength to do both at the same time, accidentally put the leapfrog method into practice for the first time.

This prompted Nimitz in July to recommend bypassing Kolombangara on purpose in favor of the next target, the island of Vella Lavella, to the northwest. Halsey and Rear Admiral Theodore S. Wilkinson, who relieved Turner as Third Fleet Amphibious Force commander in mid-July, enthusiastically took up the idea. MacArthur's later claim to have invented the leapfrog by attacking Lae instead of Salamaua in September 1943 is specious twice over, not only because SOPAC tried the technique first but also because it was Blamey, the bypassed Australian commander of 'New Guinea Force,' who chose Lae. The underlying idea in all these cases was to exploit the proliferation of Japanese garrisons instead of allowing them to achieve their objective of delaying the Allied advance on Japan. By judicious forward leaps, enemy garrisons, even quite large ones, could be left to rot behind the front, sealed off by sea and air, to be mopped up later when convenient. The Japanese, who had only a handful of troops there, were thinking of fortifying Vella Lavella themselves in a bid to delay the inevitable advance on Bougainville even further, but decided against it on 13 August. The Americans completed a thorough reconnaissance before the end of July 1943 and assembled a mixed invasion force of forty-six hundred Army troops and Marines under Brigadier General Robert McClure, USA, to be delivered by Wilkinson's amphibians from Guadalcanal. The only drawback in the Americans' plan to bypass Kolombangara was that they did not have enough ships to seal it off altogether.

The first of three waves achieved complete surprise on 15 August, but Japanese aircraft interfered with the second and third, causing no more disruption than the awkward landing conditions had already done. There was no Japanese counter-attack, as the junta had decided to strengthen its defense of Rabaul by shortening rather than extending its perimeter. The next 'decisive battle' on their right flank would be on and for Bougainville; the Kolombangara garrison would be evacuated thither. For this purpose, they decided to take a staging post for their troop barges at Horaniu, on the northeast tip of Vella Lavella, with a small force covered by four destroyers. This was done by the 19th, despite a skirmish at sea; the Americans advanced slowly across the island in vastly superior numbers (sixty-three hundred men by now) and erased Horaniu on 14 September. They were expensively and unnecessarily relieved by New Zealand troops shortly afterward; the encirclement of the last six hundred Japanese left on the island was complete by 1 October. The Japanese, meanwhile, had completed their own northward evacuation of Kolombangara by exploiting gaps in the US blockade.

Instead of abandoning their little garrison on Vella Lavella to its fate, a militarily correct decision which the Japanese, on previous form, might have taken without a second thought, they decided to evacuate it – just as the US Navy applied its mind to the removal of the New Zealanders. The result was another naval clash, the Battle of Vella Lavella. To rescue less than a battalion, Rear Admiral Ijuin Masuji brought from Rabaul nine precious destroyers of Japan's dwindling and irreplaceable stock (nearly forty lost in a year), three of them adapted as transports. Admiral Wilkinson scraped together six destroyers – all that could be spared from convoying troops – and put them under the command of Captain Frank Walker, USN. The two forces became aware of each other's presence in the Slot before midnight on 6 October. Within minutes, one Japanese and one American destroyer were sinking. USS *Chevalier* collided with her sister ship *O'Bannon* for good measure, in a confused clash. *O'Bannon* was out of the fight, and so was the third destroyer of Walker's own group of three, *Selfridge*, hit by a Long Lance at great range. Captain Harold Larson now came to the rescue from the south with the other three, just after midnight on the morning of the 7th. But the Japanese were able to lift their little garrison and sail away, abandoning their sinking destroyer *Yugumo* and notching up a clear if also sterile tactical victory in their stock-in-trade, the night action. As usual, both sides made seriously exaggerated claims. Essentially, however, the Americans had wasted five months preparing and three months executing their advance just 250 miles up a group of islands three thousand miles from Tokyo. The Japanese were now out of the central Solomons as well as the southern, but remained in considerable strength on Bougainville, the main island of the northern Solomons.

The former German colony of Bougainville, mandated to Australia in 1919, was another piece of extremely hostile territory, 130 miles by thirty, covered in dense jungle except for a coastal plain at the southern end. The Japanese had a garrison of sixty thousand men (two-thirds Army) on Bougainville and its associated islands, the Shortlands, Buka, and the Treasury Islands. There were five military airfields, and the Japanese had made a determined effort to sweep the islands clear of Allied Coastwatchers and residual Australian Army elements. A US submarine had to steal into a northern bay in March 1943 to evacuate those few who evaded these attentions. Halsey's purpose in attacking Bougainville was to seize forward air bases for fighters covering the bombing of Rabaul, still the principal Cartwheel objective. The strategic command was MacArthur's, but the work was to be done by Halsey's Third Fleet, supported by Admiral Fitch's land-based aircraft, Wilkinson's amphibious force, and Vandegrift's I Marine Corps, all controlled from a forward headquarters at Guadalcanal. Wilkinson was in tactical command of the invasion itself until Major General Roy Geiger, USMC, established himself ashore as commander of ground forces. Wilkinson's force was the minimum thought capable of doing the job, because by this time Nimitz was preparing the main body of the Pacific Fleet for his central thrust across the ocean to Japan, described in the next chapter.

Halsey's plan was to seize Empress Augusta Bay, on the south-west coast of Bougainville, bypassing Choiseul, the adjacent large island to the southeast, the Shortlands and the Japanese bases on the south coast of Bougainville itself. The landing was to be made at Cape Torokina, the only point that offered shelter for shipping on an otherwise unprotected coast open to the monsoons. While General Kenney's V Air Force struck Rabaul

(New Britain) and Kavieng (New Ireland) from New Guinea, the Solomon Islands air group ('Airsols'), based at Guadalcanal as part of Fitch's South Pacific Air command, moved up to Munda late in October to attack the five Japanese airfields on Bougainville. The Japanese, knowing something big was coming but unable to work out where it would strike, decided to build up Rabaul as their central air base. To the two hundred or so planes of the naval 11th Air Fleet still operational there after Kenney's highly overrated raids, Admiral Koga of the Combined Fleet decided to add 173 aircraft from three of his precious carriers for Operation RO – the destruction of the next Allied advance by air attack.

The first Allied move was to seize the Treasury Islands with troops from the New Zealand Third Division on 17 October 1943, partly to distract the enemy and partly to gain logistical support bases for the Bougainville operations. The Japanese garrison was tiny but resisted valiantly. On the main island of Mono, a well-dug-in heavy-machine-gun nest was giving the invaders trouble and preventing unloading. One of the first items ashore was a giant Seabee bulldozer. Captain Robert P. Briscoe, USN, who commanded the light cruiser *Denver* in the subsequent Battle of Empress Augusta Bay, recalled:

They closed the front doors and this twenty-ton bulldozer, which happened to be one of the first pieces of equipment out, this man manned his bulldozer and lifted his blade up to give him protection in his cab and trundled out, dropped his blade in front of the strongpoint and in about ten minutes had covered the whole place over to a depth of about six feet. And then, not being satisfied, he rode back and forth over it and tramped it down well and covered up the entire strongpoint, and from that time on there was no further opposition to that particular landing.

The Japanese were taken by surprise, and the Treasury Islands were under Allied control by 6 November. A Marine paratroop battalion landed on Choiseul by way of a further distraction while the Treasury operation was still going on and raided several Japanese posts before being evacuated on 4 November.

*   *   *

The first wave of the attack on Empress Augusta Bay came from the Third Marine Division (Major General A. H. Turnage, USMC) – more than fourteen thousand men on twelve transports, covered by eleven destroyers and various smaller craft as well as a strong air umbrella. Cape Torokina was defended by fewer than three hundred Japanese, so the landing on 1 November was impressive mainly for speed and excellent organization, the product of so many expensive lessons in the Solomons and North Africa. One small Japanese air raid from Rabaul during the landings was beaten off with no significant loss. Later in the day, a hundred Japanese carrier planes came in from Rabaul but again were beaten off without noteworthy damage to the invasion force. The transports withdrew, most of their work done, at nightfall. They did not return until the naval Battle of Empress Augusta Bay had been fought and won to the west of the beachhead.

The Japanese naval command at Rabaul decided to attack the landing force with two heavy and two light cruisers plus six destroyers, led by Rear Admiral Omori Sentaro from Rabaul, who came rushing in from the northwest. The American Task Force 39, four light cruisers and eight destroyers commanded by Rear Admiral Stanton Merrill, was regrouping off Vella Island after bombarding the island of Buka, north of Bougainville, on 1 November, as another distraction from the invasion. Forewarned by reconnaissance, Merrill formed a single, well-spaced line of four destroyers, four cruisers and four destroyers and sailed north up the western side of the Solomons. They met three Japanese columns, the two heavy cruisers in the center with a light cruiser and three destroyers on either side. The leading American destroyers launched a torpedo attack, and then the cruisers opened fire with their six-inch guns, quickly crippling light cruiser *Sendai*, whose sudden slowing caused the two destroyers astern of her to collide and pull out of the line. The Japanese left flank now consisted of just one destroyer. One of the three in the right flank, *Hatsukaze*, collided with heavy cruiser *Myoko*, Omori's flagship, and lost her bow. *Denver*, the rearmost American cruiser, took three direct hits from eight-inch guns and had to slow down to avoid going down by the bow until the flooded forward compartments could be sealed off. Of the

US destroyers at the rear, *Foote* had her stern blown off by a torpedo and *Spence* was damaged by a shell which contaminated her fuel.

All three damaged American ships were saved; *Sendai* and *Hatsukaze* were dispatched by US destroyers with gunfire and torpedoes. At daylight, Merrill's force managed, by violent maneuvering and much expenditure of anti-aircraft ammunition, to avoid all but minor damage from an attack by over a hundred planes from Rabaul, which brushed aside the sixteen Allied fighters sent to intercept them. Merrill's victory was marked, if not actually marred, by a stupendous expenditure of ammunition – nearly forty-six hundred rounds of six-inch alone, for a maximum twenty hits; two torpedoes out of fifty-two struck home, both on a crippled ship.

The Japanese withdrew to Rabaul, where Omori was disgraced and relieved of his command. Admiral Koga had already decided on a serious reinforcement of his surface forces in New Britain and sent no fewer than seven heavy cruisers of the Second Fleet from Truk, escorted by a light cruiser, four destroyers, and suitable auxiliaries. Halsey had no big surface ships, because Nimitz had earmarked them all for his impending attack on the Gilbert Islands. But he did have, on loan from Nimitz, the two carriers *Saratoga*, fully recovered from her two torpedo attacks, and *Princeton* (light), Task Force 38 under Rear Admiral Frederick Sherman. They had backed Merrill's prebattle bombardments and were refueling at the southern end of the Solomons. Halsey decided, *faute de mieux*, to send them in to attack the concentration at Rabaul before it could be used to dislodge the Americans at Empress Augusta Bay. 'Airsols' and Kenney's V Air Force were asked to provide cover and extra bombers. The carrier attack by ninety-seven planes was launched from about 230 miles away, and sank only the one ship, but so much damage was done to four heavy and two light cruisers plus two destroyers that Admiral Koga concluded Rabaul was too dangerous for his cruisers and withdrew them, never to return. However, the Japanese claimed to have sunk both American carriers in an air attack that caused them no damage at all.

Nimitz now lent Halsey two new fleet carriers, *Essex* and

Bunker Hill, and the new light carrier *Independence*, forming Task Group 50.3 under Rear Admiral Alfred Montgomery, to reinforce Sherman for a second strike at Rabaul on 11 November. In fact the two groups struck separately; Sherman's attack was hampered by bad weather, but Montgomery's was able to deliver nearly two hundred aircraft over the port. One heavy cruiser was crippled, her stern torn off, one destroyer was sunk and one heavily damaged, several other ships taking moderate damage from the Americans. Montgomery's second strike going in was met by a large Japanese counterattack of well over a hundred planes coming out, at noon. The American strike was canceled, and a huge mêlée blew up over the carriers and their covering ring of nine destroyers for three-quarters of an hour. The US ships maneuvered violently and put up a thick barrage of anti-aircraft fire, losing just eleven planes and ten sailors wounded. Japanese aircraft losses were distinctly higher, perhaps 20 percent. Since some of them were from carriers, the result was another useful erosion of the Japanese fleet's air arm on the eve of Nimitz's attack on the central Pacific islands.

On Bougainville, the Americans had beefed up their beachhead successfully in the first week of November but embarrassingly failed to intercept two small reinforcements landed by the Japanese from Rabaul, one very close to Torokina at night and the other at the northern end of the island. The Japanese persisted in sending air raids over but never in sufficient numbers to do any more than inconvenience American naval, air and ground forces. The main effect was yet more attrition of Japanese carrier planes, pilots and air crew. More than two-thirds of the 173 carrier aircraft sent to reinforce Rabaul, and half their pilots, had been lost for no tangible advantage and the 11th Air Fleet they had been sent to strengthen was in scarcely better case. Nevertheless, relying on their own ludicrously overblown claims of losses inflicted on the enemy, which amounted to a large battle fleet, they claimed victory in RO-Go, which was now consigned to oblivion after a word of thanks from the emperor. In the air and on the ground at the end of November, Bougainville was relatively quiet as the Japanese took a long breather.

At sea there was one more action, on 25 November, known as the Battle of Cape St George, the sixth and last in this prolonged second phase of the disproportionate struggle for the Solomons. Once again it was a fight that developed as a result of a Japanese Army attempt at penny-packet reinforcement, this time nine hundred men in three destroyer-transports escorted by two destroyers, all commanded by Captain Kagawa Kiyoto, IJN. As they approached Buka, the American Navy's finest destroyer-leader, Captain Arleigh Burke, was sent to intercept them with five vessels. In the early hours of the 25th the Americans 'got the drop' on the surprised Japanese and launched a spread of torpedoes which fatally damaged two large new Japanese destroyers, *Onami* and *Makinami*, leaving the three transports unescorted. In the ensuing gun action, one of them, *Yugiri*, was also sunk. No American ship was even hit in this swift and sure operation on Thanksgiving day. The Marines roundly defeated a Japanese attack on their enlarged beachhead on the same day. By Christmas the American force ashore behind its fortified perimeter consisted of two Army divisions (37th Infantry and Americal) in General Griswold's XIV Corps; General Geiger had handed over command when he and the Third Marine Division were relieved on 15 December. Eight battalions of Seabees and a New Zealand Army engineer unit had built first a fighter strip and then a much longer one, capable of taking heavy bombers for attacking Rabaul.

MacArthur, meanwhile, landed nearly two thousand troops of the 112th Cavalry Regiment at Arawe, on the southwest coast of New Britain, at the opposite extreme of that banana-shaped island from the hub of Cartwheel, Rabaul. The entire First Marine Division landed at Cape Gloucester ten days later, on Christmas Day, on the other side of the island and sixty miles to the north. All this was really a waste of effort, probably based on a landsman's erroneous belief that you need to occupy both sides of a strait in order to control it (an idea that never struck the British as necessary at Gibraltar, or even in the English Channel). MacArthur already had total control of the Huon Peninsula of New Guinea, on the other side of the Dampier Strait. But his Marines were absorbed into an appalling morass of rain-soaked jungle which proved a much tougher enemy than the ragbag collection of Japanese rear-echelon troops who occasionally fired on them. The veterans of Guadalcanal had never known terrain or weather to match the conditions they had to endure at steaming Cape Gloucester. Admiral Barbey's VII Amphibious Force also put a regiment ashore at Saidor, on the New Guinea coast, at New Year to tighten the Allied grip on the waters between New Guinea and New Britain. Rabaul was now a useless liability for the Japanese: of no value as a naval base, unable to launch effective air attacks on an enemy whose flying strength was much superior and growing all the time – yet defended by a mighty garrison of a hundred thousand idle troops, whose formidable fortress the enemy unaccountably saw no need to assault. Cartwheel had not altogether gone as the Americans had intended, especially in terms of time, but the desired effect had been achieved: Rabaul was neutralized.

The Japanese junta was marching to a different tune in its efforts to come to terms with an unprecedented development in modern Japanese history: defeat. The surrender of fascist Italy in September 1943 seemed to kill off residual Japanese interest in the Tripartite Pact as a military alliance, which it had hardly been in any case. On 15 September IGHQ drew up a 'New General Outline of the Future War-Direction Policy,' which was endorsed at an imperial conference on the 30th. This called for a redistribution of forces within a reduced perimeter – the 'absolute national defense sphere' – on a line from Burma and Indochina through the East Indies, across western New Guinea and the Carolines to the Marshalls. Thus the main conquests and sources of raw materials would be kept, lines of communication shortened and need for transport reduced as the Navy at last made a serious effort to set up a convoy system, based on a plan for 360 escorts and two thousand antisubmarine aircraft. This would have been all very well if the Allies had been prepared to give the Japanese a respite. As it was, the redefined 'all-costs' defense zone already meant abandoning three hundred thousand troops who could not be rescued from places like Rabaul, the Solomons and eastern New Guinea – on paper enough men for twenty divisions. Long before Hitler began to move phantom armies to a notional plan in his Berlin bunker, the Japanese junta drew up grand schemes for making merchant and naval ships in hundreds, planes in tens of thousands and artificial fuel in millions of gallons, none of which took the slightest account of the realities. The government had introduced, as long ago as 20 January 1943, an 'urgent wooden-ship building plan' because of the desperate shortage of shipping and steel and the ready availability of wood. The idea was to build standardized types of between seventy and five hundred tons; only the largest, between three and five hundred tons, would have steel frames; the rest of the steel allocated to shipbuilding was now reserved for warships.

The men at the front were not unaware of Japan's mounting difficulties. As First Lieutenant Uchimura of the Third Company, Second Battalion, 238th Infantry Regiment, 41st Division, confided to his diary at Sattelberg, New Guinea, late in 1943:

In air superiority . . . we are about a century behind America . . . This present war is termed a war of supply. Shipping is the secret of victory or defeat in this war of supply in countries thousands of miles across the sea . . . To have regular shipping lanes, air superiority is essential. Ah, if only we had air superiority . . . If only we had planes . . .
Received a report of the withdrawal from Lae and Salamaua. Ah, what tears of bitterness swallowed! At dawn on 3 September the American and Australian armies, which can boast of absolute superiority in the air in an arrogant and blatant manner, severed our main supply lines . . . They carried out a parachute landing in the vicinity of Nadzab, which was the key point of our escape route, and finally cut our supply lines . . . We were forced to withdraw from the Lae-Salamaua line which we had striven to maintain amid ever-present difficulties for half a year . . . I feel that the commanding officer [of the 18th Army] must be brokenhearted . . . Officers and men throughout the regiment must have felt they were drinking poison . . . The enemy motor roads that have been built through the jungle are something of which their materialistic civilization must be proud.

First Lieutenant Kuroki Toshiro, commanding the Third Company of the 20th Engineer Regiment attached to the 20th Division, accustomed, like all his countrymen, to a rice diet, noted with rising passion at about the same time and in the same area:

Potatoes, potatoes! The battle in the Finschhafen area was full of potatoes. It would be impossible to live without potatoes. Since our arrival on 11 November we have had hardly any rice. We added a few potatoes to what rice we have had and continued the fight. We have an army, a division and an area army, with a commander-in-chief, a divisional commander, a chief of staff, a director of intelligence and what have you, but in the front line we have to contend with a rotten supply situation and live a dog's life on potatoes.

You will not find many smiling faces among the men in the ranks in New Guinea. They are always hungry; every other word has something to do with eating. At the sight of potatoes their eyes gleam and their mouths water. The divisional commander and the staff officers do not seem to realize that the only way the men can drag out their lives from day to day is by this endless hunt for potatoes. How can they complain about slackness and expect miracles when most of our effort goes into looking for something to eat!

You would hardly think divisional HQ could be [un]informed. They are comparatively close to us. But there is no getting away from it – Army must think we have plenty of everything and are getting along all right. What a laugh! We have a perfect illustration in this theater of the good old Chinese saying:

> To hell with the boys in the firing line
> So long as the bigwigs get along fine!

The indifference and bungling of staff officers and high commanders should be a court-martial offense. Their crimes are worse than desertion, cowardice, or running away under fire, but they get off scot-free. If you ask me, the only way the Army commander and his immediate subordinates could make it all up to the fighting troops and officers is by committing hara-kiri. The men in the rear, whose biggest job is talking, know nothing about the soldiers in the hills, in the valley and in the native villages of the forward areas, dying off like flies. The stupid fools!

General Imamura's Eighth Area Army was left in no doubt from September 1943 that it was fighting no more than a series of delaying actions. The Navy uncomprehendingly defied the spirit of the new arrangement by insisting blindly on hanging on to Truk, its great base in the Carolines, whose value was severely reduced when those islands became part of the new forward-defense line. The Americans were already hammering at the 'absolute sphere,' having occupied the Ellice Islands, near the Gilberts, in August 1943, and sent carrier air raids against Wake, Marcus, the Gilberts and Nauru in the succeeding weeks.

Huge effort was expended in redistributing some forty battalions of imperial troops among all manner of garrisons; there were still forty-one thousand men defending Bougainville. Had it not been for Truk, the Japanese Army might well have reined in its southern front rather farther, in an attempt to buy time. The immediate threat of overwhelming enemy force was an argument the generals could understand, even if their capacity to follow it to its logical conclusion was restricted by fear of loss of face.

On the diplomatic front in this transitional period between the Japanese offensive and what was to be the main Allied counteroffensive, the United States had already become the locomotive of the United Nations and of the Anglo-American alliance at Casablanca, as we saw above, but still had to confer with its main allies. A conference code-named 'Trident' took place in Washington in May 1943. Churchill arrived in New York on the world's largest liner, the *Queen Elizabeth*, decked out in gray for her wartime role as uncatchably fast troopship, on the 12th – as if to make the point that Britain was still number one in some respects. The British favored mopping up in the Mediterranean before attacking in France; the Americans wanted the Pacific to benefit from any spare military capacity left over after full provision had been made for crossing the Channel. The outcome was a compromise that suited the Americans rather better than the fudged outcome of Casablanca. The British got their invasion of Italy and consent to the postponement of their promised land-offensive in Burma, but the Americans got a date – May 1944 – for the invasion of France, which was to be 'first charge' on Allied resources, and simultaneous consent to maintaining and extending the pressure on Japan. Cartwheel was to continue, and the Americans were free to attack the Marshall and Caroline mandated islands in 1943–44, when they were ready, in line with a decision by the Joint Chiefs on 8 May to unleash Nimitz. The British wondered aloud and pointedly whether such a double demand on resources could be justified. They were ignored.

The next top-level conference came only three months later.

Code-named 'Quadrant,' it took place in Quebec, Canada, and had the same old agenda. The British argued that the imminent success in Sicily should be followed by an assault on the Italian mainland to knock Mussolini out of the war, as a sound preparation for the cross-Channel invasion. The Americans wanted no more diversions from 'Overlord,' as the invasion of France was now code-named. The compromise was to do both, with priority for resources going to Overlord, which was to be commanded by an American general (Eisenhower). The planning section of the Anglo-American Combined Chiefs of Staff (CCOS) produced a highly cautious, British-accented, new strategic program for the defeat of Japan by 1948 (sic!). This included an advance across the Central Pacific islands paralleled by a landward drive through Southeast Asia and China, with a view to acquiring Chinese ports and airfields for a total blockade of Japan.

The American Joint Chiefs were appalled by the time scale and wanted Japan beaten in 1945 or within a year of the defeat of Germany, whichever was the sooner. The British informally agreed to an acceleration of the war against Japan. The Americans proposed bypassing Rabaul, which was to be left to rot under continuing Allied air raids, and mounting simultaneous advances across New Guinea and the Central Pacific, to be completed by the end of 1944. A new Supreme Allied Command, Southeast Asia, was set up, with Admiral Lord Louis Mountbatten as SACSEA to superintend the long-deferred recovery of Burma. SEAC soon became known to American cynics as 'Save England's Asiatic Colonies'; its supremely self-promoting and ambitious commander was the nearest approximation to Douglas MacArthur that Britain was able to produce.

A triple conference involving first the Anglo-Americans and China, then the 'Big Three' with the Soviet Union (the 'Big Four'), and finally the Big Three again took place in November and December 1943. The Combined Chiefs of Staff met first at Cairo at a conference code-named 'Sextant' and came to terms at last on an 'Overall Plan for the Defeat of Japan,' centered upon CINCPAC's drive across the Pacific throughout 1944 and MacArthur's conquest of the rest of New Guinea. The vexed question of who would be in overall command of the two thrusts when they merged was still left open, although it was recognized that Nimitz was more likely to make significant progress than COMSOWESPAC. The political meetings at Sextant were otherwise mainly concerned with China, in which Churchill took little interest, leaving Roosevelt and Chiang largely to make their own arrangements. From Cairo they went for their first summit meeting with Stalin at Tehran, the capital of Persia (Iran), divided at that time into British and Soviet zones of occupation. Stalin confirmed, as had been indicated by Foreign Minister Molotov in October, that the Soviet Union would declare war on Japan three months after the final defeat of Germany, a promise he was to keep to the last embarrassing letter. The rest of the proceedings were concerned with Europe and the postwar character of the United Nations, as was the resumed Cairo Conference of the Big Three early in December. The ensuing Cairo Declaration promised to liberate all Japanese conquests, to restore Manchuria, Formosa, and other Japanese-occupied territory to China, and to grant independence to Korea.

With no real military allies to consult, Japan could hardly compete in the matter of conferences, and indeed staged only one in 1943: the Greater East Asia Conference in Tokyo on 5 and 6 November. The Greater East Asia Co-Prosperity Sphere had previously been viewed as an arrangement for the greater ease of Japan, in which other Asian states and nations would take their place as allocated by Tokyo. There had been vague promises of independence for Burma, the Philippines and Indonesia if they cooperated, but none for Korea or even India. As Japan's military position deteriorated in 1943, the junta promised Wang Ching-wei's northern Chinese puppet regime, based in Nanking, eventual independence. In the middle of the year, Tokyo recognized Subhas Chandra Bose as head of a 'provisional government of Free India' and of the sizable 'Indian National Army' formed from British-Indian troops captured in Malaya and Singapore. Bose had figured in a dramatic escape from prison in 1940 and got himself to

Berlin; when Japan went to war with the West he was taken to Tokyo by U-boat. As an earnest of good intent, Bose was given the Andaman and Nicobar Islands in the Indian Ocean to administer in October 1943, which gave him a clearer idea of what 'independence' under Tokyo's tutelage meant: Japanese hegemony, as in Manchukuo and northern China.

By the time of the GEA Conference in November, the Japanese had begun to realize that they needed some active sympathizers. Burma, Siam, the Philippines, Manchukuo and Nanking China were represented, but only by puppets, and Bose was among the observers. Japan tried to present its war as one of liberation for all Asia from Western domination, after which Asian nations would live harmoniously together in accordance with the principles of justice, independence and mutually beneficial economic development.

The conference expressed its faith in the superiority of Asian culture and the contribution it could make to the welfare of all mankind, and its total opposition to racial discrimination and Western materialism. As a Japanese answer to the Anglo-American 'Atlantic Charter' of August 1941 this was not only too little, too late; it was also undermined in advance by Japan's conduct of its wars on the ground since 1931, which was irreconcilable with the joint anti-Western crusade it was now seeking to proclaim.

The Atlantic Charter was already working rather like the American Constitution, which was not expressly intended for slaves – on the contrary – but eventually became the instrument of black emancipation. The charter was conceived with unstated reservations (by the British on freedom for Europe's colonies, and by the Americans, who intended to supplant British imperialism with what the world soon identified as neocolonialism). But it was already seen, even by many British officials, as extending Roosevelt's 'Four Freedoms' to all mankind. No nation that went to war on the basis of its inherent belief in its own superiority over all others, as Japan and Germany had done, could match the universal promise, however hedged, hollow, and hypocritical it might prove to be in practice, of the Atlantic Charter. There was no matching 'Pacific Charter,' as many on both sides of the Atlantic had urged; but the Atlantic Charter's offspring, the United Nations Declaration, more than made good that deficiency in the hopeful eyes of those awaiting liberation. Next in line to receive it in the Pacific was the sparse population of the Gilbert Islands.

10
Hitting the Beaches

A microscopic examination of the globe would yield nowhere of less apparent importance than the myriad islands that are lumped together under the collective name Micronesia, and which, since the war, have formed several independent states of commensurate influence – virtually nil. Neither the Americans, the British nor the French gave a second thought to the wishes of the few inhabitants when choosing sites for testing nuclear weapons. Yet these coral atolls – whose hard, flat surfaces make such sound airfields, and whose lagoons offer so many natural harbors – played an enormous part in the strategic calculations of Japan and its enemies alike. An Allied advance from the direction of Australia through New Guinea, the Philippines and Formosa to Japan, the American Navy argued late in 1943, would be exposed to constant attack on its right flank from, successively, the Ellice Islands to the east, the Gilberts, the Marshalls, the Carolines, and the Palau group to the west, all five hundred to a thousand miles away (i.e. posing a threat to the line of advance from long-range bombers), and behind them, to the north, the Marianas. All these groups except the Gilberts had been wrested from the Germans during the First World War and mandated to Japan after it. The Japanese added the Gilberts to their collection in December 1941; as we have seen, Carlson's submarine raid on Makin in August 1942 only served to stiffen the Japanese defenses. The Americans moved into the undefended Ellice Islands, seven hundred miles southeast of the Gilberts, in October 1942, as a forward base for moving against the Gilberts when the time came. In Nimitz's plans as they developed in 1943, the Gilberts themselves were only a preliminary for the tougher proposition represented by the Marshalls.

From Micronesia, the America-Australia route and the westward march of the US fleet could also be attacked. From the Marianas, the Japanese could cover the Philippines; but if the United States Navy got hold of the Marianas, the American Army Air Force could bomb Japan with the next generation of big bomber, the B-29 (it is not entirely cynical to suspect a naval 'sweetener' for the Army here). Or so the more cautious heads on the Joint Chiefs of Staff argued; there were a few who advised ignoring the whole of Micronesia in what would have been the biggest bypass of them all. Attacking these places would take American ships beyond the reach of their own side's land-based aircraft and within range of the enemy's. But after various conferences and discussions, Nimitz drew up a definitive plan for his next move in the Pacific: an attack on Makin, which had a seaplane base, and Tarawa, also in the Gilberts, which had the only airfield in the archipelago – Operation Galvanic. In time for this effort Nimitz expected to have a huge fleet with many new ships at his disposal, including ten fast large and medium carriers, seven escort carriers and a dozen battleships. The Central Pacific force round which this rapidly growing strength gathered during 1943 was restyled the Fifth Fleet on 15 March (when Halsey's South Pacific force became the Third) and was commanded by Vice-Admiral Raymond Spruance, whom we met at Midway, from 5 August. To help the precipitately expanded fleet to shake down and gain experience, the new carriers, including USS *Essex*, the first of the new generation, a new *Yorktown* and a new *Lexington* were used in preliminary air raids on Wake and on Marcus Island, northeast of the distant Marianas, as well as Makin and Tarawa, in September. They achieved little, but the main object was practice.

Submarines sent in advance to positions near the targeted islands were used for the first time as 'lifeguards' to recover ditched air crew, an intelligent subsidiary employment for the boats in their classical but recently overshadowed role as fleet auxiliaries. The Americans also established a new forward base on their own equatorial Baker Island, six hundred miles east of the Gilberts, on 1 September. Heavy reliance would now be placed on Vice-Admiral William Calhoun's Service Force, Pacific Fleet. From this was drawn Service Squadron 4 (later subsumed into the even bigger Service Squadron 10) – basis of the famous 'Fleet Train' which enabled the Americans to keep moving at sea whether there was a harbor handy or not. It was a mobile supply-and-repair facility, ammunition dump and salvage service. Service Squadron 8 consisted of a fleet of tankers and oilers supplying ships and planes with fuel.

On the 24th the practiced Rear Admiral Richmond K. Turner, as short-tempered and salty-tongued as ever, became commander of V Amphibious Force, which divided into two for the double objective of the Gilberts operation. V Amphibious Corps consisted of one division of Marines (Major General Julian C. Smith's crack Second) and various Army and Marine units amounting to about half a division, all under the command of Major General Holland M. 'Howlin' Mad' Smith, USMC. The Second Division was to attack Tarawa, and the rest of the troops were in the Northern Attack Force against Makin. The Smiths had always to bear in mind the uncomfortable fact that the Joint Chiefs expected a much more elaborate invasion of the Marshalls only ten weeks after the assault on the Gilberts, set for 20 November 1943. The attack would be executed by an invasion fleet of two hundred ships carrying thirty-five thousand troops, 120,000 tons of supplies, and six thousand vehicles. Turner himself commanded the Northern Attack Force (Task Force 52), leaving the Southern (TF 53) to be deployed against Tarawa, by his deputy, Rear Admiral Harry Hill. TF 50, subdivided into four task groups, consisted of six fleet carriers and five light ones escorted by six battleships and assorted vessels – the mightiest assemblage of seaborne airpower yet. Rear Admiral John H. Hoover's TF 57 had several hundred land-based Navy, Army, and Marine aircraft operating from the Ellice Islands.

The senior Japanese officer on Makin was a first lieutenant called Kurokawa who had eight hundred men (including 275 laborers), no boats, and no planes – they got away – at his disposal. Admiral Turner had sixty-five hundred troops of the Army's 27th Division, a New York National Guard formation, to throw at them; they were commanded on the ground by yet another Major General Smith (Ralph C.). The first big landing on a coral atoll brought serious problems in getting so many inexperienced men and so much equipment ashore;

and the 27th Division was anything but a crack unit. It took its time doing what should have been a simple job. Hundreds of panicky men allowed themselves to be tied down by a handful of Japanese snipers. On D-plus-1, General Holland Smith, living up to his nickname, roared ashore to investigate the holdup. By the morning of 23 November the hapless 27th managed to finish its little task of occupying the main island of Butaritari with losses of sixty-four dead and 150 wounded.

The Japanese launched one serious air raid of forty-six planes from the Marshalls and damaged the new light carrier *Independence* on the evening of D-day (20 November), forcing her to withdraw all the way to Pearl Harbor. Four days later, submarine *I175* from Truk (Lieutenant Commander Tabata Sunao, IJN) sank the new escort carrier *Liscombe Bay* with a single torpedo, which sent up a pillar of burning fuel a thousand feet high. For the rest, the US carriers beat off all Japanese air attacks with relative ease, a task that involved the first night actions at sea by fighters. The Japanese lost about a hundred planes in their surprisingly weak counterattacks; the Americans lost forty-seven plus another seventy-three in accidents, reflecting the inexperience of the new carriers and crews. The Americans probably sank three Japanese submarines as well, but definitely lost two of the ten they had themselves deployed round the Gilberts, *Corvina* to *I176*, and *Sculpin* to destroyer *Yamagumo*. So much for the doings of Northern Attack Force and its escorts.

One hundred miles farther south, the main objective was the principal island of the Tarawa atoll, Betio, where the airfield lay. The Japanese had constructed elaborate and extremely strong defenses with a great deal of well-protected artillery and bunkers, hidden obstacles and barbed wire. There were some forty-five hundred fighting troops and another twenty-two hundred construction troops and Korean laborers, all commanded by Rear Admiral Shibasaki Keiji. His orders were to fight to the last man. Even though he was killed on the day after the Americans landed, they fought to the last 146; all the rest died.

After a preliminary bombardment by three battleships and associated vessels which expended much heavy ammunition to little effect against the virtually impregnable Japanese bunkers, the Marines landed one battalion in each of three places on the north side of the island of Betio, less than two miles long and up to six hundred yards wide. Much reliance was placed on 125 of the new 'amphtracs' – amphibious, tracked, armored vehicles that could ferry troops and supplies across water and land alike. Artillery and tanks were to follow. They met ferocious resistance. In the afternoon, eight hours after the dawn attack, General Julian Smith radioed Holland Smith, who was with Turner off Makin: 'Issue in doubt.' The best efforts of carrier planes and destroyer bombardments were unable to tilt the balance in favor of the Marines, who were pinned down on the beaches at nightfall. The last reserves of the Second Division were landed on the crowded, corpse-strewn shore by noon on the second day, the 21st. At this stage, the Americans were able to split the defense in two by crossing the island to the south, where units from the corps reserve landed to consolidate a new beachhead. Early on the morning of the 22nd, D-plus-2, the battered Japanese garrison sent its defiant farewell radio message, promising a last charge. It failed, but the first determinedly opposed American amphibious landing in the Pacific Campaign gave those who carried it out much food for thought when planning future operations.

The Americans lost 1,009 killed and 2,101 wounded out of the 18,000 or so engaged in ground operations, a casualty rate of more than 17 percent, pretty high by American standards of 'acceptability' and still more so for less than three square miles of coral. Ninety out of the 125 amphtracs employed at Tarawa were destroyed. Among the principal errors identified in detailed analysis, carried out with great speed for the benefit of the larger undertaking due soon in the Marshalls, were inadequate naval bombardment, too little air bombardment too late, too few amphtracs (and insufficient reliability), and poor radio equipment ashore. Above all, perhaps, as a contributory factor to the high reckoning for Tarawa was American ignorance of local waters and tides; the landing at low tide forced many Marines to wade across the reefs as the only practicable means of getting ashore and exposed them to prolonged and withering fire.

Sensational reports in the American media gave an impression of unnecessary tragedy incurred in a useless place. The most important result was that the United States forces remembered most of the lessons so painfully learned. Like the good leader he was, Nimitz had commanders study every error and go over the problems again and again, trying alternative solutions. Once more, meanwhile, vastly superior force had blundered to a sort of triumph over vastly greater commitment on the part of a garrison outnumbered four to one. Betio and the adjacent islands of Tarawa Atoll were secure within a week of the hotly contested landing, which undoubtedly came as a great shock to the Americans. Admirals Turner and Hill had their initial analyses on Nimitz's desk by 30 November. As a first taste of what it was going to cost to dislodge a well-prepared, sizable, determined, and well-led Japanese garrison from a Pacific island, it filled many Americans with foreboding. But they had won an excellent forward base for the next step against the Marshalls, and were better prepared than they could otherwise have been for what now threatened to be a rather more formidable task than Tarawa – much tougher than they had originally bargained for.

The expensive assault on the Gilbert Islands marked a seldom noticed turning point in the war against Japan. Hitherto, interservice rivalry notwithstanding, the two American Pacific commands, one Navy, one Army, had been partners in a single enterprise, the pincer movement against the strategic Japanese base of Rabaul in New Britain (Bismarck Archipelago) with the aim of breaking through the Bismarck barrier to the Philippines, which lay across Japan's access route to the Dutch East Indies and its raw-material supplies. MacArthur drove toward it on the left from eastern New Guinea over the Dampier Strait to the western end of New Britain Island; Halsey, Nimitz's South Pacific commander, pushed up the Solomons chain on the right from Guadalcanal and Tulagi to Bougainville. The two advances were under a single strategic command (MacArthur's) and could and did assist each other, notably in mutually supportive air operations. Having so successfully helped neutralize Rabaul, the Navy surely had no real need to go back to the Gilberts and start a new advance from a point fifteen hundred miles to the rear and to the northeast of the Bismarcks. Once it did so, the two commands were marching to different drummers and were too far apart, both literally and metaphorically, to help each other very much.

What might have been achieved had a single command had access, as soon as they became available, to all the assault troops, all the landing craft, the engineers and construction units, the land-based planes, the carriers, the surface ships, and the submarines which America so lavishly and brilliantly assembled and then unevenly divided, is a matter for speculation. But ordinary logic does suggest that a united thrust by concentrated forces to cut Japan off from its southern conquests would have ended the war at least a year earlier, especially if accompanied by an all-out economic blockade from sea and air, the proper business of navies in a protracted, long-range war. We have already had clear indications of Japan's special vulnerability in shipping. This should have been obvious at the time to an American Navy which, with its other hand, severely weakened by an unnecessarily spendthrift Pacific effort, was helping to save the British from the German submarine blockade.

As it was, the competing demands of the Solomons and New Guinea wings of Operation Cartwheel, at the same time as Nimitz prepared for what amounted to a second front, inevitably left all three undertakings short of essentials and undoubtedly slowed them all down. The fact that Japan's best effort to interfere in the Gilberts operation could only produce a strike by forty-six aircraft to help its garrison does rather undermine the argument that the protection of the right flank of the advance on the Philippines required the neutralization of all Micronesia by a vast fleet. But the US Navy, no more stubborn than the US Army in refusing to surrender unity of command to the other service, was, by virtue of its control of the means of delivering troops, supplies and mobile air bases, able to ensure that its flanking movement along Japan's outer island possessions became the main thrust of the Pacific Campaign. The failure to impose unity of command in the Pacific goes right to the top; only Roosevelt could have broken the stalemate on the Joint Chiefs between Marshall and King which kept the two thrusts across the Pacific apart for so long. But the president, as so often, preferred to defer a difficult decision, and in the end never took it. This must have been his most expensive equivocation. It would have meant choosing between Army and Navy, between MacArthur and Nimitz as sitting tenants, or else appointing a fresh supremo. It would also surely have meant choosing the New Guinea route to the Philippines, not only because it was much shorter but also because it gave much more scope for the use of large numbers of the Army's troops and land-based aircraft without detracting from the Navy's ability to employ its Marines and carriers. In such case, the whole of Micronesia could indeed have been bypassed and only the Palau group secured to cover western New Guinea and the Philippines.

Nimitz lost no time in beginning the process of softening up his next target, the Marshall Islands, by carrier strikes which would also yield much-needed photographic reconnaissance material. Rear Admiral Charles Pownall led Task Force 50, of four fleet and two light carriers and supporting ships, to raid the principal atoll of Kwajalein, with its airfield and large anchorage, on 3 December 1943, sailing from the southern Solomons north about the Marshalls to attack them from the east. Ignorance of ground conditions caused the first wave of over a hundred aircraft to achieve comparatively little, destroying four cargo ships and perhaps fifty planes on the ground and in the air. Pownall nevertheless canceled the planned second strike on Kwajalein and, after bombing his second target – Wotje Atoll, to the east – withdrew for fear of a Japanese counterattack from some of the many other islands in the archipelago. Inexplicably, the American carriers had no fighter pilots trained in night combat, which was unfortunate because the Japanese sent in small groups of unescorted bombers to make at least a dozen raids during the night of 3–4 December and again on the following night. Eventually the *Lexington II* took a torpedo 'up her skirt' and lost her steering; but all ships got to Pearl Harbor, claiming another twenty-nine Japanese planes shot down.

The invasion of the Marshalls, Operation Flintlock, had to be postponed by thirty days to 31 January 1944, for lack of troop transports. It was to be executed in two stages, first the southern part of the scattered group and then the smaller, northern group round Eniwetok Atoll. Turner had concluded from Tarawa that the preparatory big-gun bombardment should be three times greater and wanted more of everything: troops, landing craft, stores and ammunition. All his wishes were granted. Nimitz, his staff and his senior commanders meanwhile became immersed in a fierce, even blistering debate about where to strike. Wotje and the atoll of Maloelap had functioning airfields capable of handling bombers; Kwajalein was known to have a fighter field (only after reconnaissance was it established that it also had a bomber strip three-quarters complete); all these and Mili and Jaluit atolls, with their fighter fields, were strongly defended. Opinion was all but universal among the senior officers that the targets should be Maloelap and Wotje; Spruance, Turner and Holland Smith all pressed for this, Turner distinguishing himself by his choice of words. Nimitz listened and, relying on the new photographs, pulled rank and insisted on Kwajalein. When Turner challenged this once too often, Nimitz quietly offered him the chance to resign. The caustic amphibious commander backed down. The other four atolls with airfields were simply to be bypassed. The undefended island of Majuro, three hundred miles southeast of Kwajalein, was to be taken as a temporary forward fleet base.

Turner, commanding Task Force 51 of 297 ships, was personally to lead the Southern Attack Force (TF 52) against Kwajalein island, landing Major General C. H. Corlett's seasoned Seventh Infantry Division, USA. Rear Admiral Richard L. Connolly led the Northern (TF 53) against the smaller islands of Roi and Namur, in the same atoll, sending in the new Fourth Marine Division under Major General Harry Schmidt, USMC. Rear Admiral Harry Hill's task group (TG 51.1), acted as Reserve Force and included a detachment for Majuro (TG 51.2), to support Turner or go on to Eniwetok as appropriate. General Holland Smith was in overall command of eighty-four thousand troops (including thirty thousand rear-echelon personnel) – twice as many as for the Gilberts – and sailed with Turner aboard the attack transport *Rocky Mount*, specially adapted as a command ship and better suited for the role than even the largest warship. Backing the invaders over and above Turner's armada was the newly formed TF 58, the Fast Carrier Force, under Vice-Admiral Marc Mitscher, in four task groups of three carriers each, a total of six fleet and six light carriers escorted by eight new battleships and a plethora of heavy and light cruisers, destroyers and auxiliaries, land-based aircraft and six submarines as scouts. Such was the staggering product of two years of all-out American naval construction since Pearl Harbor.

Rear Admiral Hoover's 350 land-based combat aircraft, including the Army's VII Army Air Force and Marine and Navy planes, pounded the Marshalls from the Ellice and Gilberts for seven weeks, disregarding Japanese night raids on their new bases, which caused only minor inconvenience. But they did not establish control of airspace over the Marshalls. This was achieved in the last two days before the landings by the carriers, heavily attacking as many as four different targets simultaneously and destroying the last 150 serviceable planes left to the enemy locally.

The Japanese knew – how could it be otherwise? – that the Americans were coming. But Admiral Koga of the Combined Fleet, who was under orders to fight no more than a holding action in the Marshalls pending the reinforcement of stronger positions to the north and west, had no means of knowing where. He misread enemy intentions and was caught completely wrong-footed. He was not about to send in his reduced stock of carriers, denuded of planes and pilots as they had been for the defense of Rabaul. Without carrier cover he could not deploy his battle fleet against the American carriers either. They already comfortably outnumbered what Japan had possessed at the outset of hostilities. He deployed an unknown number of submarines as sentinels; the Americans found and almost certainly sank four. Koga reinforced Mili, Wotje and Maloelap at the expense of Kwajalein, which he assumed would be attacked last and which was left with a garrison of nine thousand troops. Captain Arima Seiho, IJN, headed about 3,750 men on the islands of Roi and Namur, which are joined together by an isthmus and a causeway, at the northern end of the triangular chain which constitutes Kwajalein, the world's biggest coral atoll; the rest were on Kwajalein island at the southern tip.

Northern Force first established itself on five neighboring islets and then landed with little initial opposition, thanks to three days' very close and heavy ship-and-air bombardment, on the south side of Roi-Namur in the early hours of 1 February. Much confusion was caused by rough weather in the lagoon, heavy rain, overcrowding of the available sea room, and inexperience among troops and the crews of landing craft and amphtracs. The Marines found a smashed landscape on Roi, almost wholly taken up by its airfield, and had little difficulty with the three hundred Japanese troops left alive after the shelling and bombing. Namur offered a little more resistance, but both islands were secure in little more than twenty-four hours. A seriously opposed invasion would probably have been a catastrophe for the Americans because of the enormous complexity and consequent serious confusion of the landing process itself, but the much heavier preliminary bombardments prevented a second Tarawa at Roi-Namur.

Forty-five miles to the south, Turner sent in against Kwajalein the Southern Force, the Army men who had taken Attu and Kiska in the Aleutians and were now stewing in temperatures half a thermometer higher. The Seventh Division's troops had to take more than two dozen islands and islets, of which as many as half might be manned by Japanese – but only after a mighty battering from seven US battleships, other vessels and planes directed against a banana-shaped island just two and a half miles long and its adjacent islets. The landing, on the western end, went much better than in the north, as if it had been an exercise, with successive waves of landing craft, some of them acting as gunboats for close fire-support, rolling in neat lines toward the chosen beaches. Within minutes, it seemed, twelve hundred troops, tanks, and specialized vehicles were ashore and proceeding smoothly about their assigned tasks amid mounds of supplies.

It took Rear Admiral Akiyama Monzo, commanding the garrison of about four thousand, two hours to shake off the effects of three days of bombardment and organize stiff resistance from a variety of strongpoints several hundred yards back from the beaches. The Americans had put eleven thousand men ashore by the afternoon of 1 February, but General Corlett's two regimental combat teams were in no hurry to advance, relying on artillery and the new rocket-equipped ground-attack fighters to prepare the ground for them. The sole tank-battalion made little progress, being cautiously led and not coordinated with the infantry. The desperate Japanese charges of the first night were easily surpassed in ferocity by those of the second; on 3 February, the third day ashore, Japanese resistance was even fiercer, making good use of extremely strong bunkers and pillboxes which had survived sixteen-inch battleship shells and thousand-pound bombs. The struggle for the atoll strongpoint became an inch-by-inch affair among the concrete emplacements; it came to an end on 6 February.

Virtually the entire Japanese garrison died (only 265 prisoners were taken), as did 372 Americans (sixteen hundred were wounded). Given the overall ratio of nine attackers to two defenders, it should have been a little easier; but the ratio of dead – twenty-one Japanese for every American (i.e. nearly a hundred to one in proportion to the forces engaged) – speaks for itself. A rational commander would have saved most of his men by a judicious surrender after putting up enough resistance to satisfy honor (as defined in the West). But to the Japanese, anything less than fighting until the last breath was dishonorable. That was the principal reason the Americans found it necessary or thought it best to overwhelm such an enemy, who usually conceded defeat only in death, with relentless firepower. And if American strategy had major faults, whether of division, diversion or disproportionate application of strength, which it certainly did, we need only remind ourselves that Japanese aggression lay at the root of it all. This time the Americans were better prepared than in the Gilberts and kept their casualties to a more-than-reasonable minimum. Here is what a friendly (and doubtless highly envious) British observer, Commander Anthony Kimmins, RN, who saw the landings at Salerno in Italy as well as Kwajalein, thought of it all:

[It is] the most brilliant success I have ever taken part in . . .
Nothing could have lived through that sea and air bombardment
. . . A staggering force built up in these various amphibious task
forces. It is something that is quite new in naval warfare . . . an
absolute treat to watch . . . Generally speaking the organization was
absolutely perfect. [The bombardment] was the most damaging thing
I have ever seen. [Ashore] I have never seen such a shambles in my
life. As you got ashore the beach was a mass of highly colored fish
that had been thrown up there by nearby explosions.

One can absolutely hear him saying it.

Thus encouraged, the Americans at once turned their attention
to capturing Eniwetok, at the northern end of the Marshalls.
This was scheduled for 1 May, but Nimitz agreed with Spruance
and the commanders on the spot that the right recourse was
to strike while the iron was hot and deny the Japanese time
to reinforce. However, Eniwetok was exposed to air and
surface attack from the greatest Japanese base in the whole
of Micronesia – Truk in the Caroline Islands, principal advance
base of the Combined Fleet and defended by four airfields.

So powerful and confident of its invulnerability was Mitscher's
mighty Task Force 58 that he saw no need to worry about
land-based aircraft, either the enemy's or his own, as he
undertook the task of neutralizing Truk, seven hundred miles
from Eniwetok, on the shortest notice. He launched thirty raids
in two days, 17 and 18 February, none of them involving fewer
than 150 planes. The Japanese lost 270 planes on the ground
or in the air, 191,000 tons of shipping including two dozen
first-class merchantmen, two light cruisers, four destroyers, two
submarines and five other vessels, either to these air attacks or
to a task group of two battleships and escort, led by Admiral
Spruance himself in his new flagship, *New Jersey*, which lay
in wait for would-be escapers. Fortunately for the Japanese,
Admiral Koga had effectively abandoned Truk as a principal
Combined Fleet base a week earlier because he feared just such
an onslaught. Most of the warships had withdrawn to Japan itself

or to Palau, which now became the main forward fleet base and flew Koga's flag. To make assurance doubly sure, Mitscher also bombed Saipan, the main Japanese bomber base in the Marianas, a thousand miles west-northwest of the Marshalls, on 23 February.

The downgrading of Truk had two immediate and momentous consequences. On 20 February Koga ordered all remaining naval aircraft based at Rabaul to move to Truk, setting the seal on the earlier Allied neutralization of New Britain despite the abandonment of the occupation plan. The Bismarck base 'withered on the vine' as intended and played no further significant role in the war except that it continued to tie down the equivalent of five Army divisions (which could not be moved anyway, for lack of ships). And, as noted above, thanks to these events, Tojo added the cap of chief of the Army General Staff to his collection of hats on 21 February; Admiral Shimada, the Navy minister, became chief of the Naval General Staff. This set the seal on the junta's complete takeover of the Japanese government; not since the creation of the General Staff in 1889 had there been such a concentration of power in the Cabinet. This arrangement was intended to improve coordination of political, military, and naval strategy, especially in the burgeoning shipping crisis, which was already the chief threat to the war effort. It looks more effective on paper than it turned out to be in practice, because individual Japanese commanders were as accustomed as ever to getting their own way (usually as interpreted by their staffs), and because Imperial General Headquarters, the epicenter of real power, was still two separate organizations, one military, the other naval, under one roof. The 'rationalization' of the leadership merely confirmed that the great schism between the services went all the way to the top, that the Army was dominant in the state, and that there was no place at all for civilians in the governance of the Japanese Empire.

The capture of Eniwetok, a scattered necklace of an atoll with some forty islands and islets offering a total of little more than two square miles of standing room, was left to Admiral Harry Hill's Reserve Attack Force in Operation Catchpole. Brigadier General T. E. Watson, USMC, commanded eight thousand men of the reinforced 22nd Marine Regiment, and two thousand in two reinforced battalions of the 106th Army Infantry Regiment. They were to be pitted against a total Japanese garrison of two thousand Marines and two thousand Army troops. The first US landing, to the north of the atoll on 17 February, after another heavy bombardment, would have made Commander Kimmins eat his words. Despite the complete absence of immediate reaction from the Japanese, the invasion itself was essentially a logistical and administrative shambles. The main objective in the north – Engebi Island, with its airfield – inevitably fell, however, on 19 February. About a thousand Japanese died, or as near 100 percent as made no difference, compared with eighty-five Marines (twice as many were wounded).

Eniwetok Island and Parry Island, to its northeast – the American objectives on the southern rim of the atoll – were a different proposition. They had been reported unoccupied and therefore were only lightly bombarded. But the defenders, Major General Nishida's First Amphibious Brigade, had eight hundred men on Eniwetok and 1,350 on Parry, under strict orders to stay in their well-concealed underground bunkers until the expected invaders started to come ashore. The Americans were lucky in having chosen to attack in the north first, because they found papers there revealing the presence of the hidden garrisons in the south. General Watson therefore decided to attack the two islands in succession with all his forces, instead of simultaneously with half each. The Marines landed on Eniwetok on the 19th, after an intense but brief and ineffectual bombardment, amid more congestion and confusion on the beaches, soon to be followed by Army troops. The island was declared secure on the afternoon of 21 February.

The exhausted infantry were left in possession, while the Marines were re-embarked with orders to make their second landing, on Parry Island, the next day, with the Marines who had already overrun the northern part of the atoll. Another landing against stiff concealed opposition ensued, but the Americans were in total control by nightfall the same day, having lost

110 killed on the two southern islands, with 355 wounded. They counted more than seventeen hundred enemy dead and captured forty-eight to add to the sixteen prisoners taken in the north. The Marshalls were now firmly held by the United States, which was already well advanced in constructing forward bases and staging posts on several of the islands for the furtherance of the great Central Pacific naval drive against Japan, now ten weeks ahead of its cautious schedule.

The next item on the agenda was the Mariana Archipelago, a thousand miles closer to Tokyo, including long-lost American Guam. The four Marshalls atolls still held by the Japanese were not invaded but just bombed, day after day, week after week, and left to wither. The defenders had huge stockpiles of supplies, which was lucky for them because they never received any more, except for the occasional night delivery by submarine. If American warships went close they were fired on by shore batteries; anti-aircraft guns managed to down a few American planes from time to time. The garrisons slowly starved, but Wotje, Maloelap, Jaluit and Mili hung on for the emperor until after the Japanese surrender, living on fish, such produce as could be grown, and eventually rats or even less appetizing items.

MacArthur, who can now be seen as the wrong man in the right place, was under instructions emanating from the Quadrant conference in Quebec in August 1943 to push onward along the northern coast of New Guinea across Japanese-held Dutch territory to the western end of the island. His Sixth Army, alias Alamo Force, had executed a ninety-mile leapfrog, an unopposed landing at Saidor, at the beginning of the year, cutting off fourteen thousand Japanese troops to the east. Four Australian divisions out of Blamey's six were at home regrouping and retraining, leaving two in New Guinea. The Americans were consolidating at Arawe and Cape Gloucester in New Britain, while General Kenney's V Air Force was battering Kavieng, the major Japanese base at the northern tip of New Ireland in the Bismarcks.

Action at sea in SOWESPAC was desultory since Kinkaid's

large miscellany of ships enjoyed local superiority, but the Japanese were as implacable as ever when it came to a fight. Before he went to Mitscher as his chief of staff in March 1944, Captain Arleigh Burke, USN, the accomplished destroyer-commander, was leading the 23rd Squadron of five ships on patrol in New Ireland waters in February when they sighted a small Japanese destroyer. He invited it to surrender but was rewarded with a salvo from the ridiculously outgunned enemy's four-inch guns. The Americans responded with a hail of five-inch from all five destroyers, which blew up the Japanese vessel in fifteen seconds. Some 150 survivors took to the water and Burke ordered their rescue, in defiance of the well-known Japanese reluctance in such circumstances already referred to. Let Burke's recollection of the ensuing incidents do duty here for all the others witnessed by Allied servicemen on land and sea. As usual the destroyer-men resisted rescue, which meant capture and dishonor, at least until one sailor had been picked up and waved an assurance that all was well, once he was made to understand that they were not about to be summarily executed in accordance with the not uncommon Japanese way of dealing with prisoners.

However not all of the Japanese wanted to be taken prisoner. There were many weird methods of committing suicide . . . Many of them cut their throats; some of them had no knives and would try to bump their heads against the wreckage on which they were resting. Eventually they would succeed and drown. Some of them could not hit themselves hard enough to kill themselves . . . They would try to dive down and drown, and surprisingly enough it seemed quite difficult for a man to drown himself voluntarily . . .

There was one man on the port side who did not want to drown; he was swimming very slowly towards the ship . . . He was about all in. We threw him a life-ring . . . and he smiled up at the bridge . . . He got one hand on the edge . . . which slipped away from him. He looked up in a sort of dismay . . . and sank. You could see him struggle slowly [and] drown. It seemed a shame that so many of those people who wanted to die could not have died easier and the man who wanted to live could not have lived.

There was one particularly gruesome suicide . . . This lad had a knife. It apparently was not very sharp. He tried to cut his throat and could not do it. He hacked at himself and he bled quite a bit but apparently he was unable to sever the proper arteries . . . Then he decided he would stab himself in the heart. He tried there and he had trouble with this too

. . . He finally stuck the blade of his knife in his mouth and hit it with his right fist. That did the trick. He killed himself satisfactorily.

Ironically, at this stage of the Pacific Campaign the United States Navy was busy executing a revised version of the old Plan Orange – an island-by-island advance across the Central Pacific to the Philippines, just as MacArthur himself had advocated readopting as a strategy in the middle of 1941. The difference was that the United States Army was not in possession of Manila Bay to welcome the fleet, as Orange envisaged. Even after Sextant in Cairo at the end of 1943, there was no word for MacArthur about pressing on to the Philippines, which was his overriding ambition: had he not said he would return? At a conference of Pacific commanders at Pearl late in January 1944, a prematurely delighted MacArthur found a surprising degree of support among naval chiefs, including his naval opposite number, Nimitz. CINCPAC preferred Truk and Palau to the distant Marianas as the next objectives after the Gilberts and the Marshalls, whereupon he would have flank protection for the single thrust at Mindanao in the southern Philippines which he too favored.

Before the attack on the Gilberts in November 1943, the Pacific Fleet had been limited by the operational range of land-based fighters for the length of each step forward; the Fleet Train and Mitscher's new Task Force 58, with its dozen carriers (four more still to join when ready) and its swarms of fighters, gave Nimitz instant air superiority wherever he wished to go. But King was the power behind the separate Central Pacific line of advance, as he soon reminded Nimitz in a harsh letter. King trotted out the familiar argument about how a single advance would leave Allied communications wide open to flank attack, conveniently forgetting his own mightiest creation, TF 58. MacArthur sent Sutherland to Washington in February to plead with the Joint Chiefs for one thrust at the Philippines, but the arrogant chief of staff was not a skilled negotiator from a position of weakness and was turned down flat.

Determined not to be relegated to manager of a sideshow, MacArthur cast about for something dramatic to do – and fastened upon the Admiralty Islands, an archipelago north of New Guinea and west of the northern Bismarcks. With these in

Allied hands, the advance across New Guinea would be safe from land-based air interference. Aerial reconnaissance showed grass growing on the main Japanese airfield on Los Negros Island, which V Air Force had been bombing frequently. In fact the Japanese garrison commander, Colonel Ezaki Yoshio, was 'doing an Eniwetok' and playing possum, such planes as he had left dispersed in camouflaged revetments and his four thousand troops hidden in the jungle with orders not to fire at enemy aircraft. MacArthur gave the new Seventh Fleet commander, Vice-Admiral Thomas Kinkaid, barely a week's notice to transport, cover and land a thousand men of the first Cavalry Division with supporting units for a reconnaissance in force of Los Negros under Brigadier General William Chase. They went ashore on Leap-Year Day and captured the airfield in two hours because the Japanese were expecting any attack to come from the other side of the island, the south, where the main anchorage lay. Chase beat off piecemeal Japanese night attacks, terrifying but small, until 2 March, when another fifteen hundred cavalrymen (who were giving a convincing imitation of infantry) arrived to support him, covering four hundred Seabees, who started at once to clear and extend the runway. By the time another brigade of the 'First Cav' arrived a week later, the Japanese resistance had dwindled; by 18 March the Admiralties were secure. MacArthur's lucky gamble – and bold or desperate opportunism, which won him complete strategic surprise – had been justified by events. A properly organized resistance by what was initially a numerically much superior defense could have given the Japanese a rare victory; but Ezaki was not up to the opportunity. The isolation of Rabaul was now hermetic, and MacArthur was master of the air and the waters of northern New Guinea all the way across. The new outer perimeter of the 'Absolute National Defense Sphere' was broken before it had hardened.

We have seen how the Japanese regarded the Solomon Islands in MacArthur's command as an area to defend in depth outside the perimeter of the Absolute National Defense Sphere. They unfortunately chose to continue this defense of Rabaul even after that strategic emplacement had lost its purpose. Bougainville, at the northern end of the chain, was the largest island, with the most room for maneuver. The irrepressible Lieutenant General Hyakutake Harukichi, unfazed by his defeats on Guadalcanal in the south, was preparing to redeem himself with an attack in divisional strength against the fortified American beachhead at Empress Augusta Bay with its three airstrips. There was to be no penny-packaging this time; all the 15,500 men he could muster, and every big gun, would attack the perimeter and overrun the air bases. Odds of two to one against him would not be allowed to make any difference; the Japanese were having to get used to the enemy's habitual numerical superiority. As it turned out, however, General Griswold had a whole reinforced corps of sixty-two thousand at his disposal, or four to one, including the 37th Infantry and Americal divisions. He also had a captured copy of the Japanese plan of attack, which began on 8 March with the heaviest artillery barrage the Japanese had ever mounted in the entire Solomons campaign. This much they had learned from their enemy. The yelling infantry then attacked in three places, at the northeast, the east and the center of the defense perimeter. With the usual one-sided traffic in casualties, the Americans took fifteen days to beat off the onslaught. From now on the Allies were secure on Bougainville. Mopping up the residual Japanese pockets of resistance – left almost exclusively to the Australians, who often wondered aloud what they were doing on such strategically dead and intrinsically horrible ground – took the rest of the war.

General Douglas MacArthur had other things on his mind, specifically his new war plan, 'Reno IV,' calling for a mighty six-hundred-mile leapfrog along the northern New Guinea coast to Hollandia, a large step in the right direction, toward the jumping-off point at the western end, for the invasion of his cherished Philippines. Hollandia looked like a good spot to build a base for long-range heavy bombers in aid of the Philippines operations. In fact he was about to advance nearly fifteen hundred miles westward from the Admiralties in three months, after taking five times as long to get control of Papua and the Bismarcks. For the first step he took the view that he needed to keep Halsey's Third (South Pacific) Fleet, with its fleet carriers, to reinforce his Seventh, which had no such luxuries. But Nimitz wanted to combine his South with his Central (Fifth) Fleet for the northerly or right-flank naval thrust against its toughest objective yet, the Marianas. Once again the Joint Chiefs were racked by the same old row about which horse to back in what should have been a one-horse race with a single jockey. And once again, on 12 March, they put their (fortunately plentiful) money on both. MacArthur was to carry on as before, with a view to landing in Mindanao in November. Nimitz was to neutralize Truk altogether, assault the Marianas in June, and the Palaus in September.

Halsey's command was effectively dismantled, with the ships and maritime planes going to Nimitz and the troops and Army air to MacArthur, who thus acquired his second air force, the VII. That at least was logical as far as it went; but not as logical as a single strategy would have been at this stage of the war, when it was abundantly clear that the United States had the initiative – and the forces to overwhelm Japan if it could only decide where to apply the pressure. The new combined Pacific Fleet now acquired a unique command system. Spruance and his staff were to alternate with Halsey and his in seagoing command; under the latter it would be called the Third Fleet and under the former the Fifth. Sadly for the crews and all but the most senior admirals, everybody else stayed in place: for many American sailors, marines and fliers, leave became a distant memory and an even more distant prospect as the bulk of the United States Navy came under one admiral's flag to form the mightiest battle fleet ever assembled.

It remains astonishing that President Roosevelt did not at this stage, if not sooner, issue a clear-cut order to the Joint Chiefs to come up with a one-thrust plan using this invincible armada and MacArthur's five divisions to finish the Pacific Campaign before the final stages of the war against Germany, before the Russians could join in the war against Japan, and (for Anglophobes everywhere) before the British intervened with their planned but unwanted and unneeded Pacific Fleet. Nimitz flew to Brisbane on 25 March to present MacArthur with an offer of twenty-four hours' carrier cover by TF 58 for the Hollandia leap, Operation Reckless, which was close to the limit of land-based range, but not a minute more, in those dangerously narrow waters. It was their first private meeting of the war; they compromised on a week's extra support from small escort carriers.

Operation Persecution was to seize Aitape, more than a hundred miles east of Hollandia, at the same time as the latter, so that fighters could fly from there to help the main landing. Lieutenant General Robert L. Eichelberger, corps commander, had two divisions, the 41st and 24th, for the undertaking. The defense was weak and unprepared for this strategic surprise on 22 April, and the Americans soon established themselves in both places. Lieutenant General Adachi Hatazo, commanding the 18th Army at Wewak, well to the east, where he had been expecting the next American blow to fall, took two months to crawl a hundred miles through the jungle to Aitape, strongly garrisoned in the meantime by the US 32nd Division. On 10 July 1944, Adachi broke through the American line and gave them a nasty shock; only late in August, when nine thousand Japanese nearly half Adachi's strength, had been killed in combat, was Aitape safe. After all that, Hollandia proved to be useless for heavy bombers because its soil was too soft. MacArthur had already moved on, ordering the invasion of Wakde, 250 miles farther west along the coast, on 17 May. The local airfield was taken in two days and was in operational use by the Americans on the third, but hard fighting developed when Lieutenant General Walter Krueger, MacArthur's Sixth Army (Alamo Force) commander, ordered an attack on the nearby town of Sarmi, which happened to be the local Japanese military headquarters of the 36th Division (Lieutenant General Tagami Hachiro). Although the Americans won control of the area, the Japanese were not cleared from Sarmi itself until the war was over.

Next came Biak, an island with three airfields dominating Geelvink Bay, north of the neck of western New Guinea, which ends in the Vogelkop ('bird's head') Peninsula and Cape Sansapor, designated jumping-off point for the Philippines. Major General Fuller's 41st Division landed on Biak in the misnamed Operation Hurricane on 27 May and ran into a very fierce defense conducted by Lieutenant Colonel Kuzume Naoyuki, complete with bunkers, a maze of pillboxes and even tanks, rarely mustered by the Japanese in the Pacific theater, especially at this stage of the war. Despite Krueger's dismissal of Fuller in deference to MacArthur's frustration, and the temporary substitution (after three weeks) of Eichelberger in the field command, the equivalent of three battalions of scratch Japanese troops held off one and a half American divisions for a whole month. The Japanese had recognized the capture of Biak as a threat to their position in Mindanao (Philippines) and their new plan for a 'decisive battle' against the American fleet in the area, and had decided to send in the Second Amphibious Brigade of the Southern Army's strategic reserve from Mindanao by Tokyo Express. This Operation Kon was to be covered by warships and naval air reinforcements from the Philippines, the Marianas, and even Japan itself, on the orders of IGHQ (Navy). Aerial fights and naval skirmishes on the Solomon Islands model ensued, badly disrupting the persistent Japanese attempts to save Biak. The third and largest reinforcement effort by transports, involving an overwhelming escort of the two super-battleships *Yamato* and *Musashi* with appropriate support, was called off at the last minute when the 'decisive naval battle' loomed, which was very lucky indeed for the Americans in New Guinea. From now on, fresh Japanese troops arrived in small numbers by barge at night, adding perhaps a thousand men piecemeal to the garrison.

The last-ditch defense of Biak was conducted in and from two caves which commanded and fired upon the captured main airfield, preventing its use by the Americans. When the eastern cave was abandoned to relentless bombing and flamethrower attacks on 28 June, Naoyuki ordered his colors burned and committed suicide. When the time came to withdraw from the western cave, a hundred wounded men gave up their lives in a final stand, which enabled the last 150 unwounded defenders to escape. All the other Japanese soldiers were dead, compared with 438 Americans killed and twenty-four hundred wounded. The Americans took even higher casualties from virulent tropical intestinal diseases. At the end of June, Krueger completed the capture of the island of Numfoor, sixty miles west of Biak, aided by massive air bombardment and a parachute drop. Within a month, the Vogelkop Peninsula was under American control. The last step by MacArthur's forces before the invasion of the Philippines was an unopposed landing in mid-September on the island of Morotai, at the northern end of the Moluccan Archipelago, halfway between New Guinea and Mindanao.

11

The Marianas and the Great Turkey Shoot

Admiral Nimitz had been preparing for Operation Forager, the attack on the Marianas, specifically Saipan, Tinian and Guam at the southern end of the group, since February. They were to provide advanced naval bases for the next move – whatever that might be – and airfields from which the new B-29 'Superfortress' Army bombers could attack Japan. Half of Task Force 58 – six carriers in two groups – made the first air raid on the islands on 23 February. It included a strong reconnaissance element, as there was no ground intelligence about any of the islands except American Guam. US submarines, whose climactic year of 1944 we shall consider in due course, did considerably more damage than aircraft in this preliminary stage. The campaign was recognized by the Japanese beforehand, at the time and afterward as the turning point in their southern war: the really, really 'decisive battle' (until the next one). Land-based bombers took over in March, flying from SOWESPAC, SOPAC and Marshalls airfields.

The overall task commander for the invasion in June 1944 was to be Admiral Spruance, which meant the Pacific Fleet would be called the Fifth. Vice-Admiral Mitscher would cover as usual with Task Force 58, on this occasion made up of fifteen carriers and suitably massive escort in four task groups. Newly promoted Vice-Admiral Turner was to head the Joint Expeditionary Force in TF 51. This had two components, of which the larger was the Northern Attack Force (TF 52), personally commanded by Turner and transporting Lieutenant General Holland M. Smith's V Amphibious Corps of two divisions against Saipan and Tinian. The other was the Southern Force (TF 53) against Guam, under Rear Admiral R. L. Connolly with Major General Roy Geiger's III Corps of one and a half divisions. Rear Admiral William Blandy's Task Group 51.1 carried Major General Ralph Smith's 27th Infantry Division as floating reserve for either corps. The

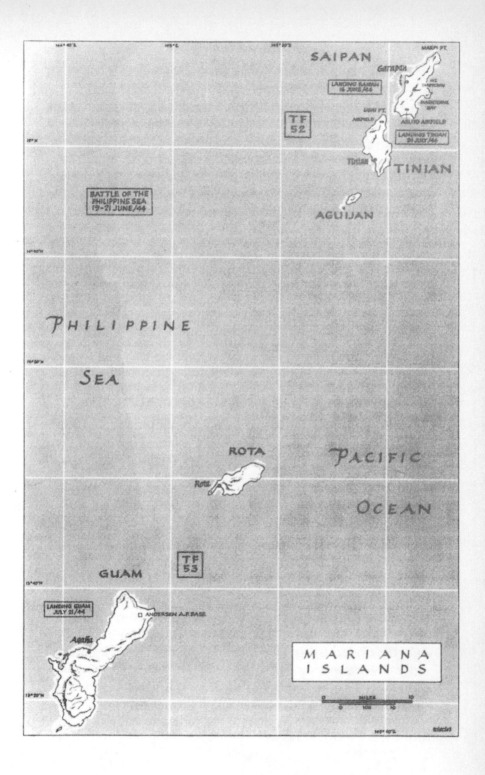

SAIPAN

MARPI PT.

Garapan

LANDING SAIPAN
15 JUNE /44

TF
52

USHI PT.
AIRFIELD

AS TANEPAG

MAGICIENNE
BAY

ASLITO AIRFIELD

LANDING TINIAN
24 JULY /44

TINIAN

TINIAN

BATTLE OF THE
PHILIPPINE SEA
19-21 JUNE /44

AGUIJAN

PHILIPPINE

SEA

ROTA

PACIFIC

Rota

OCEAN

GUAM

TF
53

LANDING GUAM
JULY 21/44

ANDERSEN A.F.BASE

Agaña

MARIANA
ISLANDS

MILES

77th Army Division was earmarked as army reserve. All in all there were nearly 130,000 troops, tying up 535 vessels for at least three months, all operating more than a thousand miles away from the nearest American base at Eniwetok, twenty-four hundred from the main intermediate base at Guadalcanal and thirty-five hundred from Nimitz's headquarters. The troops numbered only 22,500 fewer than those deployed in the all-important opening phase of Operation Overlord – the invasion of Normandy – on 6 June, nine days earlier. That the United States could mount two such vast operations on opposite sides of the world simultaneously speaks volumes about the growth of American military might in the Second World War. Not the least of many productive feats was the manufacture of 82,028 landing craft, over eighty thousand naval aircraft and 1,054 surface warships down to destroyer-escort size, in five years of naval expenditure totaling $105 billion in 1940s money.

Saipan was defended by the Japanese 31st Army, in fact a scratch corps of thirty-two thousand men commanded by Lieutenant General Saito Yoshitsugu. He came under the nominal authority of our old acquaintance Vice-Admiral Nagumo Chuichi, the former Mobile Force commander, now in charge of a ragbag local flotilla of small vessels, who knew better than to try to tell an Army officer of equal rank what to do. Saito was in charge, and he had his own division, the 43rd, one mixed brigade (the 47th Independent), and various miscellaneous Army and Navy units, not all of them front-line. But they were still twice as many as the Americans expected, giving the latter an advantage of only four to one compared with the desired eight. The fixed defenses were decidedly incomplete and much equipment and many stores intended for Saipan had never arrived, thanks to US submarine and air activity. Saito had shown little sense of urgency about using the materials he had on hand, probably because he could not bring himself to believe, until the evidence was overwhelming, that the Americans would strike so hard so soon and so far from their bases. On 10 June he pronounced himself ready for the forward defense of the island on the beaches. Five days later he got his opportunity.

* * *

For the Japanese Navy the impending attack on the Marianas was too close to home to abide by its post-Midway policy of keeping its powder dry until a chance to fight a 'decisive battle' with the US Fleet presented itself. Vice-Admiral Koga Mineichi, Yamamoto's successor as C-in-C of the Combined Fleet, had likewise been killed in an air crash, accidental this time, in the Philippines in March 1944. He was succeeded by Admiral Toyoda Soemu, who in April drew up a plan for Operation A-Go: the decisive battle was now to be in defense of the Marianas, in the sea area east of the Philippines and north of Australia. The Navy was once again very much Japan's second service; since it was unable to stop or even interrupt the inexorable advance of the hugely expanded American Navy across the Pacific toward Japan, the survival of the Empire depended on the Army, last defeated in Korea 352 years earlier – if you overlooked the Solomons and New Guinea and the Gilberts and the Marshalls . . .

Shortages of fuel and tankers limited the radius of action of the bulk of the IJN, now known as the First Mobile Fleet, to within a thousand miles of Palau. Vice-Admiral Kakuta Kakuji's new land-based First Air Fleet, formed in July 1943, was responsible for air cover, with more than five hundred planes at bases in the Marianas (172), the Carolines and the Philippines. This effort had been diluted in advance by airborne attempts to interfere with MacArthur's New Guinea operations – an unwitting, retrospective justification of King's whipsaw strategy rationalizing America's twofold offensive. The attrition of carrier planes and pilots, foolishly parceled out to reinforce Rabaul and Truk, had been so severe since November 1943 that the Japanese fleet carriers had at times all but ceased to function as effective fighting units. From February 1944 the five fleet carriers that were supposed to be the spearhead of the Navy had been in Japan for reorganization and retraining. On 1 March Vice-Admiral Ozawa Jisaburo, Nagumo's successor since November 1942 in command of the Mobile Force or Third Fleet, became C-in-C, First Mobile Fleet, formed from the previous Second and Third fleets – five battleships and four destroyer-squadrons from the Second, and from the Third nine carriers (the four large and five light ones in three divisions), with a combined total of 473 planes, protected by five destroyer-divisions. This force was assembled in the southwestern Philippines and was on maximum alert by the end of May. Operation A-Go was rehearsed on a large chart before the emperor on 2 May 1944. The change of commander had brought no fundamental change of plan: Koga had already drawn up his Plan Z for a full-dress fleet engagement, as the Americans knew from top-secret documents they found at Hollandia, Dutch New Guinea (a handy fringe benefit from MacArthur's great leapfrog).

From the Marianas, eleven hundred miles south of Tokyo, the Home Islands themselves were within range of the American Boeing B-29 bomber just coming into production. Thanks to a remarkable piece of American carelessness, the Japanese already knew something about this airborne monster. In May 1943 a B-29 on a test flight over the northern Solomons, where Japanese forces still held substantial territory, was shot down and its pilot captured. From him was extracted enough information to make it clear that everything should be done to stop the Americans using the bomber, either from bases in China or from Pacific islands. Flying at a maximum speed of 360 miles per hour with a ceiling of 36,000 feet, the Superfortress had an endurance of 3,250 miles with half a load; it could carry 20,000 pounds internally and a further 24,000 pounds externally. It also carried a crew of up to fourteen who were protected by lavish armor, two 20mm cannon and ten .50-caliber machine guns. This compared with the top speed of 287 miles per hour and the maximum payload of 12,800 pounds boasted by its Boeing predecessor, the B-17 Flying Fortress, which could fly 1,100 miles with a full load. The first B-29 raid on Kyushu, the southernmost Home Island, was misleadingly mild on 15 June 1944. It was launched from Ch'eng-tu, in southeastern China, where the planes had refueled after flying in from India. Some four dozen fleetingly bombed a steelworks without putting it out of action. But it was the first time that American bombers had been seen over Japan since the Doolittle Raid more than two years previously, an obvious portent, even if it was to be four months before the Americans were able to use the captured and greatly expanded air base on Saipan for this purpose.

The first American seaborne, pre-invasion bombardment of Saipan on 13 June by the seven fast new battleships, temporarily diverted from their main role of protecting Mitscher's carriers, was a waste of time and fifteen thousand shells. Only when the appropriately trained, older battleships of the two bombardment groups commanded by Rear Admirals Jesse B. Oldendorf and Walden Ainsworth took over the next day and hit both Saipan and Tinian was the shelling more effective – but not effective enough to wreck the well-protected Japanese shore positions. General Saito had been given due notice of imminent invasion – and, from the concentration of fire, where it was likely to strike. The Americans would obviously land in the southwest of the island. Saito planned to pin them down until Admiral Ozawa arrived with the Japanese Fleet from Tawitawi in the Sulu Archipelago to the southwest of the Philippines. As the shells roared overhead like passing locomotives, American Underwater Demolition Teams, an innovation first used in the Marshalls, reconnoitered the chosen invasion beaches, swimming in from small landing craft brought up by destroyers. The courageous frogmen looked for channels through the coral and for any hidden snares offshore, such as mines and tank traps.

It was eighty-three degrees on the Fahrenheit scale at dawn on D-day, 16 June 1944, as the forward elements of the Second and Fourth Marine divisions began to go ashore in more than seven hundred landing craft. As she had done in the early days at Eniwetok, 'Tokyo Rose,' the voice of NHK Japanese Radio's English-language service, mocked the invaders and promised them an early death (she was Mrs Iva Ikuko Togori D'Aquino, then aged thirty-eight and a nisei – an American citizen of Japanese parentage – who was visiting relatives in Japan when war broke out and was recruited as a disc jockey for the Japanese propaganda campaign). The two regimental combat teams that led the assault came under heavy fire from strongly protected and well-concealed artillery, mortars, and heavy machine guns. Some of the Japanese fire was coming from the eastern side of

Saipan, almost three miles away, and even from the adjacent
island of Tinian to the south. This added to the confusion, some
of it avoidable, most of it unavoidable, of the intricate landing
plan, which nevertheless put eight thousand men on the beach
in twenty minutes. A diversionary pass at a beach well to the
north of the landing position had not fooled the garrison for a
moment. So intense and sustained was the welcoming barrage
and lashing crossfire that it took the Americans three days to
attain their modest objective for day one, a notional line up to
one mile inland across a front of nearly four miles.

This beachhead was secured on 17 June, by which time
forty thousand assault troops had taken at least 10 percent
casualties. At dusk on the 16th the corps reserve, the Army's
27th Infantry Division, began to go ashore despite Japanese
infantry counterattacks by day and especially by night (as usual,
their effect was mitigated by the Japanese command's piecemeal
approach). Meanwhile US submarines watching the Philippines
reported two large formations of ships approaching from either
side of the island of Samar, immediately south of Luzon: the
carriers were coming from San Bernardino Strait to the north
and other major surface units from Surigao Strait to the south.
Admiral Spruance therefore called off the invasion of Guam,
at the southern end of the Marianas, scheduled for 18 June,
and prepared for a major naval action. Hence the decision to
commit the 27th so early, and also to have the Guam invasion
force act as an extra reserve for the unexpectedly difficult task of
conquering Saipan. Air cover of the Saipan operations was left
to the escort carriers, while the seven fleet and eight light carriers
of Task Force 58, with their 956 aircraft in four heavily escorted
groups, looked westward to the arrival of the Japanese fleet. The
Americans too lusted after a 'decisive battle,' a Trafalgar of the
carrier age; they correctly believed that their enemy had been
avoiding the showdown they had been waiting for. Japanese air
attacks caused the invaders little difficulty.

On the fourth day the Japanese gave up trying to halt the
Americans on the beach and moved back to exploit the
extremely tortuous terrain of inland Saipan, with its limestone
rocks and gorges. The Americans cut off the southern end of
the island by an eastward rush which left them in possession of

the lightly damaged airfield at Aslito by 20 June. On that day command of the invasion force passed from Admiral Turner to General Holland Smith – a sure sign that the Marines regarded themselves as secure ashore.

At sea, the boats of Vice-Admiral Takagi Takeo's Sixth Fleet (the submarine force) contributed nothing to Operation A-Go against the US Fifth Fleet but lost seventeen boats to destroyers or planes out of the two dozen or so deployed. American submarines, by contrast, sank three tankers and four destroyers off the Philippines. And although Spruance was aware, thanks to American submarines, that the Mobile Fleet had left Tawitawi on 13 June, heading east through the Philippines toward him, he also knew there could be no clash before the 17th, because of the distance involved. So he calmly allowed the carrier raids on Iwo Jima and Chichi Jima in the long, straggling string of islands running due south of Tokyo, to go ahead on the 16th as planned. These islands had airstrips used as staging posts for planes from Japan going south to join A-Go. By the 18th Spruance knew that Ozawa's forces were in two sections and diverted extra cruisers and destroyers from Admiral Turner off Saipan to augment the protection of TF 58. Mitscher's force now consisted of the four reunited carrier groups and Vice-Admiral Willis Lee's seven modern battleships, backed by eight heavy and thirteen light cruisers and sixty-nine destroyers.

The Japanese, who mustered eleven heavy and two light cruisers plus twenty-eight destroyers to protect their nine carriers and five battleships, had the wind advantage: on carriers it pays to have the wind against you to help launch and land planes as you approach the enemy. In that region at that season, the prevailing wind blew from west to east. On paper, Ozawa had the edge in aircraft as well, because, though his carrier planes were outnumbered two to one, he had some 540 land-based planes to back him; and nearly all his aircraft types outranged their American counterparts, which were also out of reach of their own land-based allies.

On the 18th the two converging fleets searched for one another, each desperate to get in the first air attack. The

Japanese succeeded in the afternoon, despite a false alarm which led to an abortive strike. Ozawa decided to attack from at least three hundred miles, outside the American carrier planes' maximum practicable operational range, the next morning. For his part, Spruance stuck firmly to the view that his main task was to cover the invasion force on Saipan no matter what happened. He therefore sailed west by day and east by night, back toward the Marianas, in case the Japanese attempted to get in behind him during the night and make an 'end run' against the invasion fleet, an outflanking maneuver taken from American football. Spruance still had no clear indication of his enemy's whereabouts; all he knew was that Ozawa was on his way, and there were indications that he had split his forces, suggesting a possible feint. Nonetheless, the American admiral's customary caution, which had proved so wise at Midway, irritated some of his staff, who wanted to charge straight at the enemy. Hindsight shows that, had he sailed westward without reversing course at nightfall, the Japanese would probably not have got in the first attack. But he had no idea where the enemy was on the night of 18–19 June. Such retrospective quibbles also overlook the result of the ensuing battle, which is ungrateful to say the least. The artificial controversy about Spruance's alleged overcaution reflects an immature desire not just to win the fight, which is enough for most people, but to win every round without error, which is unrealistic. Spruance's selfsame caution led him to launch a limited strike against Guam, to forestall an attack from there and to make it harder for the Japanese carriers to use the island as a supporting base or refuge. Three dozen Hellcats therefore saw off a similar number of Japanese planes from Guam at little cost to themselves in a protracted mêlée, before the first Japanese carrier planes struck the American fleet.

Ozawa had arranged his force in two groups, a hundred miles apart, as they all sailed northeast toward the Marianas in the early hours of the 19th. To the fore were three carriers, each surrounded by a ring of escorts. In the rear were the other two divisions of three carriers apiece, each group encircled by an escort of battleships, cruisers, and destroyers. The American carriers were in their four groups, three of four and one of three, each group surrounded by escorts; Lee's battleships and escort provided a powerful anti-aircraft screen between the 'flattops' and the Japanese to the west.

The first wave of sixty-four Japanese planes came over shortly after 10:00 A.M.; forty-two were shot down in exchange for one bomb-hit on battleship *South Dakota*. No aircraft got through to the US carriers, which were listening in to the directions thoughtfully provided by a Japanese scout droning overhead. He also told the American fighter-controllers exactly where to look for the next attack. The second wave consisted of 128 planes. No sooner was it airborne from the carriers of the rear than Warrant Officer Komatsu Sakio saw a torpedo track heading for fleet carrier *Taiho*. He took the extraordinary decision to dive straight at the missile, and died in the undeservedly frustrated attempt to stave off disaster. His was the first of ninety-seven Japanese planes lost in this attack, which yielded no more than a crop of spectacular near-misses. The third attack consisted of forty-seven aircraft that swung round to come in from the north; only seven of these were shot down, but they did not get as far as the US carriers. The fourth Japanese raid, by eighty-two planes, came in after 2:00 P.M. Only nine emerged unscathed; nineteen more got away but were too badly damaged for subsequent use. In these four raids, at intervals of roughly one hour, plus associated operations, Ozawa had launched a total of 373 aircraft at the American carriers, of which 240 were downed; perhaps another fifty were destroyed at or near Guam, and a further twenty-five or so were lost to other causes. Not one hit had been scored on a carrier and only very minor damage inflicted on the American force as a whole. The Americans still had no precise idea where the attacks were coming from. The havoc caused by the unprecedented ferocity of their anti-aircraft and fighter defense led the Americans to dub this Battle of the Philippine Sea 'the Great Marianas Turkey Shoot.'

But the story is not yet complete. Some of the bombers that had been cleared off Mitscher's carriers while the enemy raids were being repelled attacked Guam and Rota islands in the Marianas again and again to deter land-based attacks on their carriers. And although Spruance and Mitscher were in the extraordinary position of having won a huge defensive air battle without establishing the whereabouts of the enemy carrier force, two American submarines had found the Japanese and profited from the discovery.

We saw how Warrant Officer Komatsu died valiantly trying to save the new carrier *Taiho*, Ozawa's flagship, from a torpedo. This was one of six fired by USS *Albacore* (Commander J. W. Blanchard, USN) from periscope depth. Komatsu did manage to take it with him to the bottom, but one other missile struck home, forward on the starboard side, at 9:15 A.M., causing what appeared to be containable minor damage: a large aviation-fuel tank was ruptured but there was no fire. Next submarine on the scene was Lieutenant Commander Kossler's *Cavalla*, which fired six torpedoes at Ozawa's group of three carriers just after noon. Three struck the *Shokaku*, veteran of the Pearl Harbor raid, and she blew up and sank in minutes. Meanwhile the hull of the *Taiho* had filled with aviation-gasoline vapor because someone innocently used the ventilation system to try to get rid of the smell. It needed only a single spark to set off the thirty-three-thousand-ton carrier like a colossal bomb. The inevitable occurred at 3:32 P.M. on 19 June. Ozawa and his staff managed to transfer by lifeboat to the heavy cruiser *Haguro* as his flagship broke up and sank.

The Japanese withdrew from the waters west of the Marianas. Mitscher's scouts proved as incapable of finding the enemy after the victory as they had been before it, and Ozawa lived to fight another day (a fact he surely came to regret). The Japanese carrier force had ceased to exist as a credible threat. But if Spruance erred on the side of caution, as his critics maintain, Mitscher surely erred even more in not sending out more air searches before, during and after the battle of the 19th, especially on the night of 19–20 June, when none was flown, even though he had scores of aircraft and fresh pilots equipped for the task. Not to seek with every instrument at your disposal an enemy whom you expect to fight at dawn seems bizarre to the point of negligence. Spruance properly planned to follow up and attack the Japanese carriers if only he could find them; but they were already heading northwest under strict radio silence as the Americans turned west to chase them, and they got away – for the moment. The total of American losses in all operations on the 19th was twenty-nine aircraft and twenty-seven air crew, plus thirty-one sailors killed by the three hits on or near ships scored by the Japanese. Such was the wildly uneven outcome of what was the largest carrier-battle so far recorded.

It was only later that Ozawa discovered the true extent of his defeat on the 19th. He knew, of course, that he had lost two fleet carriers, but he had not seen a single attacking American plane. This could have meant success against the enemy's carriers, as the returning pilots claimed in their customary exaggerated way. The missing Japanese aircraft were just as likely to have landed on Guam, according to plan, as to have been shot down. Ozawa may have withdrawn from the field, but he fully intended to come back for a second round, and sent search planes from his vanguard of three light carriers to look for the Americans. And even though he was on a crowded cruiser with no sign of *Zuikaku*, the third and only surviving member of his First Carrier Division, Ozawa did not think of handing over tactical command to the leader of the vanguard, Admiral Kurita. He did not change his mind when *Zuikaku* caught up at lunchtime, not even after he transferred his flag to her and learned the extent of his losses from incoming shore-radio and returned pilots. He was down to exactly one hundred serviceable combat aircraft, so he postponed his planned attack until the next day, the 21st, intending to call up land-based reinforcements for the return match.

But in mid-afternoon the unblind eye of Lieutenant R. S. Nelson, USNR, a scouting pilot from the indestructible USS *Enterprise*, at last got a sighting of Ozawa's fleet, then in three groups sailing slowly west and refueling. Task Force 58 was 280 miles to the east, at the limit of its planes' operational range. Mitscher, so unnecessarily cautious during the previous night, now recklessly committed a full striking force of 216 planes, untrained for fighting, landing, or even flying at night, at the end of the day, while sailing in the wrong direction (with the wind behind him) for launching and recovery. Ozawa managed to get seventy-five planes into the air against them. The Americans crippled two tankers which had to be scuttled later; then fleet carrier *Hiyo* was struck by one or more torpedoes, caught fire and eventually sank after dark by the bow. *Zuikaku* was damaged but got away to Japan. One other carrier, *Chiyoda*, and battleship *Haruna* were hit, but not seriously, in the most effective torpedo-bombing action ever fought by the US Navy. This time the Japanese failed to press home an attack on the Americans. As his last throw, Ozawa, now down to thirty-five operational carrier planes, ordered Kurita to make a surface attack at night, but he too failed as completely as the Japanese pilots to get near Task Force 58. The American pilots, flying on their last ounces of fuel, returned to a carrier force lit up on Mitscher's order like a seaside resort to guide them home. His task force lost twenty planes in combat, but four times as many crashed or ditched as they tried to return; only forty-nine air crew were ultimately recorded as missing after the search-and-rescue operations of the 21st.

On the orders of Admiral Toyoda, Ozawa withdrew after dark on the 20th with six carriers, five battleships, thirteen cruisers and twenty-eight destroyers as good as intact. But only thirty-five carrier aircraft were left to him as he made for Okinawa in the East China Sea, halfway between Formosa and Japan. None of his ships was slowed down by damage, although two bore scars as we have seen. Spruance did not strain to catch them, because he was anxious to recover as many downed air crew as possible from the water. Japanese ships even this far into the war generally 'had the legs' to get away from their American opposite numbers and duly proceeded to do so. The Americans followed only in the hope of catching stragglers, of which there were none. Spruance turned down a suggestion from Mitscher to send Lee's battleships at top speed after the retreating Japanese as a waste of time, fuel and effort. Only scout planes, flying at the limit of their range, sighted the Japanese heading northwest after dawn on the 21st, 360 miles ahead and out of reach of American carrier strike-aircraft. Spruance gave up the 'chase' at dusk.

Task Group 58.1 under Rear Admiral Joseph J. Clark detached itself, with Mitscher's approval, to mount another air raid on Iwo Jima before returning to Eniwetok. The American pilots destroyed sixty-six Japanese planes sent to intercept them, dumping their bombs and abandoning the attack on the island to deal with the airborne enemy. In all, Admiral Clark subsequently managed three attacks on Iwo Jima and Chichi Jima before returning to his advance base. MacArthur's bombers attacked Yap, in the Palau Islands, and naval Liberators bombed Truk several times during the week of the Battle of the Philippine Sea, to discourage Japanese efforts to turn the course of the struggle for the Marianas. Spruance had undoubtedly won this unique maritime clash by a very large margin; but Mitscher and others ungenerously took the view that he had missed a golden opportunity to destroy the bulk of the Japanese fleet's ships. Spruance himself never forgot that he had to defend the invasion fleet, Ozawa's objective, as well as attack the enemy, a double responsibility not shared by Mitscher. He therefore waited defensively for the enemy to attack him instead of going all-out to get in the first blow. Spruance's critics maintain that the close-support forces he left with Admiral Turner would have been enough to deal with any Japanese attempt at an end run, but that smacks of hindsight, as informed by the even more remarkable naval event in the Philippines in the ensuing October. Nor was Ozawa as feeble an opponent as the result implies: he was ultimately let down by the failure of his superiors to provide him with a constant stream of properly trained pilots (for reasons already discussed). His offer to resign after the battle was rightly rejected by Toyoda.

Spruance's essentially defensive action with a force designed for attack, together with his judicious counterattack, gave the Americans indisputable command of the airspace and water round the Marianas, a strategic gain that would not have been enhanced by a huge extra effort to impose even more material destruction on a defeated enemy. Spruance himself told Morison after the war that, given a second chance, he would have been more aggressive. But that too was hindsight speaking. And his undeniable victory was attained for the loss of just 130 aircraft and seventy-six air crew in all.

Ashore on Saipan, the outcome of the protracted and bloody wrestling match could now take only one form, even if those on the ground took a long time to realize it. The Marines at first thought they had been 'abandoned by the Navy' once again, as at Guadalcanal, while the Japanese defenders deduced that Ozawa had chased the American ships off. But the three American divisions ashore, backed by constant gun bombardment from the close-support groups and air strikes from the escort carriers, inexorably took control of the island over the ensuing eighteen days, the Japanese enduring terrible losses as they retreated to the northern end of Saipan.

In the struggle for Mount Tapotchau, the highest point of the island, the 27th Division, largely composed of unreconstructed National Guard units from New York state and of a distinctly higher average age than the Marines, was pushed back while the two Marine divisions advanced on either flank, causing the American line to sag alarmingly by nearly a mile in the middle. General Holland Smith decided on 25 June to remove his namesake Major General Ralph Smith from the command of the division, replacing him with Major General Sanderford Jarman, who was on hand to command the Army garrison after the anticipated victory. It was the sixth time a divisional commander had been replaced in one or another of the Pacific commands for failure to deliver the quick results required, and there was justice in the measure. What was new about it, however, was that the general who did the sacking was a Marine while his victim was in the Army. An almighty interservice row duly followed, the facts of the case soon lost to view under the prejudices of both sides. The Army accused Marine commanders of unfamiliarity with large units, reckless impatience and disregard for casualties, while the Marines accused the Army of overcaution and incompetence. It was all highly unsavory, but the 27th held on by the skin of its teeth and eventually pushed forward to straighten the line. By 5 July the surviving Japanese defenders were cooped up in the north with their backs to the cliffs and the sea below.

Two days later, three thousand of them took part in the biggest *banzai* charge on record (the word is a battle cry meaning literally 'ten thousand years,' a declaration that the soldier is ready for eternal life on dying for his emperor). The screaming Japanese troops, armed with grenades or only bayonets, found a gap in the shaky line of the 27th Division on their right or western front and burst through it, rolling up and decimating two infantry battalions and pouring down the west coast until they came up against Marine artillerymen, who snatched up small arms and stood their ground. The ignominy of the 27th was completed by the arrival of destroyers to lift off the remnants of the broken first and second battalions of the 105th Infantry Regiment. Holland Smith ordered the entire division withdrawn. But on 9 July significant organized resistance came to an end with no fewer than 23,811 Japanese troops dead (as counted by American burial parties) and 1,780 taken prisoner. This compared with 3,426 Americans killed and 13,099 wounded, a casualty rate of all but 25 percent. Although Holland Smith took the salute at a flag-raising ceremony on 10 July and handed over command two days later to Major General Harry Schmidt, USMC, it was not until one calendar month afterward that Spruance felt able to pronounce Saipan secure in American hands.

The horror of the most intense territorial struggle of the Pacific war so far did not, unfortunately, come to an end with the suicides of General Saito and Admiral Nagumo in their respective command bunkers in the caves of northern Saipan on 6 July, or with the last mad charge by soldiers and sailors of the following day.

A curious message emanating from the Imperial Palace in Tokyo (as distinct from Hirohito himself – the identity of the originator of this idea, explosive in Japanese terms, was never established) was sent to the governor of Saipan on 22 June, against the wishes of Prime Minister Tojo. It informed the civilian population of the island, one week after the American invasion, that any civilian who died in the fighting then raging would enjoy the same glorious hereafter as a soldier who died for the emperor – also head of the state Shinto religion. There were at least twenty-two thousand nonindigenous inhabitants – mostly Japanese, with some Korean laborers – on the island. As the cries of the last *banzai* charge faded into the distance, thousands of people – men, women, children, old and young, even babies, and one mother actually in the act of giving birth – plunged off the cliffs in droves at the northern end of Saipan into the shark-ridden waters. Film taken by Marine combat photographers of this macabre act of mass hysteria survives in all its cruel clarity. The last soldiers left alive held off the encircling Americans and urged on where they did not actually threaten these human lemmings. They shot some to encourage the others. Parents dashed their babies against rocks before going over the edge; children clasped live hand grenades to their emaciated bodies as they jumped. Marine interpreters with loudspeakers called on them in Japanese to give themselves up, promising fair treatment (nearly fifteen thousand civilians had done so or were to surrender shortly afterward, a much higher proportion than usual). Nevertheless, nearly eight thousand noncombatants were lost in this pointless, suicidal stampede, taking the total of Japanese who died in vain for Saipan past the thirty-thousand mark.

The Fourth Marine Division landed on Tinian, five miles across the strait from Saipan, on 24 July, soon followed by strong elements of the Second Marine Division, and the Americans had it firmly under control by 2 August. The invasion was preceded by a furious row between Holland Smith and Richmond Kelly Turner about the choice of beaches. Only when Spruance intervened and found all his staff except Turner in agreement with Smith that a landing in the north was better than one in the south, where the main settlement and therefore the strongest defenses were, did the aggressive amphibious commander climb down. The invaders took the Ushi Point air base on 26 July, and work began at once on repairing and expanding it to take B-29s. This remote airfield would help change the course of history as it had never been changed before, one year later. More than a thousand Japanese died defending the island against overwhelmingly superior force. An attempt to repeat the civilian mass-murder-cum-suicide seen on Saipan led to only a few score deaths and was undermined by American loudspeaker persuasion from shore and ships at sea. As on Saipan, the Americans felt unable to use force because the soldiers instigating the suicides were intermingled with the civilians and often had live grenades in their hands.

*   *   *

Guam, which should have been the second, was the last of the three objectives to fall, thanks to the Battle of the Philippine Sea and the intensity of the struggle for Saipan. The 77th Army Division was shipped from Hawaii to back General Geiger's III Marine Amphibious Corps of one and a half divisions for the assault, to be delivered by Rear Admiral Richard Connolly's Southern Attack Force (Task Force 53). The admiral was known as 'Close-in Connolly' for his point-blank barrages in the Marshall Islands campaign, a habit he maintained in thirteen days of unprecedented shelling and bombing in preparation for the Guam landings. Lieutenant General Takashina commanded a garrison of nineteen thousand men ensconced in elaborately constructed defenses, including dug-in artillery, bunkers, and underwater obstacles, through which paths had to be cut by the Underwater Demolition Teams.

The Marines went ashore on 21 July in two places on the west coast of the gauntlet-shaped island, soon followed by the well-trained troops of the 77th – another largely New Yorker formation, which proved to be as sound as the 27th had been unsound. Two massive Japanese counterattacks had been beaten off by Marines and troops jammed together on the two overcrowded beachheads by 26 July, whereupon resistance ceased to be a serious threat. The Stars and Stripes went up again on the old Marine parade ground at 3:30 P.M. on 29 July 1944, in an emotional scene watched by Spruance, Holland Smith, Geiger and tearful local inhabitants. Only on 10 August, however, was Guam, the first piece of captured US external territory to be liberated from the Japanese, declared 'secure.' Of the fifty-five thousand American servicemen engaged, 1,440 were killed and 5,650 wounded. More than nineteen thousand Japanese, including noncombatants, died, and another 1,250 were captured; but the last Japanese unit did not surrender until 4 September 1945 (Lieutenant Colonel Takeda and 113 men). The last soldier of the emperor took rather longer to give himself up. Corporal Yokoi Shoichi of the 38th Infantry Regiment emerged in tatters from the Guam jungle carrying his rifle, still functioning but with its wooden stock rotted away – in January 1972. Even then he wept tears of shame on television over Japan's defeat – not, it may be noted, over Japan's aggression which invoked that defeat.

As far as the Japanese were concerned, the fall of Tinian and Guam was a mere postscript to the irretrievable national loss of face brought on not by the loss but by the mere fact of the invasion of Saipan. There need be no doubt that this was the great psychological and political turning point of the entire misbegotten southern-front campaign of the Japanese Empire. It was militarily and strategically important too, partly because it represented the first loss to the Allies of territory that had been Japanese before the war, and partly because it gave the Americans an invulnerable base for bombing Japan (that the Chinese airfields could not be regarded as safe was to be shown by Japanese attacks as late as March 1945). The associated defeat in the Philippine Sea caused a terminal loss of faith among those in the know in the Navy. The Marianas were part of the Absolute National Defense Sphere, inviolable, guaranteed by the Army above all as Japanese soil which would never be lost to an enemy. As Marquis Kido, Lord Privy Seal and key confidant of Emperor Hirohito, saw it:

The Japanese people in general had placed much expectation on Saipan. They had thought that Saipan was heavily fortified and heavily defended, but this proved otherwise, and the consequences greatly shocked the Japanese people. In order to meet [the threat to Japanese soil] General Tojo, the prime minister at the time, took upon himself another office – that of chief of the Army General Staff; and by assuming greater powers he incited a great deal of opposition against him, which was already growing in Japan and intensified from day to day to a policy of 'down with Tojo.' War production at that time was not making any headway and the people's life was becoming more difficult.
    As a consequence, the Tojo government, in order to meet the situation and to obtain results, intensified the various controls on the economic life of the people. This in itself was another point which drew strong criticism against him from the people. The opinion among the civil population was to give a wide latitude to the people themselves and to permit them to exercise greater initiative. This they believed would secure better results than otherwise. These developments culminated in the fall of the Tojo Cabinet in July 1944.

41. "Symbol": Roosevelt and Churchill at the Casablanca conference (code-named Symbol) in January 1943, where the USA superseded Great Britain and FDR unilaterally announced the war aim of unconditional surrender.

42. US troops go ashore at Leyte, Philippines, in October 1944.

43. Aboard a light carrier: shifting Hellcats on USS *Princeton* in October 1944.

44. Kamikaze: suicide attack on battleship *Missouri* in April 1945.

45. The scene on Iwo Jima shortly after the American invasion.

46. The marines crawl up Mount Suribachi, Iwo.

47. Raising the Flag on Iwo Jima (for the second time, actually)—the most famous picture of the Pacific Campaign, by AP's Joe Rosenthal

48. The end is nigh: US carrier planes raid Japan in July 1945.

49. & 50. Above and below: Tokyo after the fire bombing raids by B-29s.

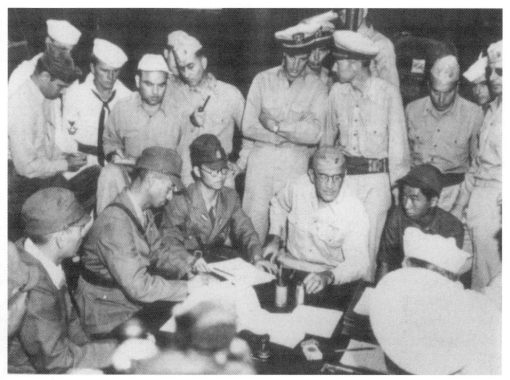

51. The Americans accept the return of Wake Island on the surrender of its Japanese occupiers in June 1945.

52. The "Big Three" at Potsdam, where Churchill, Truman, and Stalin issued their ultimatum to Japan.

53. & 54. Above and below: Hiroshima.

55. MacArthur as Supreme Commander, Allied Powers, countersigns the Japanese surrender on the USS *Missouri* on September 2, 1945.

56. The Imperial Japanese Army disbands.

57. The Emperor surveys the ruins of his capital.

The first sign of unease came from Tojo himself, only five days after the Americans invaded – on 20 June, the day the Saipan air base was captured. One of his aides told Prince Higashikuni, a member of the imperial family and an adviser to the emperor (Japan's first postwar prime minister), that the general was thinking of resigning because he felt he lacked Hirohito's confidence and himself had no confidence in victory. The discussion spread to other imperial advisers and eventually reached the emperor's ears. There was a groundswell for peace centered upon Tojo's predecessor, Prince Konoye, but there was concern to protect the imperial throne from blame for the threatened defeat and for the war itself, which could mean the end of the emperor system, since the Allies were demanding unconditional surrender. If Tojo stayed, he would be the obvious and convenient scapegoat; if he went, others would come under fire, not excluding Hirohito. Marquis Kido, effectively the emperor's alter ego, saw Tojo on 13 July and demanded he give up his Army chief-of-staff post, sack the Navy minister, Admiral Shimada, and add what we would call 'doves' to his Cabinet from the ranks of pro-negotiation ex-premiers and former ministers. This was done on the advice of the *jushin*, the group of former premiers, which was close to the throne and the main influence upon its occupant after the military. But the doves refused to serve in a Cabinet still led by Tojo. The inscrutable general now felt his loss of face to be intolerable and resigned on 18 July. His immediate reason was the general loss of confidence in himself at court and among the imperial advisers; the real cause was the Saipan disaster.

The entire Cabinet, as usual, fell with him, after thirty-two months in office. It was also four years to the day since Tojo had become Army minister in the second Konoye Cabinet. Tojo, however, retained the mantle of principal scapegoat in the eyes of Japan's enemies: having failed in a suicide attempt, he was the principal defendant at the Tokyo war-crimes trial and was hanged in December 1948. He was meanwhile succeeded on 22 July by another general, Koiso Kuniaki, the mildly dovish erstwhile 'Tiger of Korea' and governor-general of the peninsula, who brought with him the decidedly dovelike Admiral Yonai Mitsumasa as his almost equal partner, nominal deputy and Navy minister. Tojo wanted to stay in the Cabinet and revert to his old portfolio as Army minister, but the 'peace party,' if such it can be called, had the upper hand for the time being and blocked this. The new war minister, however, was the fire-eating General Sugiyama Hajime, proof that the hawks in the General Staff junta still held the reins of real power in Japan. General Umezu Yoshijiro, previously C-in-C of the Kwantung Army and key commander on Japan's other front in China, was appointed chief of the Army General Staff at the same time. Umezu was not the only one in the new Cabinet who had yet to be convinced that the game was up. The peace campaign was all but stillborn, even though the Japanese assault on India from Burma, begun in March 1944 as a last attempt to stem the tide of defeat, had finally been driven off by the British General Slim in the first week of July. The American General Stilwell was closing in on Myitkyina, in the north of the country, after a long siege. Japan fought on, preparing for the next 'decisive battle' on the southern front. Only the emperor could have kept the peace initiative alive, but at this stage he chose not to do so. His people braced themselves for the feared American air bombardment. This was a surprisingly long time coming, but all the worse when it finally came, as we shall see.

Watching it all from the sidelines was Admiral Wenneker, the German naval attaché and frustrated advocate of submarines, who noted at this time:

Saipan was really understood to be a matter of life and death. About that time they began telling the people the truth about the war. They began preparing them for whatever must happen. Before that they had been doing nothing but fooling the people.

A much greater threat to Japan than the loss of the Marianas had already been raging at sea for the whole of 1944 – the American submarine campaign, which had got into its stride in 1943 and was set to strangle the Japanese war effort long before the first B-29 appeared over Tokyo.

The new and more effective Torpex explosive (first used

in February 1943), the improvements to torpedoes and their detonators by October 1943, and the more favorable operational climate created by the US Navy as a whole enabled the American submarine service to make an important contribution to the war of attrition against Japan in that year. A total of 350 patrols produced 335 sinkings of shipping, totaling about 1.5 million tons, of which about four hundred thousand were oil tankers. Even so, the Japanese managed to build enough new tankers, arguably their most important category of vessel, naval or mercantile, to raise their net total at the end of the year to 863,000 tons compared with the 686,000 tons they had at the end of 1942 (by which time the Americans had sunk 725,000 tons, or 180 ships of all kinds). But the nontanker merchant fleet decreased by 1.1 million tons to 4.1 million. Nimitz's submarine operational plan issued at the end of June 1943 listed carriers as the prime targets, followed by battleships and light carriers. A proper understanding of the potential of submarines and the enemy's Achilles' heel would surely have put tankers at the top of the list, an omission all the more surprising given the US Atlantic Fleet's detailed, first-hand knowledge of the desperate struggle in the other ocean, in which the economy of another island-empire had almost been strangled by submarines.

The latter part of the year saw some exceptional patrols by individual boats. The American love of heroes and some unusual exploits combined to produce much propaganda for home consumption. USS *Wahoo*, under Lieutenant Commander Dudley W. 'Mush' Morton, USN, was the most famous submarine in American service. Initially operating from Brisbane, Australia, it came to fame after its third patrol, during which it stole into Wewak harbor in New Guinea and crippled a Japanese destroyer in January 1943. After withdrawing, Morton, on his first patrol as skipper, attacked a convoy, damaging a tanker and sinking two freighters and a transport. The boat then surfaced to machine-gun the troops milling about in the water from the latter. Morton's reward for all this was the Navy Cross and the Distinguished Service Cross. When he brought his boat into Pearl Harbor on 7 February, he and his crew were lionized by the publicity machine, not least in order to inspire a service whose leadership was still marked by caution and whose morale had been damaged by the torpedo crisis and other problems. Morton's second patrol, in the spring, yielded a numerical record of nine ships sunk, in the shallow and therefore exceptionally dangerous waters of the Yellow Sea, between Korea and China.

In May the *Wahoo* was in the northwestern Pacific and sank three ships. Yet Morton felt he would have sunk six but for his faulty torpedoes. When he got back to Pearl Harbor, where he was now based, he stormed into the office of COMSUBPAC, Vice-Admiral Charles A. Lockwood, and spoke his mind as only the fleet's star skipper could. He convinced Lockwood at last that the magnetic exploder in the Mark VI detonator was faulty; in June 1943 the admiral ordered his captains not to use it, with Nimitz's approval. But the submarine command of the Seventh Fleet, at Fremantle, Australia, continued to use it, which meant that captains 'commuting' between the two fleets followed the opinion of the flag officer under whose command they happened to be at the time – or quietly did as they thought fit. It is not unreasonable to describe this state of affairs as ludicrous. When Admiral Kinkaid took over the Seventh Fleet in November, he ordered Rear Admiral Ralph Christie, COMSUBSOWESPAC, to deactivate the unreliable pistol. By that time, USS *Wahoo* was lost. On Morton's fifth patrol – in the Sea of Japan, west of the main island of Honshu – he sank four ships, totaling 13,300 tons, including a transport with 544 men aboard, with the new Mark XVIII torpedo. On 11 October a Japanese aircraft depth-charged a submarine to destruction near La Pérouse Strait, in all probability the *Wahoo*. At any rate, it was never heard from again, one of fifty-two American submarines lost in the war to all causes.

In response to the belated Japanese discovery of the value of convoy at about this time, the American submarine command at Pearl Harbor introduced what it called the 'wolf pack.' This was of course an imitation of the German example – yet not as set in this world war but in the First, when the introduction of convoy by the Allies prompted the Germans to put two or more boats under a flotilla commander who used one of the group as his 'flagship.' In the Second World War, Admiral Dönitz introduced packs as soon as he had enough boats, but gave up the idea of on-the-spot control at sea because submarines could not communicate except at or on the surface – radio does not function underwater. The Germans assembled packs on an *ad hoc* basis; when one boat saw a convoy it reported to headquarters, which then radioed all available boats, sometimes as many as twenty or more, to the area to attack individually. But against the poor antisubmarine defenses of the Japanese Navy, the Americans found their small-flotilla tactics sufficiently fruitful to persevere with them (official title: Co-ordinated Submarine Attack Groups).

Thus Captain John Philip Cromwell, USN, leading Submarine Division 43, was aboard USS *Sculpin* (Commander Fred Connaway, USN) on 19 November 1943, with a view to forming the first pack on his first war-patrol. The boat was on station east of Truk when it found a Japanese convoy. Depth-charged by the destroyer *Yamagumo*, *Sculpin* surfaced but inevitably lost the ensuing exchange of gunfire. Connaway decided to abandon and scuttle. Captain Cromwell, who knew all about Ultra and Operation Galvanic, the American assault on Tarawa and Makin in the Gilbert Islands, identified himself – rather belatedly, one may think – as a security risk who could not afford exposure to the ministrations of Japanese 'interrogators.' He therefore coolly decided to go down with the boat. When the survivors returned from prison after the war, Cromwell's unique sacrifice was recognized with a posthumous Congressional Medal of Honor.

So the first wolf pack was formed by Captain Charles B. Momsen, commanding Submarine Squadron 2, from USSs *Cero*, *Shad* and *Grayback*, at the end of the year. They practiced working together using 'TBS' (talk between ships), short-range, high-frequency radio and cipher to communicate. Tactics were for the center boat to attack a convoy and withdraw to reload, whereupon the two on either side would attack in turn, each hoping to drive the targets onto the torpedoes of the other. The joint patrol sank three ships of 23,500 tons, and Momsen was awarded the Navy Cross. But his conclusions were that the Dönitz system of centralized control would probably be better, that the main problem was inevitably communication between boats at sea, and that no separate flotilla-commander was needed: the senior skipper present could do the job, saving precious space and unnecessary complications.

Submarines could and did usually fire and forget and flee the scene. One of the most successful American boats of the war, USS *Barb* (seventeen ships, including two naval, of 96,628 tons, sunk), was on patrol in the Kurile Islands, between northern Japan and the Soviet Union, in the middle of 1944. This is how the captain, Commander Eugene B. Fluckey, USN, remembered an attack on a small Japanese convoy in the area.

The first ship that came toward us was a large transport. We submerged upon sighting her, made our attack and sank her with three torpedo-hits. . . . As this ship sank in a very graceful fashion, just slowly settling, still going ahead under full power, some landing craft floated off her which were filled full of Japanese Army personnel [who fired on *Barb* with machine guns when it surfaced; it made off and sank another ship].

We returned to the scene of our crime, the last sinking, to pick up a prisoner to see if he could give us any information. It was a very gruesome picture. This was the first time that I had ever returned to the scene of a sinking and it was rather an unholy sight. The clouds at that time were settling down close to the water, it was just getting dark, the atmosphere was much like one you'd expect from Frankenstein. The people were screaming and groaning in the water. There were several survivors on rafts. The water at that time was very cold, about twenty-seven degrees. These people were gradually freezing and dying. We took the most lively-looking specimen aboard, who was very anxious to come aboard . . .

Given the lethal threat presented by the American submarine campaign, the Japanese response was not just too little too late but positively pathetic. Like the British Admiralty in the First World War, the Imperial Japanese Navy either misunderstood or actively disdained convoy as out of place in a war of aggressive aggrandizement. Some admirals doubtless thought that creeping about the ocean protecting merchant ships was effeminate or cowardly or quite possibly both. They were not trained to see this as playing ducks and drakes with a dwindling national treasure and the entire economy of the Empire, upon which the ability to wage war depended. Given the alleged importance of the economic lessons of the 1914–18 war, it is more than merely amazing that the Japanese did not absorb the lesson of the great British naval change of heart in April 1917, when convoys were introduced in the nick of time to prevent defeat. It was this decision that ensured it was the Germans and not the British who were decisively starved of food and raw materials. Had the Japanese looked back to an earlier British example, when the Royal Navy unquestionably ruled the waves – the Napoleonic Wars – they would have noted that toward the end of that global conflict, which was in naval terms the first real world war, the British Parliament actually forbade merchant ships to sail unescorted. In those days it was clearly understood that the Navy had two tasks, to attack the enemy fleet and to protect the nation's commerce, of which the latter was the more important. Indeed, the use of convoy brought the two roles together, in that men-of-war often found their enemy counterparts covering merchantmen.

In the beginning of the war [antisubmarine warfare – ASW] was not considered of great importance [Captain Kamide, IJN, a naval ASW pilot and staff officer, recollected]. In December 1943, prompted by serious shipping losses, we quickly increased the strength and enlarged the organization of the 901st Squadron . . . The escort fleet also operated escort carriers for protection of convoys.

The IJN set up its First Escort Squadron only at the end of 1943, under Vice-Admiral Nakajima Torohiko. It consisted initially of ten destroyers, eight coastal-defense ships, three minesweepers, and a few miscellaneous craft such as submarine-chasers and armed transports (called 'special gunboats' and roughly equivalent to the unsuccessful British 'armed merchant cruisers'). This modest ASW outfit covered convoys traveling the routes between southern Japan, Formosa, the Philippines, Singapore, North Borneo, and Palau. Local naval commands were expected to provide protection for local convoys: we have already seen how many naval battles developed in the classic manner over Tokyo Expresses – troop and/or supply convoys with warship escort. Indeed, the destroyer and cruiser transports used by both sides were another throwback to the days of sail, when merchantmen carried their own guns and it was often difficult to tell them apart from warships. The first convoys on these principal interior lines of communication consisted of just one, two or three merchantmen; the senior captain present took command of convoy and escort. Later, about fifteen captains were detailed as specialist convoy-commanders.

When the convoy-protection formation was raised to the status of the First Escort Fleet in November 1944, under the flag of Vice-Admiral Kishi Fukuji, four rear admirals were retained to command the largest or most important convoys. The fleet started with about sixty escorts, including four fleet destroyers (now extremely scarce and irreplaceable), forty-five coastal-defense vessels, four minesweepers, two sub-chasers and miscellaneous converted transports. Some 220 aircraft (on paper; actually never more than 170, because of competing needs for planes) of the 901st Naval Air Flotilla, based at Takao, Formosa, were also attached to Kishi's flag. Founded as a squadron in December 1943, this unit reached its maximum real strength in January 1945 with a scratch force of eighty patrol aircraft, fifty medium bombers (thirty 'Kates' and twenty 'Bettys'), thirty Zero fighters and ten flying boats. Aircraft were assigned on the basis of availability, not suitability. 'Bettys' and flying boats proved the most useful, not least because they alone could function at night. In the last months of the war some twenty specially adapted 'Bettys,' code-named 'Lornas' by the Americans, joined the 901st as the most effective Japanese airborne escorts. They were equipped with radar and a 'magnetic anomaly detector,' which discerned a submarine's presence underwater from the air by the steel of its hull to a depth as great as nine hundred feet. Other technical aids for antisubmarine warfare developed at the naval-aviation research facility at Yokosuka included two types of depth charge of 250 and sixty kilograms (550 and 132 pounds), adapted from conventional bombs. They had time fuses rather than the more sensitive but more complicated hydrostatic (pressure-operated) fuses favored by the Allies and the Germans. The maximum setting of sixteen seconds would cause the depth charge to go off at eighty meters (over 250 feet). Originally they were set to go off sooner, but Congressman Andrew Jackson May, a member of the Military Affairs Committee of the US House of

Representatives, changed all that. After a tour of the Pacific war zone, he gave a press conference at which he said that American submarines were surviving well because the Japanese were setting their depth charges to go off too soon. This was a leak of which, unlike those about American cryptanalytic successes, the Japanese took full cognizance. Admiral Lockwood was furious, reckoning (somehow) that this unpardonably stupid indiscretion cost the US Navy eight hundred men and ten boats.

There was a Grand Escort Fleet headquarters under IGHQ (Navy) in Tokyo, which superintended escorts everywhere, including those organized by local naval commands and those provided to cover the separate Army convoys. It kept a plot of estimated Allied submarine positions, fed by observations and radio direction-finding reports. But the Army and the Ministry of Transport each had a separate shipping-control agency as well as the Navy, which led to pointless duplication. The Navy's effort reached its peak of defensive efficiency, such as it was, in May 1944, when only one ship was lost while in convoy (the rest of the major losses at this period were among unescorted ships, as in both world wars in the Atlantic). The maximum offensive success was achieved in the autumn of 1944, when escorting forces claimed four or five US submarines destroyed out of ten operating between Formosa and the Philippines. The 901st may have claimed twenty submarines in all, although its pilots put the figure at five hundred!

In the spring of 1944 the airmen tried to kill submarines with torpedoes, which had worked so well against battleships; but submarines were able to avoid them by crash-diving, and the experiment was abandoned after five months. Even a 'curly' torpedo with a contact pistol, a weapon that did four complete, downward-spiraling circles underwater, ideally dropped two hundred meters ahead of where the submarine went down, proved ineffective. Only an advanced magnetic detonator might have made this idea work. The Japanese came to radar late, as we have observed, and used it for sub-hunting without regard to the fact that its emissions could and did warn the intended prey. By the end of the war, when it was academic, Japanese radar could detect a submarine at ranges between twelve miles and five hundred meters. Sub-hunting aircraft had no guns but only depth charges, and their tactics on finding a boat were to attack, report the contact, mark the spot where the boat dived and circle overhead. No proper coordination was developed between air and surface escorts: if a ship responded to an aircraft sighting, the plane would indicate visually the last known position of the enemy submarine. Nobody systematically sought German advice on submarine or antisubmarine warfare. The Japanese ASW commanders regarded submarines as the main threat to their shipping until October 1944, when they were supplanted by the ubiquitous carrier aircraft, and, from the beginning of 1945, land-based planes.

The American Submarine Operational History may describe the US Navy's underwater arm as 'the world's greatest submarine force,' a piece of hyperbole Grand-Admiral Dönitz would not be the only man to dispute. But it is also honest enough to concede:

It would do very well however for all submariners to humbly ponder the fact that Japanese anti-submarine defenses were not of the best. If our submarines had been confronted with Allied anti-submarine measures, the casualty list of the submarine force would have been much larger and the accomplishment of Allied submarines much less impressive.

The curiously ineffectual Japanese submarine arm, the Sixth Fleet, seems to have virtually lost heart and given up the struggle when the Americans invaded the Marianas. 'Our war was lost with the loss of Saipan,' said Vice-Admiral Miwa Shigeyoshi, the last commander of the Sixth Fleet. 'The loss of Saipan meant [the Americans] could cut off our shipping and attack our homeland, and our submarine operations were completely shut out.' The Japanese tried to counter American radar by coating their boats with rubber and giving them conning towers which sloped inboard from the top, so that they would reflect a beam downward toward the water rather than back to the enemy. They also consulted the Germans to some extent in the spring of 1944 about antiradar precautions. They were advised to minimize the boat's profile on the surface and fit the snorkel (an underwater breathing tube which enabled a boat to stay under at periscope depth for very long periods). But the Germans proved at best indifferent at radar detection and radar avoidance themselves, and Japanese boats were never anywhere near as good as theirs. Japanese radar countermeasures therefore had little or no effect overall, although they did develop adequate radar detectors and fitted their boats with search radar from June 1944.

The principal failing was tactical, as we have observed. The Germans strongly advised their ally to attack American merchant shipping. But 'We wanted to attack the American fleet,' said Miwa simply. After 1942 the Japanese submarines gave the US fleet no serious trouble as they concentrated on helping the Army feed and supply its sporadic collection of island garrisons and bringing back the sick and wounded. Three or four managed to get to Germany and back in a generally abortive trade in vital war materials and technical items. Japan's force of operational boats never rose beyond a net total of sixty throughout the war. To communicate with their own planes and surface ships, Japanese boats were best advised to contact headquarters and have the message relayed. Little thought was given to cooperation between submarines and aircraft. By the end, the Japanese were building submarine aircraft-carriers of over five thousand tons in order to launch fighters at US carriers; the idea was Yamamoto's, and there were to have been eight. The requirement was cut to four, but even so none became operational: the superstructure was so large that they would have been sitting ducks for advanced American seaborne and airborne radar. The only one to put to sea was lost off Guadalcanal. In the dying throes of the war, the Japanese Army was trying to build its own submarines to supply island garrisons – without consulting the Navy or even showing the designs to naval experts – yet another example of interservice rivalry hampering the war effort.

Captain Genda Minoru, Yamamoto's aviation staff-planner, saw this Army-Navy rivalry as terminally damaging to Japan's entire war effort:

There was no single agency which could coordinate both the Army and Navy. Had there been someone between the emperor and the two branches of service who had the power to coordinate, I feel things would have gone much better. As it was, each branch tried to carry out operations of their [sic] own with insufficient understanding of the other branch. The war was primarily a naval war, I personally feel, but as far as the IGHQ is concerned that was not necessarily the case.

How different things were in the United States! There they could afford two separate lines of attack against the enemy. Because their two services could not agree on one or the other – still less on a supreme commander for the Pacific, even though they insisted on one (an American, naturally) for Europe – they did both. Thus the Americans took thirty-eight months to undo what it had taken the Japanese only 180 days to do by way of conquest. This has to be seen not only as evidence of the unique stubbornness of the Japanese fighting man but also as the result of Admiral King's whipsaw strategy – switching the focus from front to front to keep the enemy confused, which it unquestionably did. King argued that the double advance pre-vented the Japanese from concentrating their forces (which they could not move about very much anyway, because of American control of air and sea). But it also prevented the Americans from concentrating theirs, and they had the choice, whereas their enemy did not. In terms of miles on the map, Nimitz advanced more than twice as far in eight months – three thousand miles, from the Gilberts via the Marshalls to the Marianas, within striking distance of Japan – as MacArthur's thirteen hundred miles. The comparison is, however, unfair, because of the fundamental difference between the two campaigns. Nimitz applied immense force against a series of small islands and atolls, whereas MacArthur hacked his way across New Guinea in leaps that could be no larger than the range of his land-based fighters. Nevertheless, the US Navy had won the race in the sense that it was the first to gain a base, the Marianas, from which the Home Islands could be hit hard.

The next question was, inevitably, 'Where do we go from here?' MacArthur by this time was positively obsessed with fulfilling his promise to the people of the Philippines of March 1942: 'I shall return.' The Navy, specifically King and to a lesser degree Nimitz, wanted to go for Formosa, which was rather closer to Japan than the Philippines, and even slightly closer than the Marianas. The Joint Chiefs had been thinking for some time of successive landings on Luzon in the Philippines, then Formosa and finally the east coast of China, where they could set up any number of airfields to bomb Japan. In so doing they were behaving remarkably like the British, whose long shopping list of Mediterranean objectives looked to the Americans like a sustained attempt to avoid a frontal assault on the Germans in Fortress Europe. But as the enemy is wont to do, the Japanese had their own ideas, and in mid-April 1944 launched Operation Ichi-Go to capture the airfields already being used by Major General Claire Chennault's 14th US Army Air Force in eastern China, forcing an Allied retreat westward. This made Saipan all the more important when the time came to deploy the B-29s. It also prompted the thought that, if eastern China was perforce going to be bypassed, and if an invasion of Formosa, a very big island, looked like too large a diversion of effort after the Marianas when the campaign in Europe would be in full spate, then why not bypass the Philippines as well and go straight for the Home Islands, as General Marshall himself asked at one stage?

A Pacific strategy conference was called for 26 July 1944, at Pearl Harbor. The self-absorbed MacArthur, reluctant to admit that his constitutional commander-in-chief had every reason to take a hand in such an expensive dispute over strategy and supreme command, thought Roosevelt was merely election-eering just after his record fourth nomination. MacArthur therefore, in order to upstage Roosevelt (no mean feat), arrived late and last, after the president, in an immensely long open limousine with motorcycle escort. He strode alone up the gangplank of the president's cruiser *Baltimore* with his crumpled Philippine field-marshal's cap at the customary jaunty angle and his beat-up airman's leather jacket (had he not just flown in from Australia?), graciously acknowledging the cheers of the crowd.

There is no contemporaneous record of their meeting. MacArthur seems to have won over his chief with a political rather than a military argument: how could the United States leave a vast territory like the Philippines, for which it still had constitutional responsibility, in the hands of the enemy when it was in a position to liberate it without further delay? How would that sound on the hustings? At any rate, at the conference Nimitz argued for Formosa and MacArthur for the Philippines. MacArthur won on the 27th. He would command an invasion of Luzon, backed by the entire Third Fleet under the flag of his old friend and partner, 'Bull' Halsey. It was to be a mighty struggle which would cost the Japanese their best army and their battle fleet. As the next chapter will show, it turned out after all to be a campaign that conformed with Clausewitz's dictum about using your main strength to attack the enemy's – but only because the Japanese were as irrationally obsessed about the Philippines as was MacArthur.

The Americans were now in a position to go straight for Japan itself with their all-conquering carriers and their fleet train, their submarines, and their virtually invincible B-29s. But they chose to do it the hard way, and to do it twice. The invasion of the Philippines, which MacArthur wastefully insisted on clearing from cellar to attic, was to be followed by other huge and bloody naval diversions to the Bonin Islands and then to the Ryukyu Islands.

MacArthur also got his own selfish way in July 1944 with the Australians. Still enjoying the confidence of Prime Minister John Curtin to an extent never achieved by the unfortunate General Blamey, MacArthur issued his orders on the future deployment of Australian troops. They were to clear the Japanese out of their remaining pockets in the Solomons, New Britain and Australian New Guinea and to provide one division to reinforce the Americans after they had invaded the Philippines, with a second to follow on somewhat later. Blamey wanted these two divisions to stay together as a corps, which would have meant using them at the beginning; MacArthur therefore struck them out of his Philippine plans altogether, on the grounds that the US public would not understand a foreign army helping to liberate 'American' soil. By failing to hang together, Curtin and Blamey laid themselves wide open to being hanged separately by a foreign commander who did not scruple to misinform them, to tell them different stories or none at all. So much for one's loyal if not always uncomplaining ally. The superb, battle-hardened Australian Army was sidelined, often in futile mopping-up operations in highly unpleasant places, for the rest of the war, to the chagrin of many of its officers and men, and to the shame of a feeble government which allowed MacArthur to walk all over it. The Royal Australian Air Force and Navy were only marginally involved in the principal operations and have therefore made their last appearance in these pages.

12

The Philippines and Leyte Gulf

Having settled their internal differences about the future course of the Pacific Campaign in the usual way – a fudge which once again deferred the decision on unity of command – the American chiefs of staff accompanied the president to the second Anglo-American Quebec conference, code-named 'Octagon,' which lasted from 11 to 16 September. There another serious row broke out between Churchill and King about the British plan to return to the Pacific. A year earlier, the Americans had been pressing for more British involvement in the war against Japan (they lent Nimitz's command the new fleet carrier HMS *Victorious* in May 1943, when the US Navy was very short of such ships, a useful but hardly decisive contribution). Mountbatten had not been able to do much in the Indian Ocean region, and the planned offensive in Burma was repeatedly postponed. But what would have been welcome in 1943 was decidedly unwanted late in 1944, when Japan was manifestly on the way to defeat and the war with Germany might well end by Christmas (or so it seemed at the time). A brief British flirtation with the idea of a new Anglo-Australian command in the Pacific came to nothing. The British had powerful political reasons for wishing to be in at the death: they wanted to reclaim their Empire on the basis of participation in the final victory, even if the enemy was all but beaten already. The Americans did not want to be seen helping the Europeans recover their colonial empires and needed no help now from the Royal Navy, thank you very much. The British would actually have been wiser to concentrate in the Indian Ocean for the recovery of Burma, Malaya, Singapore and Sumatra, all in their Southeast Asia command area. The Japanese, after all, had moved a large part of their surface fleet to Singapore in February 1944 in order to be near their fuel supply in the East Indies.

King took the view that the British, having spent three years

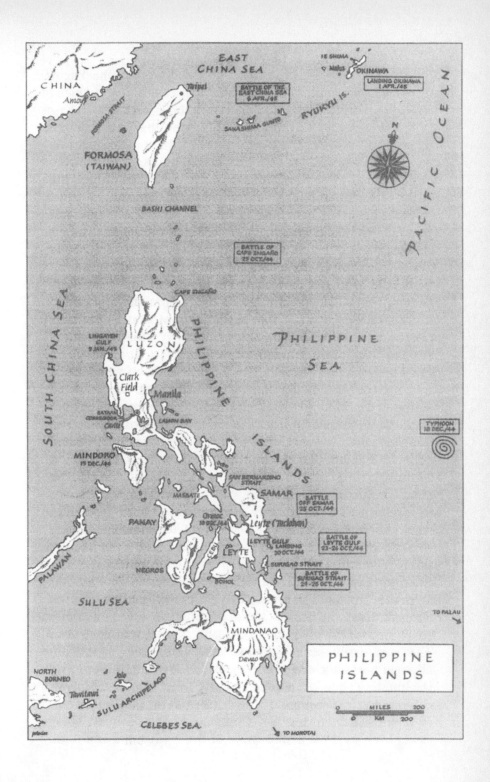

EAST
CHINA SEA

IE SHIMA

Naha OKINAWA

LANDING OKINAWA
1 APR./45

CHINA

Amoy

Taipei

BATTLE OF THE
EAST CHINA SEA
6 APR./45

FORMOSA
STRAIT

SAKISHIMA GUNTO

RYUKYU IS.

N

PACIFIC OCEAN

FORMOSA
(TAIWAN)

BASHI CHANNEL

BATTLE OF
CAPE ENGAÑO
25 OCT./44

SOUTH CHINA SEA

CAPE ENGAÑO

PHILIPPINE SEA

PHILIPPINE

LINGAYEN
GULF
9 JAN./45

LUZON

Clark
Field

Manila

ISLANDS

TYPHOON
18 DEC./44

BATAAN
CORREGIDOR
Cavite

LAMON BAY

MINDORO
15 DEC./44

SAN BERNARDINO
STRAIT

MASBATE

SAMAR

BATTLE
OFF SAMAR
25 OCT./44

PANAY

CEBU

Ormoc
10 DEC./44

Leyte (Tacloban)

NEGROS

LEYTE

LEYTE GULF
LANDING
20 OCT./44

BATTLE OF
LEYTE GULF
23-26 OCT./44

PALAWAN

BOHOL

SURIGAO STRAIT

BATTLE OF
SURIGAO STRAIT
24-25 OCT./44

SULU SEA

TO PALAU

MINDANAO

Davao

NORTH
BORNEO

Jolo

PHILIPPINE
ISLANDS

Tawitawi

SULU ARCHIPELAGO

MILES   200

Sibutu

CELEBES SEA

KM   200

TO MOROTAI

objecting to the scale of American operations in the Pacific, were now trying to muscle in on an American triumph (which of course they were). He pointed out that the British had no bases of their own nearer to the scene of action than Australia and had no fleet train; the US Navy was not about to start carrying passengers. Churchill pressed the matter, and King said it would be considered. The prime minister seethed. He had offered the British Fleet, no less, and challenged the Americans to reject it. Once again Roosevelt had to step in with a soothing word, overruling his fleet commander for political rather than strategic reasons.

But King was able to stick to his guns on the matter of logistics. The Royal Navy would have to support itself. Curiously enough, King had actually asked the Admiralty in London for help as recently as the beginning of July – specifically for the loan of six 'landing ships, infantry,' for MacArthur's Seventh Fleet by the middle of August, for eventual use in the Philippines operations. The British fell over themselves to comply, even though they were short of such ships in Europe. This led to one of the most ignominious episodes in the Royal Navy's history. The six tatterdemalion ships and one headquarters vessel of 'Force X' set out from England under the flag of Rear Admiral A. G. Talbot, DSO, RN, on 25 July 1944, across the Atlantic to the Panama Canal. The flotilla had no air conditioning, no tropical kit and precious little comfort for the crew or the troops they were expected to carry. There was a mutiny at Balboa, when the ships, already weeks behind schedule, were about to leave the Panama Canal on 1 September. Seventeen men were court-martialed when the flotilla got to the French Society Islands in mid-month. When they got to Australia, the US Seventh Fleet took one look at them, classified them as unfit for use in combat zones and decided to use them only in the rearmost areas. They eventually went to Manus, in the Admiralty Islands, the abysmal port allocated to the British as a forward base, to join the hundred ships of the British Pacific Fleet Train. They were earmarked to support the British contribution to the invasion of Japan scheduled for 1946. There can have been few cheers as heartfelt as those of the crews of these wretched and totally unsuitable ships when the news came of the Japanese surrender in August 1945.

King's attitude to British participation is fully revealed in this sharp if barely literate reply to a message from his opposite number in London, the First Sea Lord (Admiral Sir Andrew Cunningham) on 12 November 1944 (it has to be said that King used similarly withering language in exchanges with his compatriots when he was crossed):

I note your [message] designates Admiral [Sir Bruce] Fraser as 'commander-in-chief of the British Pacific Fleet.' This action sets up two naval C-in-Cs in the Pacific, which is an action not carried out in any other area or theater of action [sic], notably in the Mediterranean and in British home waters . . .

The method of employment [of the BPF] that appears to be contemplated by the command set up in your dispatch is, in my opinion, not workable . . .

I contemplate initial employment of British Fleet units under Vice-Admiral Kinkaid, the Allied Commander of Naval Forces in the South-West Pacific Area. Subsequently employment to be under Admiral Nimitz and/or Vice-Admiral Kinkaid as operations may indicate . . . I agree that a Commander (not a commander-in-chief) of the British Fleet can be set up in Eastern Australia from which British units can pass for operational control to Nimitz or Kinkaid when and if the use of forward bases can be made available. We have not as yet arranged for the establishment of such forward bases. The designation of Admiral Fraser for the duty as given in this paragraph is wholly acceptable to me.

This had the effect of marooning the highly proficient Admiral Fraser, victor in Britain's last-ever battleship clash at the Battle of the North Cape in December 1943, in Sydney, unable to exercise command at sea because he outranked Kinkaid and virtually every other American admiral he was likely to meet. Vice-Admiral Sir Bernard Rawlings therefore got this assignment, with Rear Admiral Philip Vian commanding the four carriers. Thus it came to pass that the mightiest concentration of firepower ever assembled at sea by the United Kingdom – four fleet carriers escorted by two modern battleships, five cruisers, and fifteen destroyers – became Task Force 57 of the United States Pacific Fleet. The fleet train amounted to 293,000 tons. But we are getting ahead of ourselves: the Philippines were about to be invaded, an undertaking from which both the Australians and the British, apart from a handful of their ships, were excluded in advance. As far as MacArthur (and most other American leaders) were concerned, this was a private fight, and nobody else was going to join in.

Admiral Halsey personally led Task Force 38, tactically commanded by Mitscher, from the Marshalls at the end of August to bomb the Palau Islands, which Nimitz planned to invade in mid-September; Mindanao, where MacArthur was planning to begin his recovery of the Philippines; Yap Island, northeast of Palau; and the Bonins, due south of Japan. These various raids would keep the Japanese guessing about the next American objective and weaken their airpower. Struck by the almost total lack of air opposition from Mindanao, Halsey on 13 September suggested bypassing that very large island and invading Leyte, an island between Mindanao and the main Philippine island of Luzon. As it happened, the Japanese were keeping their powder dry and had dispersed their precious planes in the jungle against the day when the Americans invaded. But the idea was put to the Pacific and South-West Pacific commands, which approved, and then to the Octagon conference, still in session in Quebec. The go-ahead was transmitted by return: Leyte was to be invaded on 20 October; Yap and Mindanao were to be bypassed. Octagon did not resolve the American dilemma between Luzon and Formosa. That was achieved at another Pacific conference in San Francisco at the end of September, which chose Luzon as the next target (for December after Leyte) and decided to bypass Formosa. Instead the Navy would deliver attacks first against Iwo Jima in the Bonins in January 1945 and finally against Okinawa in the Ryukyus in March as the last stepping-stone for the assault on Japan proper.

But before that program was agreed, MacArthur went ahead with his plan to take Morotai in the Moluccas, a group of Dutch East Indies islands between New Guinea and Mindanao. It was chosen in order to bypass the bigger, neighboring island of Halmahera, which had a large Japanese garrison and several airfields: this conformed with 'hitting 'em where they ain't,' as

MacArthur picturesquely described the process of leapfrogging. The Americans planned to put forty thousand troops ashore, two-thirds for combat and one-third logistical to build up the air base. The transport convoys sailed under heavy escort from Biak and Hollandia. The main invasion force was the 31st Infantry Division, led by Major General John Persons, from New Guinea; Rear Admiral Daniel Barbey commanded VII Amphibious Force which was to put them ashore, where there were at least five hundred Japanese troops. The landing on 15 September was unopposed, which was fortunate as the swarm of vehicles brought ashore by the troops nearly all bogged down in the exceptionally viscous mud at the southern end of the island. This almost vitiated the whole exercise because it proved impossible even for the brilliant improvisers of the US Army's Corps of Engineers to build an airstrip there. A better site was chosen in October, operational just in time to play a key tactical role in the Leyte operations and for the Australian invasion of Borneo in April 1945. Prime casualty of the Morotai operation was the submarine USS *Seawolf*, attacked and sunk in error, with the loss of all aboard, on 3 October by American escort forces, after one of their number had been sunk by Japanese submarine *RO 41*, which got away.

Some five hundred miles to the northeast of Morotai are the Palaus, where Nimitz had decided as long ago as May to occupy the islands of Peleliu and Angaur. Rear Admiral Theodore Wilkinson of III 'Phib was to deliver troops from General Geiger's III Marine Corps, also on 15 September. Nimitz turned down Halsey's proposal to bypass Palau: the expeditionary force was only two days away at the time, and Nimitz thought the bases could be useful for Leyte. In fact the invasion cost nearly two thousand American lives and was altogether superfluous. By this time the Americans were overconfident after so many successes, and Major General William Rupertus of the First Marine Division forecast victory in four days. The invaders had no knowledge of the appalling terrain, and it took two and a half months to wrest control of the island from about fifteen thousand Japanese defenders, twelve thousand of whom died in the process. The Americans were securely established ashore in a week, despite strong resistance, but it took much longer to winkle the stubborn defenders out of their network of man-made caves dug into the soft coral. Angaur, the southernmost islet of the Palaus, was invaded on the 17th; two-thirds of the 81st Army Infantry Division took just three days with only 237 men killed to overcome the resistance to the invasion itself, from sixteen hundred Japanese troops, some of whom held out in caves until late in October. By then, when the last defenders were killed, an instant American airfield had already been in operation for more than a week. On 21 September Halsey sent other troops of the 81st Division to capture Ulithi, northeast of Palau, which was wanted as an advanced naval base against the Philippines in place of Eniwetok, considered to be too far to the rear by now. There was no resistance.

Preparations for the Philippines assault also included carrier bombings of Okinawa and Formosa, aimed at local Japanese airpower. TF 38 (Mitscher in fleet carrier *Lexington II*) now consisted of nine fleet and eight light carriers escorted by a total of six new battleships, three heavy and ten light cruisers and fifty-eight destroyers, including Halsey's Third Fleet flagship, battleship *New Jersey*. Okinawa was bombed on 10 October, prompting Admiral Toyoda of the Combined Fleet, visiting Formosa, to activate the multifarious Operation Sho-Go ('victory') for the defense of the home islands; part one covered the Philippines, part two Formosa and the Ryukyus, part three southern Japan and part four northern Japan plus the Kurile Islands.

On his own initiative his chief of staff, Rear Admiral Kusaka, unwisely ordered the fleet's six carriers to send their planes ashore to help in the coming air battle. On the 11th Halsey struck northern Luzon; on the 12th he initiated three days of attacks on Formosa, where Vice-Admiral Fukudome Shigeru's Sixth Base Air Force had 230 fighters. By the end of the third attack there were none operational. Forty-two 'Bettys' from Kyushu, southern Japan, were shot down by the Americans as well, for the loss of forty-eight carrier planes on the first day. After the third day, the Americans had destroyed more than five hundred Japanese planes and done much damage to airfields and all manner of shore installations on Formosa. The Australian heavy cruiser *Canberra* and the American light cruiser *Houston II* were badly damaged by aerial torpedoes, and the fleet carrier *Franklin* was lightly damaged by a crashing Japanese bomber; all three survived. The carriers were aided in their labors by XX Army Air Force, consisting of the 20th and 21st Bomber Commands – two hundred of the new B-29s commanded by Major General Curtis LeMay, USAAF, flying from Chinese bases. Medium bombers from Chennault's XIV Army Air Force in China also joined the Philippine preparations by bombing the Hong Kong area. Several large-scale spoiling raids were executed from the carriers against targets in Luzon, the Dutch East Indies and Mindanao. The Americans lost eighty-nine planes in all, the Japanese probably a thousand – including a large proportion of carrier planes and pilots foolishly committed to the wrong battle in the wrong place at the wrong time. The prolonged and mishandled Operation Sho-II, for the aerial defenses of Formosa, was the real swansong of the Japanese carrier force, even though the ships remaining to it had not been scratched. Their turn would come shortly.

The numerical inflation indulged in by pilots of both sides may have reached its peak during the air battle, when Japanese airmen claimed to have sunk eleven carriers, two battleships, three cruisers and one destroyer, and to have damaged almost as many. There was jubilation in Japan and even a word of thanks from the emperor. Halsey was furious and sent the following message to Nimitz, which immediately went into naval lore as one of the all-time great dispatches from a commander at sea: 'All Third Fleet ships recently reported sunk by Radio Tokyo have been salvaged and are retiring at high speed toward the Japanese Fleet.' The text was promptly released in an inspired piece of PR work by Nimitz. But the Japanese preferred to believe their own propaganda. Even the Army believed the Navy enough to abandon its intelligent plan to concentrate the defenses of the Philippines on Luzon, which offered much room for maneuver, in favor of forward defense at Leyte, which it proceeded to reinforce in readiness for a 'decisive battle' against the 'weakened' US Fleet. Seldom can a belligerent's propaganda have backfired so devastatingly.

The Armada that was to return Douglas MacArthur to the Philippines was the strongest in the history of the world, with 157 combat ships and 581 other vessels of the Seventh Fleet protected by the 106 warships of Halsey's Third Fleet. The Seventh Fleet assembled at the sweltering ports of Hollandia, New Guinea (Northern Transport Group), and Manus in the Admiralties (Southern). Admiral Wilkinson's III 'Phib carried XXIV Corps (Southern Attack Force), and Admiral Barbey's VII 'Phib was to land X Corps (Northern), all Army troops from the four divisions in Lieutenant General Walter Krueger's Sixth Army. With two more divisions of reserves, garrison and logistical troops, there were 160,000 men to be landed, more than were put ashore in the opening phase of the Normandy invasion. Rear Admiral Jesse B. Oldendorf was to command all naval forces in Leyte Gulf, including the bombarding battleships and cruisers, a dozen escort carriers, the minesweepers and the survey ships, until 20 October, when Vice-Admiral Kinkaid arrived with the bulk of his Seventh Fleet. Both attack forces were to go ashore on that day at 10:00 A.M. some fifteen miles apart in the northwestern corner of Leyte Gulf, after the customary preparatory ministrations of bombardment, reconnaissance by frogmen and minesweeping. Because Halsey was lying off the northeast of Luzon hoping for a clash with the Japanese Fleet, which he tried to lure out from Japan by using his two crippled cruisers as bait, Rear Admiral Thomas L. Sprague's twelve small escort carriers had to do the bulk of the preliminary aerial bombardment from stations east of Leyte. The landings went without a single serious hitch apart from a few delays, and without meeting much resistance apart from some mortar bombs lobbed from a distance by the Japanese defenders, who at this stage comprised just one division, the 16th.

General Douglas MacArthur returned to his beloved Philippines on the light cruiser USS *Nashville*, from which he watched the first landings. After lunch he went ashore in an admiral's barge with senior commanders and the new Philippines President, Sergio Osmeña, choosing to wade ashore for the cameras even though he could easily have stepped over the bow without getting his feet wet. 'I have returned,' he told the people of the Philippines in an orotund, designer-historic speech. The old ham must have been genuinely moved, because observers said his hands were shaking. The text is too embarrassing to quote further. The US Sixth Army and its commander, General Krueger, were well established ashore, with vehicles and supplies unloaded by the afternoon of 24 October. Only a few desultory air attacks had been made during the landing operations, a torpedo-bomber badly damaging the veteran light cruiser *Honolulu*. At dawn on 21 October, however, a Japanese aircraft deliberately flew straight at the foremast of the heavy cruiser HMAS *Australia* and exploded there. The flaming debris hit the bridge, killing the captain and nineteen others and wounding fifty-four men. The kamikaze pilot had joined battle for the first time: suicide for the emperor had become a weapon of war.

The 'Kamikaze Special Attack Units' were included in the planning of Sho-Go and were a measure of desperation clearly indicating that the Japanese junta anticipated an invasion of Japan. As mentioned earlier, the word 'kamikaze' means 'divine wind,' a reference to the two storms that drove the would-be invasion fleet of Kublai Khan away from Japan in the thirteenth century. Pilots on both sides who knew they were going to die in a crippled aircraft often tried to crash their machine into an enemy plane, ship, or ground target on the principle of 'taking one with you.' The idea of individual suicide attacks had been floated in both Navy and Army from mid-1943 and had actually been spontaneously carried out by a few fanatics in 1944. It was no great departure from the medieval Bushido code. As Captain Genda Minoru confirmed:

Younger officers and enlisted fliers in operational areas, after the losses in the Marianas, tried to get IGHQ to start the use of suicide planes, saying they were ready to make the crashes. IGHQ said, however, that they should wait until a heavy enough bomb was developed . . . Vice-Admiral Onishi finally gave the first order in October 1944 to start the suicide attacks and he merely reported to IGHQ that they had been started. HQ did not know about it until then. They did not object, however.

Admiral Onishi Takijiro commanded the First Air Fleet in the Philippines at this time and had been one of the staff architects, with Genda, of the Pearl Harbor attack. He was the first formally to incorporate suicide missions in the plans to defend the Philippines and the remains of the Absolute National Defense Sphere. On 19 October, two days after taking up his new command, Onishi visited Clark Field, near Manila, and asked for volunteers among the Zero pilots to fly planes, each with a 550-pound bomb attached and set to explode on impact, directly at enemy ships. They volunteered to a man, according to Japanese sources. Nor was Onishi one of those commanders who asked his men to do something he would not contemplate for himself: he committed hara-kiri, or ritual self-disembowelment, with his samurai sword when the emperor announced Japan's surrender.

The Army developed the idea at the same time, and its first organized kamikaze attack was carried out in the Philippines on 6 November 1944, by the first of 220 Army pilots who volunteered for these terrifying flights in the second Philippine campaign. The Navy deployed twice as many there and claimed two hundred hits. Both parts of IGHQ fully supported the new death-and-glory tactic, which soon became a major nuisance for the Americans (British carriers, when they came, coped better than American because they had armored decks: carriers were naturally the priority target of the suicide pilots). Toward the bitter end of the war, crude wooden planes specially built for these one-way missions were fitted with a new one-ton bomb, and the kamikaze principle was extended to suicide boats and human torpedoes. This is what it was like to be on the receiving end of a suicidal air attack, which could be countered only by shooting the incoming plane to pieces in midair with airborne or surface gunfire:

These suicide people coming in on us are quite a problem [said Captain Thomas H. Winters, Jr., USN, of Air Group 19 aboard *Lexington II*, off the Philippines, in a contemporary interview]. They come in in a big bunch usually and then split up quite a distance out or they split up as soon as our fighters get to them, if the weather is good. They particularly like to come in on cloudy days when [there are] broken or even solid clouds, where they can duck in and out and get in close before our combat air patrols can hit them . . .

We had hit this monkey before he reached a ship and he was starting to burn but he maneuvered on down and he hit on the after end of the island . . . The personnel casualties were very heavy . . . We buried in all around fifty people from this action that day and one or two every night for a few days afterward.

We have seen how the false claim of victory in the air over Formosa led the Japanese Army to make the critical decision to try for a 'decisive battle' for the Philippines at Leyte rather than Luzon, because they thought that it would be easier to move troops about with the enemy Navy so drastically 'weakened.' Indeed the generals were puzzled by the American decision not to save a 'reduced' fleet to support an attack on the main Japanese garrison on Luzon! Field Marshal Terauchi, commanding the Southern Army (group) said on hearing of the Leyte invasion: 'The best opportunity to destroy the conceited enemy has come.' He ordered two divisions to be transferred from Luzon – the 35th Army's First and 26th Divisions, plus the 68th Independent Brigade, to aid the 16th Division already there – despite the express misgivings of Lieutenant General Yamashita, commanding the 14th Area Army in Luzon. Once again the principle of concentration of force was to be thrown away in favor of the penny-packet approach: a better move would have been to remove the 16th Division to Luzon. As far as airpower went, some 450 planes were still available on 22 October to Vice-Admiral Mikawa's Third Base (Philippines) Air Force at Manila, Fukudome's Second Air Fleet and Sixth Base Air Force, plus Onishi's First Air Fleet and Fifth Base Air Force, operating from Clark: they and Army air units frittered away their strength in a week instead of concentrating for mass attacks against a plethora of targets. Individual pilot performance was but a shadow of what it had been in the first six months of Japan's Pacific Campaign.

*   *   *

Sho-I required the Navy to come down upon the invaders of the Philippines in an 'unprecedented concentration' of Japanese seapower. The game was worth the candle: once the Americans got a firm foothold in the Philippines, the Navy would be cut off from its fuel supply in the East Indies. The consequence was the largest naval encounter in the history of the world, a triple clash which lasted three days from 23 October and goes under the name of the Battle of Leyte Gulf. The Combined Fleet had long since been a shore command like Nimitz's, so Admiral Toyoda was in remote control in Tokyo. In tactical command was Vice-Admiral Ozawa Jisaburo, the far-from-discredited loser of the Battle of the Philippine Sea, who also directly commanded his Mobile Force from carrier *Zuikaku* – the main body and Northern Force of the fleet. Under him were two striking forces specially assembled for the attempt to break up the American invasion of Leyte. The First, or Force A (or Center Force), was led by Vice-Admiral Kurita Takeo, who also directly commanded its first section from battleship *Yamato*; the second section was led by Vice-Admiral Suzuki Yoshio from battleship *Kongo*. The Southern Force consisted of two elements, each commanded by a vice-admiral: Force C, in the van, under the flag of Nishimura Shoji on battleship *Yamishiro*; in the rear was the Second Striking Force (alias the South-West Area Force or Philippine detachment, normally commanded by Admiral Mikawa at Manila) under Shima Kiyohide aboard heavy cruiser *Nachi*; with him was Vice-Admiral Sakonju Naomasa's Troop Transport Unit, led by cruiser *Aoba*. In addition to the land-based airpower already mentioned, Vice-Admiral Miwa's Sixth (submarine) Fleet deployed seventeen boats in four groups. This great fleet represented a total of seven carriers (though with only 116 planes), seven battleships, fourteen heavy and seven light cruisers, thirty-five destroyers – seventy front-line ships plus four destroyer-transports, six escort vessels, and two oilers. That it also represented rather less than a third of the tonnage of the combatant ships in the US Third and Seventh fleets reflects the way the balance of power in the Pacific had swung so decisively in the Americans' favor. As usual, the organization of this reduced but still-formidable bulk of the Combined Fleet was both *ad hoc* and unnecessarily intricate, due in part to the different starting points of the main components and perhaps also reflecting the *amour propre* of the inordinate number of underemployed vice-admirals on hand (only some of whom have been mentioned here).

Ozawa knew his carriers were no match for the American, ones not even ship for ship. Their air groups were a scratch residue, and their pilots barely half trained. But that was not the point. His plan was to use the Northern Force, coming down from Japan, as a large and suitably convincing decoy to draw the enemy's main fighting strength, Halsey with Task Force 38, away to the northeast of the Philippines. The Japanese Center and Southern forces, coming from the Linga Roads, the waters off Singapore (except for Shima's Second Striking Force, from Chinese waters), would then make two separate but coordinated and simultaneous attacks on the American invasion fleet as it lay off the east coast of Leyte from north and south of the island of Samar (of which Leyte is the southerly neighbor). In linking up with Nishimura's Force C to form the Southern Force, Shima's Second Striking Force would cross the track of Kurita's Center or First Striking Force. The two Japanese groups would approach from the western flank of the Philippines, the Americans' 'blind side.' Kurita was to pass through the San Bernardino Strait, separating Luzon from northern Samar, and Nishimura plus Shima, much the smaller group, through the Surigao Strait, between Leyte and Mindanao to the south. Their two attacks would take the Americans from the rear by embracing Leyte like an attacker applying a bear hug from behind. That was the theory; and that is how it very nearly worked out in practice. The severely outnumbered Ozawa was to prove a most worthy opponent for Halsey (especially), Kinkaid and their assorted (and providentially resourceful) rear admirals.

Kurita's five battleships and powerful supporting forces were sighted off the northwestern tip of Borneo, heading northeast at their economical cruising speed of sixteen knots, by the US submarines *Dace* and *Darter* soon after midnight on the morning of 23 October. After sending a sighting report to Kinkaid's

Seventh Fleet, *Darter* (Commander D. H. McClintock) sank the heavy cruiser *Atago*, and *Dace* (Commander B. D. Claggett) dispatched the *Maya* (proof surely that American torpedoes and technique had improved); the *Takao* was badly damaged by *Darter* and had to withdraw. Twenty-four hours later, American carrier planes on night patrol saw Kurita's ships in mid-Philippine waters west of Samar. At roughly the same time, land-based Japanese aircraft from Luzon found Rear Admiral Frederick C. Sherman's Task Group 38.3, two fleet and two light carriers, east of the central Philippines. They bombed and set on fire the light carrier *Princeton*, which finally exploded and had to be scuttled that afternoon. The cruiser USS *Birmingham*, which was trying to save her, was damaged and took heavy casualties, but the group's fighters decimated the Japanese attackers; Commander David McCampbell, flying a Hellcat from USS *Essex*, shot down nine, a world record for a single engagement.

In mid-morning on the 24th, Rear Admiral Gerald F. Bogan's TG 38.2 of three large and two medium carriers, which was not involved in this dogfight, flew his planes across Leyte to attack Kurita's ships. These had been allotted only the lightest of fighter screens by Admiral Fukudome's Second Air Fleet, which was concentrating on TG 38.3. Bogan's planes were later reinforced by Sherman's, and they proceeded to demolish the *Musashi*, 72,809 tons all up, one of the two largest battleships ever built (Kurita's *Yamato* was the other). It took all day and nearly forty direct hits by torpedo and bomb, but she turned turtle at dusk and went under. The heavy cruiser *Myoko* was driven back to Borneo with damage but survived. Kurita decided to withdraw westward, out of range of the US carriers, and regroup; he was not in retreat but only biding his time. Nishimura's Force C had also been sighted by the Americans on the 24th as it headed east toward the Surigao Strait, but none of its ships, including two battleships and a heavy cruiser, had been hit effectively by Rear Admiral Ralph Davison's TG 38.4 of two large and two medium carriers. Considering the balance of forces in the area, this was not a dazzling performance by the American carrier planes.

As he planned to step up the attacks on Kurita and Nishimura, Halsey had been wondering where the Japanese carriers were.

Meanwhile he earmarked four battleships, five cruisers and fourteen destroyers from the massive escort of his four task groups to be put under the flag of Vice-Admiral Willis Lee as a new Task Force 34. This heavy squadron would cover the eastern exit of the San Bernardino Strait if needed and, according to a follow-up signal, would constitute itself on Halsey's order. In the late afternoon of the 24th his search planes at last found Ozawa's carriers heading south from northeast of Luzon – but only after the desperate Japanese commander had sent two of them ahead to trail their coats before the American scouts. The latter reported three carriers (there were six, including two hybrids converted from battleships) and up to four battleships (there were none apart from the hybrids, which had sacrificed their after turrets for a short flight deck capable of launching but not recovering planes). The scouts' report was the 'view-halloo' for which the eager and impulsive Halsey had been waiting. Summoning all sixteen surviving carriers and all their escorts against Ozawa's seventeen ships, he led the bulk of TF 38 north in battleship *New Jersey*.

COMINCH in Washington, CINCPAC at Pearl, and Admiral Kinkaid of the Seventh Fleet aboard USS *Wasatch*, his amphibious command ship east of Leyte, had received Halsey's two messages about 'Task Force 34,' and they and Halsey's task-group commanders alike independently and unanimously assumed from them that the San Bernardino Strait was firmly closed to enemy ships by Lee's battleships. But Halsey had taken these with him to deal with Ozawa's phantom dreadnoughts. San Bernardino was wide open, and the battered but unbowed, and doubtless astonished, Kurita sailed serenely through into the Philippine Sea. Lee and Bogan and perhaps other officers had seen the danger; the two admirals made so bold as to query Halsey's decision but were ignored. Vice-Admiral Mitscher, the commander of TF 38, who often had nothing to do when Halsey was in command, was alerted by his chief of staff, Captain Arleigh Burke. On being told that Halsey had indeed been advised that Kurita was once again heading east after dark – and also that the Japanese occupation forces had put on the navigation lights in the strait – Mitscher took no action and firmly adhered to his plan for an early night. Task Force 38's

Groups 2, 3, and 4, sixty-five ships in all, were heading north at a stately sixteen knots; Vice-Admiral John McCain's Group 38.1 was under orders to join them as soon as he had finished refueling at Ulithi, to the east. This time there was to be no nuance of Spruance and his practice of overcautiously passing up golden opportunities to smash the main enemy force. That the Japanese carriers were but a shadow of their former selves after such recent disasters as the Great Marianas Turkey Shoot and the Formosa imbroglio obviously did not occur to Halsey, who could easily have afforded to leave one of his task groups behind as insurance for the invasion fleet he was supposed to be protecting. Ozawa's ruse was working like a charm.

Admiral Kinkaid, secure in the confidence that Halsey was looking after his northern flank for him, had made due and proper provision for guarding the southern approach to Leyte via the Surigao Strait. A screen of thirty-nine PT boats patrolled some sixty miles to the west of the waterway. Behind them were Rear Admiral Jesse Oldendorf's twenty-eight destroyers, with four heavy and four light cruisers, escorting the six veteran battleships, five salvaged from Pearl Harbor, of the Battle Line, commanded under him by Rear Admiral G. L. Weyler. Here, on the night of 24–25 October 1944, was a remarkable throwback to the beginning of the century, with not a carrier in sight. The approaching enemy, Vice-Admiral Nishimura's Force C, had four destroyers, two battleships and one heavy cruiser sailing in line in that order. They brushed aside the PT boats, whose contribution was to alert Oldendorf. The American destroyers attacked in waves with torpedoes, sinking two destroyers and damaging a third, which sank later, and scoring hits on both battleships. Then Oldendorf's heavy ships acted out the dream of line-of-battle gunnery admirals down the ages by 'crossing the T' of the already battered little Japanese line. The American battleships and cruisers were thus able to concentrate their broadsides on the head of the Japanese line, which could only use such guns as it could fire over the bow. Nishimura's Force C was annihilated in a quarter of an hour of shelling; only one destroyer got away. The one cruiser, *Mogami*, left the scene seriously damaged. If the Battle of Surigao Strait is seen as a separate action rather than as a part of the multiple clash of Leyte Gulf between the main bodies of the American and Japanese fleets, then it stands as the last of its kind, a fight between line-of-battle ships decided by heavy guns. One US destroyer was damaged.

The alert reader will have been wondering what happened to Admiral Shima's Second Striking Force, which was supposed to join Nishimura. It chose now, the earliest hour of 25 October, to arrive west of Surigao Strait – whereupon Shima's flagship, heavy cruiser *Nachi*, collided with the retreating *Mogami*. The vice-admiral proved to be a dud; he withdrew and took no further initiative. The *Mogami* ran out of luck altogether when American air attacks resumed in the area and sent her to the bottom. The *Nachi* met a similar fate. Kinkaid was well pleased at next morning's staff meeting. His delight turned to horror when he heard from Halsey, in reply to a cautionary query about the whereabouts of Task Force 34, that it was up north with him in the hunt for the Japanese carriers. Having digested this with his breakfast, Kinkaid was told half an hour later that the screen of little escort carriers and destroyers covering the Leyte beachheads were under attack from battleships and cruisers. Kurita had arrived. It was 7:20 A.M. on 25 October.

The brunt of Kurita's attack was borne by the Seventh Fleet's Task Group 77.4, which consisted of three squadrons, each of six escort carriers (merchant ships converted to carry about thirty planes) in two divisions of four and two, protected by a screen of seven or eight destroyers – a total of eighteen small carriers and twenty-three destroyers or destroyer escorts. The group was commanded by Rear Admiral Thomas L. Sprague; the first squadron, known by its radio call sign as 'Taffy 1,' was under his direct command, the second was under Rear Admiral Felix B. Stump, and Taffy 3 was led by Rear Admiral Clifton A. F. Sprague (no relation, strange to relate). The group carried fighters and short-range bombers and was in Leyte Gulf to provide close support for the invading troops. It was Clifton Sprague's Taffy 3 – four carriers under his flag and two more under Rear Admiral R. A. Ofstie, protected by Commander W. D. Thomas's screen of three fleet and four escort destroyers – which came under attack from Vice-Admiral Kurita's First Striking Force. This comprised Center Force A – super-battleship *Yamato*, battleship *Nagato*, three heavy cruisers, one light cruiser, and nine destroyers – and a second section of two battleships, four heavy cruisers, one light cruiser, and six destroyers led by Vice-Admiral Suzuki Yoshio. This considerable force had swept through the San Bernardino Strait and turned south, sighting Clifton Sprague's Taffy 3 just as the latter sighted Kurita's ships twenty miles to the north.

Clifton Sprague withdrew to the south, launching all available planes 'over his shoulder' to attack Kurita, who thought he was faced with fleet carriers and therefore sent in his battleships in an unnecessarily desperate 'general attack.' The Japanese fleet was thus reduced from a formation under united command to a collection of individual ships worth much less than the sum of its parts. Had Kurita kept his head and formed up his line of battle to exploit his locally overwhelming gunpower against the light forces arrayed against him, he could have got into the gulf, beaten up the invasion fleet and seriously embarrassed MacArthur, whose huge forces needed constant and massive supplies.

The Taffy 3 planes were armed only for ground-support work, yet this did not deter their pilots from attacking with everything they had, most of which bounced off. But Admiral Stump was able to send in aircraft armed with torpedoes from Taffy 2, the neighboring squadron to the south, to help his hard-pressed colleague eventually to sink three Japanese cruisers. The Taffy 3 destroyer screen boldly mounted three torpedo attacks to disrupt the Japanese battle fleet, which had nearly twice the speed of the 'Woodworth carriers' and was trying to hit them with shells of calibers up to 18.1 inches (*Yamato*'s). Destroyers *Hoel* and *Johnston* and destroyer escort USS *Samuel B. Roberts* were sunk. To avoid the torpedoes in the mêlée, the *Yamato* reversed course and disappeared to the north, removing the commanding admiral from the battle. Taffy 3's carrier *Gambier Bay* was set on fire and sunk by gunfire. As if this were not enough for the hard-pressed squadron, it was chosen at the same time to be the first target of a massed kamikaze attack, which scored six hits on Clifton Sprague's ships. One suicide pilot put his Zero into a dive on the escort carrier *St Lô* and plunged through her flimsy flight deck into the holds below, igniting her fuel tanks and torpedo store. She succumbed to the tearing explosions and sank in half an hour, the first total loss to kamikazes.

Kinkaid, meanwhile, was trying to get help to the Spragues from the scattered forces of the United States Navy, spread over more than three thousand square miles of the Philippine Sea. Oldendorf's elderly battleships were more than three hours away to the south and could not make it in time, although they tried. Halsey, meanwhile, had restored tactical command to Mitscher, who was busy launching his attack on Ozawa just as Kinkaid first queried Halsey on the position of 'Task Force 34.' Admiral McCain's Group 38.1 had completed refueling at Ulithi and was told to head due west, toward the Philippines, instead of racing northwest to join the rest against Ozawa. Later Halsey detached Bogan's Group 38.2, which reversed course and headed south toward Leyte. So did a cobbled-together and reduced Task Force 34 – Task Group 34.5 – of two battleships, three light cruisers, and eight destroyers, which arrived the best part of a day after the battle. Small wonder: Halsey completely misread the relative significance of the Japanese formations and had wanted to keep all his battleships to shoot up whatever of Ozawa's force his carrier planes might leave afloat. He reacted to Kinkaid's calls for help only when the Seventh Fleet commander sent a signal of apparent desperation in plain English at 8:22 A.M. As intended, this was picked up and exploited with remarkable speed by Tokyo Rose. Kinkaid's message was a ruse to alarm Kurita as much as to inspire Halsey with a manifestly lacking sense of urgency.

Half an hour later, Halsey diverted McCain's group. But it was only one full hour after getting a message from Nimitz himself that Halsey at last detached Willis Lee and Task Force 34 was born. Even then the line of battle thus formed headed south at a less-than-urgent twenty knots and then slowed down to a crawl to refuel destroyers in the afternoon, showing that Halsey's orders were not altogether pressing. Only after that was TG 34.5

detached by Lee to 'rush' ahead on a course parallel to Bogan's carriers to the east. America's new battleships never fought a gun action against the Japanese Fleet, which by lunchtime on 25 October 1944 was effectively a spent force.

Neither carrier group influenced the outcome at Samar – Bogan's because it was far too late, and McCain's because, to save time, he launched his strike at the extreme range of 335 miles, which meant his planes could not carry torpedoes or full bomb-loads. Their arrival over the scene of what became known as the Battle off Samar convinced Kurita that worse would soon follow; but he had already ordered a general withdrawal northward, back to the San Bernardino Strait, three-quarters of an hour earlier, at about half past noon. In any case, he had believed all along that he was embroiled with the fleet carriers.

Two and a half hours into the Battle off Samar, Admiral Nimitz at Pearl Harbor, goaded beyond endurance by the disjointed scraps of information coming from what was obviously a major action in Philippine waters, and suspecting the worst, ordered the dispatch to Halsey of the following simple message: 'Where is Task Force 34?' Unfortunately, sending such a straightforward signal was complicated. It had to be correctly addressed and then enciphered, and to protect it from possible enemy penetration it had to be topped and tailed by a few words of nonsense, known as 'padding,' which were not supposed to have any connection at all with the real message between them. The signal in its final form therefore opened with the words 'Turkey trots to water,' gave its origin as CINCPAC and the addressee as COM THIRD FLEET (Halsey). The actual message asked, 'Where is, repeat where is, Task Force 34?' The yeoman of signals who took down the text from Nimitz had detected a certain asperity in his voice and inserted the repetition on his own initiative, sharpening the impact considerably. And as a literate man he was aware that 25 October was Balaclava Day. That reminded him of the gallantly futile death-charge of those unwilling kamikazes on horseback, the British Light Brigade in the Crimean War, nearly ninety years earlier, as inaccurately extolled by the purple poetry of Alfred Lord Tennyson:

When can their glory fade?
O the wild charge they made!
All the world wonder'd.

So, to complete the confusion of any enemy eavesdropper, he added the required second piece of padding to the message: 'The world wonders.' The top and tail would surely be recognized as the usual camouflage and discarded at the receiving end. The top was – but the tail, appearing syntactically to be part of the message itself, which should never have been allowed to happen, was not. The communications staff on USS *New Jersey* overlooked the pair of letters signifying the end of the real text and the start of the nonsense. The signal handed to Halsey at 10:00 A.M. therefore read: 'Where is, rpt where is, Task Force 34, the world wonders?' Thus was Nimitz's straightforward query transmogrified into what Halsey took as a monstrous insult and never forgave. In fact the very ferocity of his reaction can be seen as betraying his subconscious or unacknowledged realization that he had made an almighty and unforgivable blunder, which a more determined Kurita could well have exploited to disastrous effect against the exposed American invasion fleet. Ozawa's ruse had worked better than he could possibly have imagined. It is time to consider the price he was made to pay for his success, even though it was so miserably underexploited by his hesitant colleagues to the south: the Battle of Cape Engaño, third act of the great naval Battle of Leyte Gulf.

Mitscher's night fliers had rediscovered Ozawa's seventeen retreating combatant ships as early as 2:00 A.M. on the 25th, about two hundred miles east of Cape Engaño, the northeastern tip of Luzon Island. The first American wave was assembled by dawn at 6:00 A.M., circling over the carriers while the scouts were away picking up the contact of four hours earlier. They found it 150 miles to the north just after 7:00 A.M., minutes before Kurita's and Clifton Sprague's forces sighted each other five hundred miles to the south and began the most intense action of the three. Battle was joined by 8:00 A.M., and thoroughly one-sided it proved to be. Ozawa knew he was going to be beaten; he had only a handful of planes left with which to oppose 787 American aircraft (of which 527 were actually committed). In the first American strike one carrier was sunk and one damaged. In the succeeding five – each less ferocious than the last as the unseen American fleet began sending reinforcements to Samar and Leyte – Ozawa lost all three remaining carriers and one destroyer, including his sole fleet carrier, *Zuikaku*, last survivor of the six Pearl Harbor veterans. Submarine *Tama*, on her maiden patrol, one of several US boats in the area, sank a light cruiser, and TF 38's cruisers added two destroyers to the tally, which, given the overwhelming preponderance of the Americans, seems unsurprising. The two hybrid battleship-carriers were among the nine ships that got away, but they made no further contribution to the Japanese war effort. Ozawa got away too, in a light cruiser, after what he permitted himself in his action report to describe as 'a bitter experience.' In May 1945 he became the last C-in-C of the residual Imperial Japanese Fleet, by then strategically worth no more than any other large collection of scrap metal. The most talented of Hirohito's admirals died in 1966, aged eighty.

The sheer numbers of ships engaged in the three-stage Battle of Leyte Gulf tend to obscure the quality of the result which, given the opportunities presented and created by both sides, seems curiously anticlimactic. Though much larger than Midway or indeed any other naval battle, it was tactically nothing like as decisive, because so many Japanese ships got away when they should not have done. Strategically the outcome only confirmed, and not as powerfully as it might have done, what everybody knew already: that the US Navy was much stronger than the Japanese, and that the Imperial Navy no longer presented a serious threat because its air arm had withered. In the three main bouts and associated actions, the Americans lost one light and two escort carriers, three destroyers, three submarines (including *Darter*) and a few dozen planes, in depriving the Japanese of one large and three medium carriers, three battleships, six heavy and four light cruisers, eleven destroyers, four submarines, and 116 carrier aircraft. The Japanese lost ten thousand men to the Americans' fifteen hundred. Victory therefore went to the US Navy, superior in every department except strategy, but the brunt was borne by a collection of auxiliary ships, Clifton Sprague's escort carriers and destroyers, the true heroes of Leyte Gulf.

Had Kurita properly exploited Ozawa's success, one wonders what would have become of Halsey, who insisted to his dying day that he had done the right thing. Luckily he had an order from Nimitz to make the destruction of the enemy's main body his chief objective if the chance presented itself, and the main body was always regarded as the Mobile Force of carriers. He could never bring himself to concede that this once-mighty striking force had fallen on hard times and been thrown away – successfully – as a lure, still less that by leaving the San Bernardino Strait unguarded he had compounded Ozawa's intellectual victory over him. He insisted that Kurita's force (the real main body on this occasion, though Halsey would not see it as such) had been sufficiently damaged on its approach to ensure that Kinkaid's forces could handle it. It is no exaggeration to conclude that Clifton Sprague and his men saved Halsey's reputation. We may leave the last word on the Battle of Leyte Gulf to the victor of act one, the newly promoted Vice-Admiral Jesse B. Oldendorf. Asked by the press whether a victory by six battleships with strong support over two with weak support was really all that remarkable, Oldendorf said that the enemy may have been foolish to attack him with an inferior force but he 'used the old gambler's adage with regard to the Japanese. That is, if you have the strength, use it,' he said. 'Never give a sucker an even break.'

Buoyed up by the wonderful news from the Philippines, President Roosevelt did something that he had never done before: he called an impromptu press conference at the White House in the late afternoon of 25 October, Washington time, to announce the 'rout' of the Japanese fleet. On 11 November he was re-elected for a record, unique fourth term, comfortably defeating his Republican opponent, Governor Thomas E. Dewey.

During the campaign Dewey had seriously considered attacking

Roosevelt over the Pearl Harbor debacle, which still rankled despite the subsequent accumulation of victories. It was, after all, the only election to occur during American's belligerency, and no such chance was likely to recur. Dewey thought Roosevelt vulnerable on the issue and planned to exploit the 'common knowledge' that America had broken Japanese codes and ciphers to accuse his opponent of foreknowledge and criminal negligence. Getting wind of this, General George Marshall, Army chief of staff, also did something he had never done before: he interfered in party politics. He sent an aide, Colonel Carter Clarke, to Dewey with a letter he had personally written and signed. It underlined the crucial importance of Ultra and Magic to the results of the battles of the Coral Sea, Midway, the Aleutians, submarine operations and the general conduct of the war in the Pacific. Marshall denied that there had been anything pointing to Pearl except for the low-grade Honolulu consulate message, deciphered only after the event. Clarke told Dewey that Churchill believed Ultra had saved Britain and that the Japanese had not substantially altered their codes and ciphers, so that the cryptanalysis operation remained as important as ever to the war effort. Dewey took the statesmanlike course and backed off; the greatest intelligence secret survived the war by a narrow margin. Roosevelt had been lucky in his wartime opponents; in 1940 Wendell Willkie had similarly forborne to make an issue of the president's foreign and defense policies when the isolationists were still strong.

Having decided to fight the decisive battle for the Philippines on Leyte, the Japanese Army, undeflected by the failure of the fleet to cut off the American beachhead, took advantage of the geography of the archipelago to reinforce their 16th Division. From Luzon and other islands and from farther afield, including China, they sent in the entire First and 26th divisions, which, with miscellaneous supporting units, brought the strength of the corps defending Leyte to about sixty-five thousand by November – against the advice of General Yamashita in Manila. This 35th Army was commanded by Lieutenant General Suzuki Sasaki. Even so, General Krueger and the Sixth Army's two corps (four divisions) numbered twice as many, with two divisions in reserve. Weaving in and out of the islands, convoys of transports and trains of barges landed Japanese troops almost at will, usually at Ormoc, which stands on a southward-facing bay on the western side of Leyte.

The Americans were unable to halt this quite remarkable effort because the soil of Leyte made it particularly difficult to build airstrips for land-based planes and the Navy's airpower was either tired or overstretched or both. The Battle off Samar and the continuing attentions of the kamikazes markedly reduced the effectiveness of the escort carriers. The fleet train was having difficulty satisfying TF 38's unquenchable thirst for fuel and other supplies at such great distances from major American bases, Ulithi being a thousand, New Guinea fifteen hundred, Eniwetok twenty-five hundred, and Hawaii five thousand miles away. The Japanese threw in aircraft from Luzon, Formosa and Japan itself to fight for the local airspace from landward bases already built. And instead of focusing all its strength on the major objective of the Philippines, the divided American high command wanted to launch large new invasions of Iwo Jima and Okinawa islands at the same time as pressing ahead with the Philippines campaign, which would require nearly fifty amphibious landings by Admiral Barbey's VII 'Phib before it was over.

The struggle for Leyte soon boiled down to a battle for Ormoc, in the mountains between it and Leyte Gulf to the east. Overcaution by Krueger and intelligent opportunism by Suzuki allowed the Japanese to turn these into an impassable, heavily fortified barrier. The hardest struggle of the unexpectedly prolonged fight for Leyte developed in November between the Japanese First and the American 24th divisions at what the Americans named 'Breakneck Ridge.' Only when MacArthur postponed the invasion of the island of Mindoro, northwest of Leyte and south of the Luzon mainland, by ten days was it possible to outflank the Japanese by sending the 77th Division to invade Ormoc Bay, to their rear, at the beginning of December. Because the Japanese had mounted a determined but ill-coordinated series of attacks, including parachute drops and commando raids, on American bases in eastern Leyte, the

77th landed unopposed and marched inland to take Ormoc itself on 10 December. This soon put the Americans in full control of an east-west barrier across Leyte, which cut the Japanese defense in half and doomed it to piecemeal defeat.

The Japanese decision to fight a decisive battle for the Philippines on Leyte – an ill-considered forward-defense strategy which once again frittered away formidable strength – delayed MacArthur's invasion of Luzon by twenty days. This breathing space was bought at the expense of half the Imperial Fleet, three thousand aircraft, tens of thousands of fatal casualties – and the ability to mount a successful defense of Luzon and Manila itself. Sho-I was not looking good, which is hardly surprising, because the Navy had not told the Army of the plan that led to the Battle of Leyte Gulf. Further, the two services had a furious standing dispute about the allocation of the dwindling stock of tankers, and within each service there were rows between staff and field over all categories of supplies. When they learned the truth about the Navy's losses, the generals left the fleet, except for its land-based aircraft, out of their plans: the defense of the homeland, the real 'decisive battle,' would fall on the shoulders of the Army, as was only right and proper.

The junta deluded itself into thinking that Japan could hold out indefinitely by hanging on to the Home Islands, Korea, Manchukuo and adjacent China. In January 1945 the Cabinet drew up a plan for 'redistribution of industrial capacity to meet changes in the war situation' – the first concrete result of which was the laborious transfer from Japan to Manchukuo of two blast furnaces, each with a capacity of 350 tons, to facilitate local manufacture of iron and steel. Underlying such schemes was the need to reduce transport requirements to a minimum. One can only marvel at a cast of mind so rigid as to think that such sticking-plaster measures could affect the intentions of the two-headed juggernaut lumbering so invincibly yet cautiously toward the heart of the Empire. The junta, noting with satisfaction the undoubted effect of the new kamikaze campaign, persisted in its belief that the Americans would not have the stomach for the kind of hard fighting that would follow an invasion of Japan. The generals were more right than they knew, and their country was to pay a terrible price for this indomitable obduracy. Faced with such an immovable object, the Americans would cut the Gordian knot and resort to irresistible force.

Mindoro was invaded from Leyte on 15 December to provide forward naval, land, and air bases for the assault on Luzon, now rescheduled for 9 January 1945. Yamashita in Manila seized on this as justification to write off Leyte and to plan for a prolonged holding operation on Luzon: with his best troops now marooned on Leyte, he could not hope to defeat an invasion but he could draw very large forces into the interior of his big island and tie them down for a long time – as it turned out, nearly six months. IGHQ confirmed his decision on the 18th. Air cover for the Mindoro landing, virtually unopposed on land, was provided by the Seventh Fleet's overworked escort carriers, better suited for operations in confined waters than TF 38's big fleet carriers, which continued their general attacks on Japanese air and ground targets in the central and northern Philippines. The main threat to the invasion fleet came from the kamikazes, which knocked out but did not sink several warships, including the flagship, cruiser *Nashville*, and destroyed two landing ships altogether. Two fighter bases were ready for use by 28 December.

On 18 December 1944, another, older enemy put in an appearance east of the Philippines and defeated Admiral Halsey, caught unawares as he tried to meet the desperate need of his destroyers for fuel, which they had to draw from the larger ships they were there to protect. A typhoon blew up very suddenly out of the previous day's high winds and heavy swell, which had already made fueling first difficult, then impossible. TF 38 was busy supporting the Mindoro operation and bombing Luzon, which was due for another heavy visitation on the 19th. The forecasters, not so well equipped as their modern successors, knew the storm was in the offing but misread its track. In trying to sail his fleet southward out of danger overnight, Halsey passed straight into its path, his ships scattered across three thousand square miles. At the height of the storm, in the middle of the day, he issued a typhoon warning. It was 1:45 P.M., and three destroyers – *Spence*,

Hull, and *Monaghan* – had already been lost; seven other ships were seriously damaged, and the carriers lost 146 planes as they pitched and rolled like dinghies in the giant waves. Those ships that tried to fight the storm seem to have fared worst, and some captains lost sight, or were ignorant, of the rules for survival in such atrocious conditions. (A typhoon, whichever direction it follows, is a circular system of winds which in the Northern Hemisphere blow anticlockwise round the eye of the storm. This means that, if you face the wind, the eye is always about ten points of the compass to your right: which is to say, if you look into the wind and point your arm to the right straight out from the shoulder and as far back as it will go, you are probably pointing at the center. The alternative to avoidance is to heave to and ride it out.) Water Tender, Second Class, Joseph C. McCrane was one of just six survivors of USS *Monaghan*, a ship that had lived through Pearl Harbor and everything else that the Japanese could throw at her.

The storm broke in all its fury. We started to roll, heaving to the starboard, and everyone was holding on to something and praying as hard as he could. We knew that we had lost our power and were dead in the water . . . It was suggested that someone try to make their way up to the bridge but none of us had the courage to go out into the storm . . . We must have taken about seven or eight rolls to the starboard before she went over on her side. [McCrane managed to cling to the one raft of survivors that was found on 20 December.]

Every time we opened a can of Spam more sharks would appear . . . Toward evening some of the boys began to crack under the strain. One fellow insisted on biting on the shoulder of another . . . That [second] night most of the fellows had really lost their heads; they thought they saw land and houses. [One man tried Spam as fish bait.] He tried again and let the line down about ten feet below the surface of the water. A shark about five feet long came up slowly for it and as it did, [he] pulled the line in slowly and the shark followed until [he] had the bait out of the water and the shark had its head right up on the side of our raft. [Another man] took his penknife and plunged it into its head . . . We finally gave the fishing up as a bad job.

A court of inquiry sitting at Ulithi in the last week of the year blamed Halsey for the damage and the losses. The three admirals, however, found no negligence but only 'stress of war operations' and 'a commendable desire to meet military requirements.' Having studied all the evidence, particularly the weather information available to Halsey, Morison concluded that this was unfair to the 'Bull.' The oddest fact seems to this author to be that Halsey was not aware that he had lost any ships until a destroyer reported picking up survivors at 2:25 A.M. on the 19th – twelve hours after the typhoon had passed. The Navy improved its weather service, and Nimitz sent a Letter to the Fleet full of good advice on safety at sea. The three destroyers were probably lost because their oil and ballast were out of trim as a result of the interrupted refueling process. TF 38 went back to work pounding Luzon.

MacArthur's plan for the invasion of Luzon on 9 January at Lingayen Gulf, on the 'thumb joint' halfway down the western side of the mitten-shaped island, involved more troops than were put ashore in the first Normandy landing. It was the biggest amphibious operation of the Pacific Campaign, and the largest purely American one of the entire war, involving 175,000 men to be put ashore in a few days on a beachhead twenty miles wide. The first attack was made by two corps, I and XIV; these four divisions eventually rose to ten plus ancillary units – two armies in an army group – in the campaign to recapture Luzon, which was not declared secure until 30 June 1945.

But they had to get ashore first, and the kamikazes did their best to make this as unpleasant a task as possible for the naval delivery service, covered on this occasion by Admiral Oldendorf's supporting force of bombarding battleships, escort carriers providing fighter cover and a large protective screen of cruisers and destroyers. As ever on both sides when there was a choice, men-of-war were seen as more important targets than what they were protecting – vulnerable transports and supply ships. So escort carrier *Ommaney Bay* blew up and sank when a Zero was flown through her flight deck, a destroyer was lost similarly, and two dozen other warships were damaged by suicide attacks, including another 'Jeep' carrier, two battleships, three cruisers and smaller vessels. Oldendorf asked for protection by the Third Fleet, with its large numbers of more modern fighters, but by the end of the first week after the invasion of Lingayen

Bay he had lost a total of two dozen vessels with sixty-seven of all types damaged. Only then did the kamikaze onslaught ease off, a victim of its own 'success': it had run out of planes. Since the US Navy stayed put and the invasion itself went remarkably smoothly, meeting little opposition until it began to push inland, the suicide pilots had no strategic success at all.

When MacArthur ordered Krueger to press on south to Manila, regardless of the threat to his left flank posed by Japanese units well dug in in the mountains, the latter assumed his boss wanted to be there in time for his sixty-fifth birthday on 26 January. XIV Corps got to Clark Field, north of the capital, three days before that, but was then held up for a week by a stubborn rearguard action. The 37th Division left the 40th behind and made haste slowly southward. Impatient with Krueger and ever-ready to interfere in the chain of command, MacArthur personally ordered the freshly landed, light First Cavalry Division to race ahead to Manila. The 11th Airborne Division from the US Eighth Army (led by Lieutenant General Robert L. Eichelberger, MacArthur's best general) landed southwest of Manila and also dashed for the city in what had become a three-horse race – until they came up against a Japanese Maginot-style defense line four miles out, where they were halted and pinned down. A flying column of the First Cavalry won the race and became the first US unit to reach the heart of the city, on 3 February. But Manila still had to be cleared of its Japanese defenders.

The Imperial Navy, which had the equivalent of a division of sailors equipped as infantry in the city, had independently decided to fight for what the Army was wisely inclined to abandon, so the Army made the best of a bad job by joining in (General Yamashita might otherwise have abandoned the capital to get on with the main business of holding out in the north of Luzon, where he even planned to grow his own crops to feed the troops). In battering down the resistance, MacArthur's troops came alarmingly close to killing the patient in order to effect a cure. House-by-house street-fighting ensued for nearly a month, with plenty (in fact, rather too much) tank and artillery support for the Americans. Thus one hundred thousand civilians out of a total population of eight hundred thousand were killed in the process of being saved from Japanese oppression. Their relatives could be forgiven for regarding the liberation of their smashed city as a mixed blessing: it looked no more substantial afterward than Caen – the capital of Normandy, on the other side of the world – where the Germans had staged their toughest initial resistance against the Allied invaders. Manila was declared secure only on 3 March 1945, five days after Corregidor, of bitter and glorious memory, had fallen back into the hands of the United States Army. On 25 February MacArthur briefly reclaimed his penthouse suite in the Manila Hotel, which had been pillaged but not destroyed by the Japanese occupiers.

MacArthur now proved that, when he had said 'I shall return,' he meant it to apply to every nook and cranny of the huge Philippine archipelago. His troops had secured almost every strategically relevant point and could have been diverted to more important matters, such as the planned invasion of Japan. But General Eichelberger's Eighth Army spent the rest of the war jumping in and out of landing craft, making one amphibious assault after another, thirty-eight in all, each more remote than the last, recapturing 'bypassed' islands and tying down enormous military and naval forces that SOWESPAC might otherwise have been forced to disgorge. MacArthur's defenders argued that it was America's duty to relieve the suffering of those still oppressed by Japanese garrisons, however isolated; and that his double campaign to the north and the south of Manila led to the destruction not only of the Japanese 35th Army but of the entire 14th Area Army as well, a third of a million troops in all. But the Americans had told the British on the Combined Chiefs of Staff that local liberation would be left to the considerably swollen ranks of Philippine guerrillas; nor did the US Joint Chiefs order or even authorize this backward-looking conquest, which added 1,910 dead and 5,970 wounded to the totals given below. And when Australia's General Blamey used the same argument to defend his painstaking liberation of remote areas of the Solomons, the Bismarcks and New Guinea, MacArthur dismissed it as a waste of effort which might have been better used in Borneo, the last notable and separate military operation of the South-West Pacific command. But then the US general was particularly partial to having things both ways. Only General Yamashita, who was to the Army as Admiral Ozawa was to the Navy – Japan's best – caused a serious problem in the Philippines with his stubborn and skilled defense of northern Luzon, which triumphantly achieved – for what it was worth – its limited objective of tying down disproportionate US forces. Since these were led by the plodding General Krueger with the Sixth Army rather than the dashing Eichelberger, the delaying tactics worked even better.

A proper sense of priorities would have put the better general in command of the much tougher northern front, and would have combined rather than divided the available US forces to impose an early resolution of a strategically pointless fight for northern Luzon, which officially came to an end only on the last day of June. To the south, Mindanao, the largest remaining objective in the Philippines, finally fell on 27 July. The entire land campaign for the recapture of the Philippines cost the United States 10,380 killed, 36,631 wounded – and 93,400 sick or accidentally injured, a sickness rate unmatched on the Allied side in the Pacific, not even on Guadalcanal or in New Guinea.

As the first American troops established themselves in Manila at the beginning of February, the second summit meeting of the 'Big Three' – Roosevelt, Churchill, and Stalin – was taking place at Yalta, in the Soviet Crimea. Germany was obviously about to go down to defeat, although its Army, whose last offensive narrowly failed to break through the American line in the Ardennes in December 1944, was still fighting hard on both fronts for what remained of the fatherland. Roosevelt pressed Stalin to join in against Japan, but the Soviet dictator stuck to his previously stated intention of declaring war three months after the defeat of Hitler. Churchill was virtually a spectator, with few cards to play in this bargaining session between the postwar superpowers. An acquisitive Stalin wrung territorial concessions in China and Eastern Europe from an exhausted and unwell Roosevelt. Because American strategy was committed to the great maritime thrust toward Japan across the Central Pacific, the Philippines became an expensive sideshow after the capture of Manila and need concern us no further.

*   *   *

Douglas MacArthur would have played no part at all in the war but for Roosevelt's eleventh-hour resurrection and subsequent nurture of his career, a progress to unearned glory which even Hollywood in its heyday would not have dared to invent. He was the president's chosen political lightning-conductor against attack from the always powerful ultraconservative element in the United States. It is Roosevelt who has to bear the historical responsibility for choosing a mountebank to lead what should have been the main thrust against Japan, and then seeing to it that this became a campaign of secondary importance to his beloved Navy's separate, wasteful, atoll-hopping counterattack across the broad Pacific. It was not done to slake MacArthur's insatiable thirst for glory, although that was the secondary consideration cunningly disguised as primary; it was done to protect Roosevelt himself from more dangerous right-wing enemies by giving them their 'man on horseback'; they did not notice that the horse had a long but unbreakable leading rein. The general who did not hesitate to manipulate and undermine superiors and subordinates, foreign governments, the American Navy, facts and figures and his own carefully fostered image was being manipulated himself by a much cleverer operator all along. How the president must have laughed when Governor Dewey claimed during the 1944 campaign that MacArthur had been starved of resources – only days before the colossal invasion of Luzon. Had MacArthur been a real threat he could have been destroyed by leaking the staggering story of President Quezon's $500,000 gift (MacArthur's naïve acceptance surely proves what an unequal opponent he was for Roosevelt in the political field).

But monuments to glory, even if they have feet of clay, cannot be casually demolished without danger from unpredictable falls of debris. The charade, once begun, had to continue, and therefore Douglas MacArthur, showman *sans pareil* among military leaders, will, inevitably, make his most majestic appearance in the final chapter of our story.

13
'. . . Not Necessarily to Japan's Advantage . . .'

Before becoming a belligerent, the United States, like Britain, had condemned the bombing of any targets other than the strictly military, whether in the Spanish Civil War, Abyssinia or in Japan's war in China beginning in 1937. This attitude changed dramatically after the Anglo-American powers went to war. When 'precision' bombing proved usually to be a contradiction in terms and an unattainable chimera, Germany was subjected to what was euphemistically known as 'area bombing' by the RAF at night and the US VIII Army Air Force by day, with thousand-bomber raids becoming almost commonplace. Little time was wasted regretting the indiscriminate killing of civilians resulting from this policy of what the Germans not incorrectly came to describe as 'terror-bombing,' not even when incendiaries were used on big cities like Hamburg and Dresden. The atrocities committed by the Axis powers made this generalized barbarism all the easier to rationalize. Aerial bombardment was a substitute for the invasion which the British could not undertake against the Germans, or the Americans against Japanese home territory, until late in the war: it was the only way to strike back against the aggressor's territory before grasping the nettle of attacking his army. Democratic politicians needed to be seen to hit back as soon as possible, for reasons of morale: hence the Doolittle Raid, which was not followed up for more than two years.

The revolutionary B-29, whose specifications we have already examined, first flew in September 1942, and the US Army immediately ordered 250. They were not used in Europe but were reserved exclusively for Japan, which was to be bombed from bases in China by the specially formed 20th Bomber Command from the summer of 1944. First used against Bangkok, the B-29s made their opening attack on Japan on 14 June 1944. Teething troubles, poor early results and Japanese attacks on the Ch'eng-tu region, where the bases were, all held up the sustained offensive. The campaign intensified after thirty-nine-year-old Major General Curtis LeMay took over command, but at first it had no significant effect on the war effort in Japan itself, any more than the generalized bombing of Germany had much of an effect on its war production except at the very end (the lesson had still not been learned by the mid-seventies, when the USAF was carpet-bombing in Indochina). The B-29s were used to support Allied and disrupt Japanese offensives late in 1944 in Formosa, the Philippines and China. In a remarkable but sadly ignored demonstration of the suppressed art of low-level precision-bombing, two B-29s came down from an area raid and in a low-level run neatly knocked out the main dry dock at Singapore to close it to the Japanese fleet. The new American M-69 napalm incendiary bomb was first tried out on Hankow, in China, against a Japanese advance in December 1944. Large numbers of friendly Chinese citizens were incinerated.

The first B-29 raid on Japan from Saipan in the Marianas, by Brigadier General Haywood Hansell's new 21st Bomber Command, took place on 28 November, just as Halsey's carriers began launching raids on the Home Islands. Hansell's men, bombing from maximum altitude and encountering stratospheric winds for the first time, often had to rely on radar to 'find' their objective and missed more often than they hit the aircraft factories which were their first targets. Precision-bombing was not possible in such conditions. LeMay took over the 20th and 21st, now combined into XX Air Force, in January 1945. The long haul from Saipan to Japan – 1,350 miles, not counting diversions to avoid the Nampo Shoto chain of islands running south from Tokyo, which meant reduced payloads – was an unacceptable limiting factor. The US Joint Chiefs decided late in 1944 to remedy this by capturing the best island–air base in the chain: Iwo Jima, about halfway between the Marianas and Tokyo. The 'solution' to the problem of unsatisfactory precision-bombing was to switch to incendiaries, which seemed especially promising because so many Japanese dwellings were made of wood and paper: if the houses round a factory went up, the factory would be all the more likely to catch fire. As an experiment, the city of Kobe was partly set ablaze on 4 February, and Tokyo got its first fire-bombing on the 25th, when a square mile of housing was destroyed.

LeMay adopted area fire-bombing by night, at relatively low altitude for greater concentration, as his new strategy, from 9 March, when over three hundred B-29s struck the Japanese capital. Sixteen square miles were burned out in one of the very worst bombardments in history; at least eighty thousand (probably far more; nobody knows) were killed and a million made homeless. Worse was to follow as up to eight hundred bombers pounded all the main Japanese industrial and urban centers – except for a few, spared for no reason the Japanese could guess. These included Kyoto, Japan's former capital, whose significance was mainly cultural and whose destruction, it was suggested in Washington without a trace of irony, might open the way to a charge of gratuitous vandalism; Hiroshima, at the southern end of the main island of Honshu; and Nagasaki, at the western extremity of Kyushu. The B-29s were not seriously opposed and could do as they wished; the fliers salved their consciences by first dropping leaflets recommending evacuation. In March they started dropping mines in offshore waters to compound the ravages imposed on the remains of the Japanese merchant fleet by other Allied aircraft and the submarines – the last nail in Japan's economic coffin, built by blockade. Mr Akabane Yutaka, a senior civil servant who was executive officer at Air Raid Precautions headquarters in Tokyo, observed:

It was the raids on the medium and smaller cities which had the worst effects and really brought home to the people the experience of bombing and a demoralization of faith in the outcome of the war . . . It was bad enough in so large a city as Tokyo, but much worse in the smaller cities, where most of the city would be wiped out. Through May and June [1945] the spirit of the people was crushed. [When the B-29s dropped warning pamphlets,] the morale of the people sank terrifically, reaching a low point in July, at which time there was no longer hope of victory or draw but merely desire for ending the war.

Speaking of morale in general, Mr Akabane said that Saipan had been the decisive turning point. The standard of living

went down and down du.ing the course of the war. The chief trouble was not so much in rice but in secondary articles of diet – vegetables, fish, particularly in the cities. The situation was not so bad in clothing generally, and did not become acute in any particular until the last, as it held up fairly well for the first two years. The clothing situation affected chiefly the lower strata, but in the end the food situation affected all classes.

As Japan slid towards defeat, the intellectually bankrupt military junta blamed the people (just as Hitler blamed the Germans for not having the will or moral fiber to live up to his leadership). The Japanese people heartily reciprocated this contempt. Mr Akabane detected resentment in rural areas toward evacuated urban people, who were an obvious threat to local food supplies. He organized the orderly reduction of Tokyo's population from seven to two million between mid-1943 and the end of 1944 and beyond. In the country the local price of rice was multiplied by as much as a hundred. In the factories, as about a hundred cities and towns were ruined, the raids caused a widespread collapse of morale. This was reflected in absenteeism, go-slows and even brief strikes in defiance of the Kempei-tai, the military police, whose duties included keeping the civil population in line and obedient, and even running dispersed underground aircraft factories and other vital plants. Industrial managements often upheld the Japanese paternalist tradition by providing their workers with extra rations of food and clothing when they could find any. In these conditions a vast black market throve against a background of widespread corruption and abuse of power. But against the sixty-five-ton B-29s, which drove civilian morale to new depths, Japanese fighters could do very little. Not even kamikazes had an appreciable effect on the bombing. LeMay's losses per mission from all causes did reach 6 percent for a while in January 1945 (small change in European terms), but Iwo Jima altered all that. In mid-February 1945 the American carriers brought off their first thousand-plane raids to destroy Japanese aircraft, airfields, military installations, factories and other targets of relevance to the planned first stage of the invasion of Japan in the autumn of 1945.

There can be no denying that these huge air raids, though not as massive or sustained as the bombing of Germany, had a serious, if not decisive, effect on morale and, at the very end, Japan's economic capacity to sustain the war effort. At least two hundred thousand civilians were killed and as many injured. But it was the civilians who were demoralized; the Army had a different view, as well as 3.15 million men, including all reserves, under arms, backed by 1.5 million naval personnel, for the real and last 'decisive battle' – the one for Japan itself. The last-ditch defense of the Imperial Palace in Tokyo alone was scheduled to go on for a year against all comers; and after setbacks in the field there would still be guerrilla warfare. There was to be no defeat except in death.

The American submarine campaign collided with the law of diminishing returns around the end of 1944. In that year the boats sank 603 ships totaling 2.7 million tons, compared with 515 of 2.2 million from December 1941 to the end of 1943. Japanese and American contemporary experts and subsequent historians agree that the economy of the Empire, strangled at sea and battered from the air, would have ground to a halt by the spring of 1946. Japanese bulk imports of raw materials fell from 16.4 million tons in 1943 to ten million in 1944; a merchant fleet that stood at 4.1 million tons at the beginning of the year mustered two million at the end. The submarines also sank one battleship, seven carriers, two heavy and seven light cruisers, about thirty destroyers and seven submarines, inflicting severe damage on one more carrier and four cruisers.

Their biggest and most spectacular victim of the war was HIJMS *Shinano*, which was laid down as the third *Yamato*-class battleship in 1940 but was redesigned as an armored carrier, the world's largest, after Midway. She was intended to carry few planes (up to forty-seven) for her size but large stocks of fuel, munitions, and spares to replenish other carriers and the naval air effort in general. She was completed on 19 November 1944, and undertook her maiden voyage from Yokosuka on the 29th. Only a few miles out, she was sighted by USS *Archerfish* (Commander J. F. Enright, USN) and hit by four out of the six torpedoes fired. Seven hours later, the 'unsinkable' carrier sank, taking the first consignment of *marudai* glide bombs – piloted flying bombs – with her and thus postponing another unpleasant experience for those already exposed to kamikaze attacks. Only eight days earlier, Commander Eli T. Reich in *Sealion II* had sunk the one and only battleship to fall victim to the submarine campaign, the old but nimble *Kongo*, thirty-six thousand tons, which had escaped from the Battle off Samar a month earlier.

After the invasion of Mindoro in the Philippines in December, the flow of oil to Japan was cut to a trickle; in September 1944 there had been seven hundred thousand tons of tankers ferrying oil from the East Indies to the Home Islands, and at the end of the year there were two hundred thousand. By then the paucity of targets alone reduced the Americans' performance. The boats had been coursing about in the various waters round the Home Islands patroling on an intricate system designed to give all submarines a fair crack of the whip in quiet, average and busy zones. Different areas of sea were given very American code names such as 'Convoy College' (each zone within it bearing the name of an academic degree) or 'Maru Morgue' or 'Hit Parade,' which were patrolled by packs with such breezy self-selected nicknames as 'Ed's Eradicators' (commanded by Captain Edward R. Swinburne) or 'Blakely's Behemoths.' The peak of the underwater onslaught was attained in November 1944, when 340,400 tons of shipping, or seventy-four vessels, were sent to the bottom (including eighteen warships of 125,900 tons). This amounted to less than half the damage inflicted on Allied shipping in each of several peak months of the struggle in the Atlantic by German submarines in 1942–43; but the Japanese had begun the war with a merchant fleet rather less than half the size of the British, and Japan did not enjoy the reassurance provided by the incredible productive capacity of US shipyards, which kept turning out an endless stream of replacements.

The American submarine campaign against Japan was therefore more than sufficient to prove a major strategic factor in the outcome of the war. Pressed home with more fervor and larger numbers of boats (the US shipbuilding effort, naval and mercantile, was so vast that a hundred more submarines would have been merely a minor extra imposition), it might have rendered the terrible instrument actually used to end the war superfluous. As it was, carefully checked postwar calculations revealed that the American boats accounted for 1,314 ships sunk, totaling 5.3 million tons, or 57 percent of all Japan's shipping losses; aircraft took a third, overtaking the submarines in the last months of the war, and mining and surface warships took the remaining 10 percent. This total includes 201 warships.

The top 'ace' was Commander Richard H. O'Kane, USN, in USS *Tang* (twenty-four ships of 93,824 tons). *Tang* was cruelly sunk off Formosa on 25 October 1944, by a treacherous torpedo that it had fired at a transport but which ran wild and boomeranged. Nine men, including O'Kane, out of eighty-seven aboard survived and went into captivity after being beaten up by the Japanese survivors of their last attack. They all came home, and O'Kane was awarded the Medal of Honor. His was one of fifty-two American boats (3,506 submariners) lost to all causes, forty-five to enemy action, in the war against Japan. The US Navy started its war with 112 boats and added 203; losses amounted to less than 16 percent, the lowest loss-rate of the major navies (Germany lost 781 in action out of 1,162 boats built, Japan 130, Italy eighty-five and Britain seventy-five). Some fifty thousand men served in the US submarine arm, which expended 14,748 torpedoes – eleven per sinking. The poor quality of Japanese antisubmarine warfare has already been recorded; it enabled the high-quality American fleet submarines, thanks also to their uniquely advanced search-and-warning radar, to continue to behave like torpedo boats, attacking at speed on the surface, a luxury the Germans had been forced by Allied airpower and ASW technology to give up in the latter part of the Atlantic Campaign. Once they had more effective torpedoes and reliable detonators, the American boats came into their own, contributing far more strategically than the 1.6 percent of the Navy's total wartime strength which their modest numbers represented. Only sixteen thousand men went on war patrols, and their fatal-casualty rate of 22 percent was the highest of all arms in the American services, as was also the case in the German submarine branch (some 66 percent). The US Navy achieved the very high ratio of almost twenty-five ships sunk for every submarine lost, far and away the world record; the British were second best, with 9.3:1.

The Japanese Merchant Navy started the war with 116,000

men and lost 120,000 at sea during the Pacific Campaign, by far the highest such casualty rate: twenty-seven thousand died and eighty-nine thousand were wounded, sometimes horribly, by such special factors as burning fuel oil.

But Japan fought on as these losses mounted – and no more fiercely than in defense of the little island of Iwo Jima.

First Lieutenant Barber W. Conable, USMC, twenty-one-year-old executive officer of 'Love' Battery, 13th Marine Artillery, Fifth Marine Division, rubbed his eyes in disbelief when he woke up late on the morning of 23 February 1945, and cautiously raised his helmeted head over the lip of his foxhole. He could see the Stars and Stripes flying over the peak of Mount Suribachi, the dormant volcano which dominates Iwo Jima, eight hundred miles due south of Tokyo – even though he knew that between the 550-foot summit and sea level there were strongly entrenched Japanese defensive positions.

I went ashore on D-day, in the afternoon, in a two-and-a-half-ton DUKW [amphibious vehicle]. The beach was soft, black volcanic ash and very steep [Mr Conable told me in Washington at the World Bank, whose President he became in 1986]. Bulldozers had to pull the DUKWs and the guns up the beach. Our battery of four 105-millimeter howitzers was ashore fifteen minutes before dark and we dug them in right away, though we had little enough ammunition at that stage, only what we had brought ashore with us. We dug ourselves in as well, over on the far side of the island with our guns, beyond the infantry. More Marines came through our position the next morning, but our ammunition was slow to arrive.

It was my first time in battle and we were all terrified. Someone jumped into my foxhole and swore: 'it wasn't like this on Bougainville.' The officer I admired the most, the man in the next foxhole, a sergeant I knew – they were all killed. My hearing is impaired to this day – I can't hear the upper sound-registers or distinguish voices. A major came over looking for a site for a cemetery and was shot by a sniper. I guess he was the first customer. I was lucky. I didn't get wounded, but I had to give a direct blood transfusion to a wounded man on the way back to the ship when we were relieved after about four weeks.

There was humor on the battlefield as well. We were using passwords like 'Luther to Lulu,' relying on the letter L to fox the Japanese, but when one of the black Army men driving the DUKWs was challenged one night he didn't bother with the password: he called out, 'I'se

100 percent American, boss.' We had Navajo Indians on voice-radio communications so the enemy couldn't understand; I heard they turned a tank in their language into 'fort that walks' . . .

When she heard about it, Tokyo Rose said the flag on the mountain would be thrown into the sea. I hadn't had any sleep for more than sixty hours, so I didn't see them raise it, and it was wonderful to wake up to. I must say I got a little weepy when I saw it.

Pharmacist's Mate, Second Class, John H. Bradley, USN, was attached to a forward platoon of E Company, Second Battalion, 28th Regiment, Fifth Marine Division on Iwo Jima. The unit was on patrol on the morning of the 23rd under the command of the company's executive officer, First Lieutenant H. G. Schreier, who had a small Stars and Stripes in his personal kit. Bradley is the man second from the right in one of the most famous photographs ever taken, on or off the battlefield: the raising of the flag on Iwo Jima, by Joseph Rosenthal of Associated Press, the model for the Marine War Memorial outside Arlington National Cemetery, just over the Virginia border from Washington, D.C. This wondrous, too-perfect tableau is worth lingering over because it says so much about America and its most popular war: the genuine heroism, the romanticism, the will to win, the Hollywood hype and the media manipulation. The scene it represented was real; yet the photo itself was in every other sense a put-up job.

We started up the mountain immediately after the naval barrage and plane-strafing was over [said Bradley]. The Japs were still in their caves waiting for the bombardment to be lifted. When we reached the top we formed our battle line and we all went over the top together, and much to our surprise we didn't find a Jap in sight. If one Jap had been up there manning one of his guns I think he could have pretty well taken care of our forty-man patrol . . .

We used this Jap pipe and we attached the American flag on there and we put it up. And Joe Rosenthal happened to be there at the right time.

There can be no doubting the personal bravery or probity of this newsman, lugging his heavy camera into the front line with the leading troops of an invasion force attacking one of the most robustly defended of all Japanese strongpoints in the Pacific. He obviously had the luck and the news sense of the good journalist too, arriving at the right moment to see a handful of Marines justifiably cock-a-hoop over what they had done and instantly understanding the potential impact of a picture that would 'tell a story' as few others ever had.

He asked the Marines to do it again; only this time, could somebody please get hold of a much bigger flag so that it could be seen from below? A Marine borrowed the ensign of Landing Ship (Tank) *779*, beached nearby.

The first flag was a smaller flag, and it was put up by Platoon Sergeant Ernest I. Thomas of Tallahassee, Florida [said Bradley]. He put up that flag about one half-hour before this larger one was put up. It was so small that it couldn't be seen from down below, so our battalion commander, Lieutenant Colonel Chandler W. Johnson, sent a four-man patrol up with this larger flag which is the flag you see on the poster . . .

and in the photograph that went round the world, in the history books, on the bronze memorial at Arlington and on countless postcards sent home by tourists over the succeeding decades. Rosenthal won a Pulitzer Prize for his picture. Our eyewitnesses confirm its truth: the Marines did walk up the Japanese-held mountain and plant a flag on it. It was their genuine, spontaneous inspiration, and it deserves a prominent and eternal place in their country's military annals. It is no less true to say that, when asked to do it again, only bigger and better, for the camera which never lies, they did it with no less bravery and bravura than the first time round. And yet . . .

Pharmacist's Mate Bradley needed treatment himself a few days later, when he was wounded in both legs by a Japanese mortar bomb. After immediate first aid, the corpsmen took him back to the battalion aid station, and from there he was transferred within forty-five minutes to a field hospital for surgery. The next day, he was flown to a rear-area hospital on Guam, eight hundred miles to the south. Then he went by hospital ship to Pearl Harbor, 3,750 miles east. From there he was taken to a base hospital at Oakland, California, two thousand miles farther east, and finally to the National Naval Medical Center at Bethesda, Maryland, three thousand miles farther east still. A nation that could and would do all that for a wounded sailor surely had a moral advantage as crushing as its massive material superiority – over an enemy who was once again putting up a fearsome resistance unto death in the fight for Iwo Jima.

Iwo Jima is shaped like a ham on the bone, four and a half miles long and at its broader, northeastern end two and a half miles wide. Mount Suribachi is at the southern tip. Its sole purpose in the American scheme of things was to serve as a back-up airfield, long-range fighter-base and forward staging post for the B-29 bomber offensive against Japan. Commanding the defense was Lieutenant General Kuribayashi Tadamichi, who set about his task with unprecedented thoroughness and plenty of time. He had nearly twenty-four thousand men, and every civilian was evacuated in preparation for total resistance. He started building up the defenses on his arrival in September 1944, precisely when a decisive leapfrog by the US Navy from the newly conquered Marianas could have secured the island relatively cheaply. But the Joint Chiefs had their eye on Morotai and Peleliu in the Palaus at that time, and then the carriers had to cover the invasion of the Philippines. Delays there caused the original Iwo D-day of 20 January to be put back to 19 February, by which time Iwo Jima was seething with concealed redoubts, pillboxes, bunkers and tiers of caves linked by tunnels. Never was the shaky logic behind the whipsaw strategy of alternating – some would say vacillating – between two alternative fronts because of a fundamental inability or lack of will to choose between them, more devastatingly shown up. What is more, preparations were simultaneously going ahead for the much larger undertaking against Okinawa.

The shriveled Japanese Navy could only contribute submarines carrying the new *kaiten* manned torpedoes to the defense of the Bonin Islands, of which Iwo Jima is one. They failed to make any impression. Japanese airpower was battered by carrier attacks on Japan. The two functioning Japanese airfields on Iwo

Jima, one in the south and one in the center, were abandoned as the Americans dropped sixty-eight hundred tons of bombs and fired twenty-two thousand heavy shells in the biggest and heaviest pre-invasion bombardment of the Pacific Campaign. A few kamikazes, operating at extreme range, inflicted relatively minor damage on the Pacific Fleet; the escort carrier *Bismarck Sea* was sunk in the now dreadfully familiar way, a dive through the flight deck followed by rending explosions. The defenders ashore kept their heads down and survived everything except a direct hit, which came seldom.

The last bombardment before landing was the heaviest of the entire war, east or west. The invading troops, three Marine divisions (Third, Fourth, and Fifth) of V Amphibious Corps with an Army division in reserve, were commanded by Major General Harry Schmidt, USMC; they were delivered and protected by the Fifth Fleet (Spruance commanding, Mitscher with the carriers, Turner with V 'Phib). The troops landed behind a superbly precise rolling barrage which moved two hundred yards ahead of them. Each ground unit brought its own light artillery and air-spotters, and could call on tightly controlled covering fire from destroyers or air strikes from the escort carriers just over the horizon. The logistics were of a unique complexity and brilliantly organized, but the intense bombardment simply failed, even though Task Force 58's carriers added their contribution in the final stage after returning from bombing Tokyo. The ash and cinders on the steeply sloping beaches caused tremendous congestion as the Japanese built up their fire from hidden positions on the ridge dominating the shoreline on the southeastern coast of the island, where the landing took place on a two-mile front.

The three American divisions advanced in line abreast on the hinge of their right flank, the left swinging slowly north and then northeast, until the entire, uniquely dense human bulldozer was pressing northeastward yard by yard to the broad end of the island. Schmidt pronounced Iwo Jima secure on 16 March. Five days later, Kuribayashi reported by radio: 'We have not eaten or drunk for five days, but our fighting spirit remains high.' General Schmidt shut down his corps command on the island on 27 March and the assault troops began their withdrawal. The attack on Iwo

Jima, which blunted three first-class assault divisions in the final phase of the Pacific Campaign, when the daunting invasion of Japan still lay ahead, did enable the Seabees to have the first airfield, in the south of the island, operating on 4 March. The first B-29 made an emergency landing that very day, on the way back to Saipan from Japan. By war's end, twenty-four hundred B-29 landings took place on Iwo Jima for one reason or another, and it is just possible that more aviators were saved by the availability of the airstrips than Marines were lost capturing them. But this is a distorted calculation, because many of the landings were made by crews and planes that had done it once or several times before.

If the capture of Iwo Jima was necessary, some Americans surely had to suffer and die. But casualties need not have amounted to 30 percent among the landing forces, to no less than 75 percent in the infantry units of the Fourth and Fifth Marine divisions, to 4,900 killed on the island and 1,900 missing or deceased later from wounds, and to 19,200 wounded American survivors. Iwo Jima was the worst ordeal of the United States Marine Corps and the only major battle in the war in which the American casualties exceeded the Japanese dead (22,300). All this – and twenty-seven Medals of Honor – for less than ten square miles of rock and ash. Turner came under attack in the US press for allegedly wasting the lives of his men.

Okinawa, at the far end of the Ryukyu chain between Japan and Formosa, was to be invaded in Operation Iceberg, with Admiral Turner once again in charge of the amphibious phase. The British Pacific Fleet, Task Force 57, was on hand for the first time, contributing four carriers plus full escort to the vast naval forces providing distant cover. After the usual close attention from air raids and ship bombardments, Okinawa was to be assaulted on Easter Day, 1 April 1945, by the US Tenth Army, consisting of the III Marine Corps of three divisions under Major General Geiger and XXIV Army Corps (four divisions led by Major General John Hodge, USA). The Okinawa assault therefore brought the ground forces of Nimitz and MacArthur together for the first time in the second-largest amphibious assault of the campaign and of the war. The overall land commander was the resoundingly named Lieutenant General Simon Bolivar Buckner, USA, and his 172,000 fighting Marines and soldiers were backed by 115,000 technical and logistical troops – only marginally fewer than had assaulted Luzon.

The hundred thousand Japanese defenders on an island with a dense population of half a million were led by Lieutenant General Ushijima Mitsuru of the 32nd Army, who planned to repeat the tactics used on Iwo Jima – but in depth this time, on a much larger island honeycombed with gorges and caves in soft limestone. As many as three thousand Japanese planes could reach Okinawa from Formosa, Japan and many other islands in the East China Sea. As we have seen, the B-29s were by now battering and burning Japan itself with their incendiaries; the carriers had joined in too, until five fleet carriers with their wood-lined decks were damaged by kamikazes, three badly enough to require Stateside evacuation, forcing a reorganization of Task Force 58. The British were safely out of the way, battering the islands of the Sakashima Gunto, between Okinawa and Formosa, and watching the kamikazes bounce off their armored steel flight decks; one of their destroyers had to be towed to safety. They would have been more useful had they shared the copious fuel capacity of their American counterparts. But the US carrier raids in particular significantly reduced the number of aircraft thrown into the fight for Okinawa by the Japanese. Yet the closer the Allies got to Japan, the more airfields and planes there were, and not all of them could be knocked out.

The last notable amphibious operation of the Second World War got off to an excellent start with a landing on day one by fifty thousand troops from two Marine and two Army divisions, to north and south respectively, on beaches across a six-mile front in the middle section of the west coast of Okinawa, a long, carbine-shaped island pointing northeast. The Americans were expecting another, even bigger Iwo Jima and were prepared for fierce resistance. The beaches were guarded by a high sea wall against storms, and behind that was a cliff face riddled with caves largely invulnerable to bombardment. The landing place was chosen for being close to two airfields which could be used to land more troops once captured and repaired. As the invasion went ahead, it was the Japanese defenders who achieved complete surprise. They let it happen. Ushijima's plan was to keep his powder dry and fight only after the biggest kamikaze operation ever (Ten-Go) had so damaged the twelve hundred ships of the attacking fleet that the nascent invasion would effectively be cut off from its seaborne placenta. Six American vessels, including their two main ammunition-ships, were sunk, and ten driven off with damage, in the first of ten attacks over the eleven weeks from 6 April to the fall of Okinawa. The final score in ships was thirty-four sunk and 368 damaged from the air, with nearly five thousand sailors killed and almost the same number wounded – a ship casualty rate of one in four. But for years of intense training in damage control, the score would have been significantly higher and would have included at least two and possibly three fleet carriers sunk. It was the worst and most sustained ordeal of the US Fleet in the entire war. For a while morale was severely dented. Nothing Nimitz could think of, including virtually hijacking B-29s from the 'strategic bombing' of Japan to use against the myriad enemy airfields, made any appreciable difference: the kamikazes kept on coming, even if their numbers were ground down by attrition.

Having done nothing to help the defense of Iwo Jima, the Japanese Navy decided upon a last throw for Okinawa. Out came the world's largest battleship, His Imperial Japanese Majesty's *Yamato*, carrying little food for the crew and a minimum of fuel, but with her magazines crammed to bursting with ammunition. She was escorted by light cruiser *Yahagi* and eight destroyers, all under the flag of Vice-Admiral Ito Seiichi. The group was styled the 'Surface Special Attack Force' – in other words, it was on a suicide mission. There was not enough oil to get *Yamato* home. The fag-end of the beggared Combined Fleet sailed proudly out of the Bungo Suido, southern exit of the Inland Sea, on 6 April. Submarines alerted Spruance, who ordered Mitscher to attack, keeping the old battleships of the bombardment force in reserve as a last resort. Their services were not required. The carrier planes found the force in circular formation, *Yamato* in the middle, northwest of Okinawa. They mounted relentless attacks by small groups of planes from all directions. Nearly three hundred aircraft hammered away for more than a hundred minutes. Six were shot down. Only five Japanese planes turned up to help, and they were swiftly destroyed by the swarm of Americans. A dozen bombs, a dozen torpedoes struck home, and attackers and defenders lost count of the direct hits before the monster's looming superstructure swung slowly, then faster, to port and *Yamato* slowed to twenty knots, then slower. No enemy warship was to be seen, and the planes buzzed about like insects over a dying man, his great arms swiping more and more feebly at his tormentors. The nine awesome, lumbering 18.1-inch guns fired and missed, fired raggedly and missed, fired stertorously, fired singly, fired not at all. Then she turned away to die.

With the turn to starboard, the port list increased greatly [Commander Miyamoto, gunnery officer, recalled]. Efforts to right the ship . . . proved unavailing. It was extremely difficult to attempt any gunfire. Sometime between 1400 and 1430 the *Yamato* listed heavily to port and sank. . . . I was sucked down below the surface and as I came up the first time there was great disturbance in the water which may have been fragments from explosions. I don't remember for several minutes after that because I was hit on the back of the head.

Yamato rolled over and disappeared, sending gargantuan bubbles to the surface amid a cacophony of underwater explosions, crunches and clangs from vast loosened plates of armor which boomed dully against decks and bulkheads. Of her crew of 2,767 men, 2,488 were lost. Of her escort, *Yahagi* and four destroyers (and another 1,167 sailors) went down with her. The other four were all damaged but got home. A lone kamikaze attack on carrier *Hancock* killed eighty-four Americans and destroyed nine planes in this Battle of the East China Sea, the final obsequy of the line-of-battle ship. Nobody ever ordered another dreadnought. Halsey's flagship, *New Jersey*, and three of her sisters are still in service with the US Navy at time of writing, complete with sixteen-inch guns – and much more terrible nuclear weapons tucked away below deck, the only strategic justification for keeping them in commission. Two played a significant supporting role in the 1991 Gulf War.

*   *   *

To their own amazement, the Americans got the two air-fields at the trigger end of the Okinawa rifle butt inside a week, virtually cutting the island in two at the breech. But as they turned southwest, they ran up against the first of three well-prepared defensive lines, which held off three US Army divisions for two weeks. The first line finally withdrew onto the second on or about 24 April. Having been diverted to overrun Ie Shima, a small island to the northwest of the center of Okinawa, the 77th Army Division returned to join the Seventh Army and First Marine divisions for a fresh assault. Buckner rejected the idea of an amphibious landing behind the enemy line in favor of a frontal assault and was soon made to pay for it. A Japanese counterattack was bloodily repulsed, and the fighting continued, with the Americans advancing agonizingly slowly.

All American Pacific commanders were impatient with the slow pace of the Army's advance on Okinawa. The Marines, having landed to the north, had for once had the easier passage in the opening phase. A familiar dispute between Navy and Army men about the different methods of troops and Marines – methodical, meticulously prepared progress relying heavily on artillery as against the dash claimed by the Navy's soldiers – threatened to break out again, even as Navy and Army argued fiercely over the deployment of the B-29s. But the gnarled and fissured surface of Okinawa, and the fiendishly determined defenders, seemed to leave little choice. It does, however, seem astonishing that a vast expeditionary force with absolute control of the sea and, kamikazes notwithstanding, substantive control of the air, and so much experience in mounting big and complicated amphibious landings, could not think of something a little more original than hammering away at the armored front door of a singularly determined defense. This was to re-create the western front of the previous war all over again – but on a small island. Whatever happened to the flanking attack? Superior firepower, numbers, food, medical facilities and communications seemed to make no difference except to the inevitability of the final outcome – when it came.

Attrition and safety-first were the ground rules for the ground fighting in the Pacific Campaign, which produced no really great general in either command or either service. The by now stupendously overwhelming American superiority had become not an inspiration but a substitute for thought. Commendable and highly prized though concern for minimizing losses may be, especially in a democracy, there are times when caution leads to false economy – especially in a campaign where disease was almost as important a cause of casualties as enemy action. This dearth of ideas, this pedestrian lack of élan, and the inability to agree on a single thrust and a single commander against Japan combined to ensure that the American riposte to Japanese aggression was both mighty and slow – mighty slow.

Halsey, dashing to a fault but not a land commander, relieved Spruance at the head of the Pacific Fleet, which thereupon once again became the Third, on 27 May. The ships, the planes and the lower ranks fought on. By then the capital of the island, Naha, had fallen, and total victory was but a matter of time: officially twenty-five days, in that Tenth Army reached the southern extremity of the island on 21 June. General Simon Bolivar Buckner was not there to claim the title of liberator of Okinawa, having been killed by a Japanese shell at the start of the last big push. His redoubtable opponent, General Ushijima, committed ritual suicide the morning after his defeat. He had lost seventy thousand troops and eighty thousand civilians to the Americans' seventy-seven hundred killed ashore and 31,800 wounded. We have already noted the naval casualties. The United States now had an advance base for the invasion of Japan. The Seabees moved in to build a military city of dazzling proportions at even more dazzling speed.

Within a week of each other, over the first weekend of the struggle for Okinawa, the political leadership of both main belligerents in the Pacific underwent dramatic change. On 5 April General Koiso Kuniaki, who had supplanted Tojo when Saipan fell, resigned as prime minister after barely six months. He had borne the main burden of imperial displeasure over the loss of the 'decisive battle' on sea and land at Leyte. Although he had supported his own service's choice of Leyte rather than Luzon as the place to stand and fight, the decision had been made not by him or his Cabinet but by the junta generals on the active list. Koiso (1880–1950) came under more pressure after the fall of Iwo Jima, which he had helped to oversell as an invincible fortress. The incendiary-bombing campaign and the collapse of Germany did nothing to enhance Koiso's position, which was increasingly one of responsibility without real power. Since August 1944 the Supreme War Guidance Council of senior ministers and chiefs of staff had been meeting twice a week in Hirohito's presence at his bomb shelter under the palace to discuss strategy. Hirohito as good as told his close confidant, Marquis Kido, in January 1945 that he had no confidence in Koiso. Kido advised consulting the *jushin* (all former premiers). Tojo, of course, urged war to the death, and Prince Konoye advised suing for peace; the rest sat on the fence. Koiso's chief offense was, however, his last: unbeknown to the emperor, he put out feelers for a negotiated peace with Chiang Kai-shek in China on almost any terms, in order to free troops for the defense of the Home Islands. Hirohito forbade any further contact, and Koiso felt obliged to resign.

He was succeeded by the seventy-eight-year-old Admiral Suzuki Kantaro, who had narrowly escaped assassination in the abortive coup of 1936 for his moderate stance against militarism. A former Combined Fleet C-in-C and naval chief of staff, he had held several court appointments – in August 1944 he became chairman of the Privy Council – and was known and trusted by Hirohito. His appointment was a clear signal that the emperor wanted peace, but the active generals in the junta would have none of it, advocating the decisive battle to end all decisive battles in and for Japan itself. Even the 'peace party' and the emperor himself yearned for one last victory over the Americans – perhaps on Formosa – to improve Japan's bargaining position. When Hitler committed suicide on 30 April and Germany surrendered unconditionally on 8 May, the Supreme War Guidance Council seriously considered getting the Soviet Union to mediate with its Western allies: feelers were put out (and firmly rebuffed). The effect was rather spoiled shortly afterward when Suzuki torpedoed his moderate image by speaking publicly of war to the death rather than surrender. He was old, tired, frail, overtaxed, hard of hearing and probably thoroughly confused by the unprecedented situation of his country, staring defeat in the face for the first time in centuries, and for the first time ever at home. The Guidance Council as late as 8 June adopted a new program, 'Fundamental Policy on the Conduct of the War,' which looked forward to an 'honorable death of one hundred million' and no surrender. Only when Okinawa fell did Hirohito wake up and urge Suzuki, on Kido's advice, to seek a way out. The Soviet initiative was revived, and Prince Konoye was told to go to Moscow as plenipotentiary. Stalin and Molotov stalled this unrealistic maneuver to death: they were due, in keeping with their secret undertaking to their allies, to go to war against Japan on 8 August, three months to the day after Germany gave up, and they scented political and territorial profit in adhering to this promise.

Franklin Delano Roosevelt had been feeling ill since the Tehran Conference and had gone into hospital at the end of March 1944 with high blood pressure and heart trouble. Even so, he won in November, if by the narrowest margin of his four victories – 53.4 percent of the popular vote. Immediately after his inauguration on 20 January 1945, he went to Malta for the preliminary meeting with Churchill in preparation for the Yalta 'Big Three' Conference with Stalin. He was obviously not well. At Yalta, Stalin presented his shopping list for entering the war against Japan, which included the return of the southern half of the island of Sakhalin, taken by the Japanese in the war of 1904–5; the Kurile Islands, between Sakhalin and Hokkaido (ceded by Russia to Japan in 1875); and the restoration of all Russian concessions in Manchuria and northern China lost to Japan since 1904. It was to be remembered that, in agreeing to these things – subject to negotiated consent from the absent Chiang – Roosevelt not only was physically ill and mentally exhausted but, more important, believed like everybody else that Japan would have to be invaded to end the war. Roosevelt had lost none of his political skill or instinct, and in the main Yalta only 'conceded' to the Russians what their armies had already conquered in Europe. There are many grounds for concluding that Yalta was not a Western sellout but, rather, a reasonable settlement subsequently distorted and undermined by Stalin, who went back on many of the concessions he had made.

Roosevelt was back in the White House at the end of February. On 1 March he addressed a joint session of Congress from a wheelchair, something he had never used in public before. The right-handed president was also using his left hand to turn over the pages of his speech. By 30 March Roosevelt was in Warm Springs, Georgia, the home where he had recovered as much as he was going to from infantile paralysis, for a short break. The scene is most graphically described by FDR's biographer Ted Morgan. The president was desperately tired, but still working, *inter alia*, on speeches for Jefferson Day on 13 April and for the United Nations Charter conference in San Francisco on 25 April. At noon on 12 April, still at Warm Springs, he resumed sitting for the portrait artist Elizabeth Shoumatoff. Various friends and staff wandered in and out of the room, where a relaxed atmosphere prevailed, with Roosevelt signing a few documents, making a joke, and allowing the artist to measure his nose. Then he complained of a sudden and terrible headache. Taken to his bed in a coma, Roosevelt never regained consciousness. He died at 3:25 P.M. of a cerebral hemorrhage. He was sixty-three years old. It was already Friday the 13th when the troops and sailors on and around Okinawa were dumbstruck by the news, which Tokyo Rose and Hitler alike greeted with hysterical enthusiasm as an omen of final victory for their side. The Shoumatoff portrait was never finished, but its subject died secure in the knowledge that the victory for which he had schemed and worked so hard was just around the corner.

The United States Constitution functioned as smoothly as it is meant to when a president dies in harness. Harry S. Truman, former Missouri senator from Kansas City, vice-president since January in place of the strange and mistrusted Henry Wallace, was a very fit sixty and took to the most important political post in the world as if born to it from the moment he was sworn in. He had made his name as chairman of the Truman Committee (the Senate Special Committee Investigating National Defense), which kept a beady eye on the expansion of what Eisenhower taught us to call the 'military-industrial complex,' born in the First

World War and grown gigantic in the Second. Despite his lack of experience in foreign affairs, Truman, advised by secretary of state Edward R. Stettinius, Jr., plunged into the diplomatic pool with vigor and immediately took a markedly tougher line with the Soviet Union. Molotov, on a goodwill visit to Washington at the end of April, complained that he had never been given such a talking-to in his life. 'Carry out your agreements,' said Truman, 'and you won't get talked to like that.'

Two months later, Harry Truman and his aides were hard at work preparing themselves for the last of the great wartime conferences – at Potsdam, west of Berlin in newly occupied and prostrate Germany. It was due to open on 16 July 1945, and it was code-named 'Terminal.' Delayed for a day by the late arrival of Stalin, Truman was given a tour of the shattered remains of Berlin. On his return to the undamaged Cecilienhof Palace, a huge mock-Tudor 'English country house' built by the last Kaiser for his daughter and chosen as venue for the conference, an excited Henry L. Stimson, still secretary of war at seventy-eight, was waiting with a coded message. It revealed that $2 billion had been successfully spent on the largest, most intense and expensive scientific development program in history. The 'Manhattan Project' had achieved success on schedule that very day with Operation Trinity – an almighty flash and bang in the desert at Alamogordo, New Mexico. This was what the American delegation was waiting to hear: indeed, Washington had delayed Potsdam in hope of this very piece of news, whereby hangs the following tale, definitively and absorbingly told by Richard Rhodes in *The Making of the Atomic Bomb*.

The most famous scientist in the world, Professor Albert Einstein, a German Jew who, like so many other exceptionally gifted compatriots, had gone into exile in the United States before the war to escape Nazi anti-Semitism, twice warned President Roosevelt that the Germans were working on the atomic bomb. In October 1939 and, on getting no reaction, again in April 1940, the burned-out inventor of the theory of relativity wrote from his academic refuge at Princeton University to Alexander Sachs, an economist who had entrée to the White

House. Einstein feared that the Germans might get their hands on the uranium ore in the Belgian Congo (in fact, after the occupation of Belgium in 1940, the Congo mining company Union Minière presented the US with its entire remaining stock of ore, although the Germans had acquired some in Belgium). As it happened, the Germans, who did some work with uranium oxide, went up a scientific blind alley by centering their hopes on heavy water, and when their Norwegian supplies of this expensive, man-made commodity were destroyed by bombing and sabotage, the Nazi atomic-bomb effort, already hamstrung by economic considerations, finally collapsed in 1944. Having forced nearly all their best scientists to flee, the Nazis ensured in advance that they would lose the race.

Japanese, British, and American scientists all independently regarded uranium initially as the most promising element from which to garner applied nuclear energy. The Japanese had uranium-ore deposits in northern Korea, and in 1940 Lieutenant General Yasuda Takeo, director of the Aviation Technology Research Institute of the Japanese Army, initiated an investigation of the possibilities of nuclear energy. Japan's leading nuclear physicist, Nishina Yoshio, builder of the country's first cyclotrons, was co-opted, and Army aviation ordered the development of an atomic bomb in April 1941. Experiments went ahead at Rikken, the institute for physics and chemistry research. A large and mysterious nuclear installation was built at Hungnam in Korea which no Westerner ever saw: the Russians dismantled and removed it after the war and have never said what use they made of it. Rumors persist of a Japanese experimental nuclear explosion (or accident) off the coast near Hungnam on 10 August 1945, four days after the Americans dropped their first bomb. Japanese naval interest in nuclear energy focused on powering ships. The Navy had its own theoretical-research program which got nowhere and was abandoned in 1943, turning its collective attention to neglected radar. The Navy's advisers concluded that neither Germany nor the Allies would have a bomb in under ten years. But the Navy did fund a cyclotron (particle accelerator) at Kyoto University. The Rikken facility was bombed out by B-29s on 13 April 1945, marking the end of Nishina's promising but inconclusive experiments with gaseous diffusion.

On 14 May 1945, six days after the war ended in Europe, Captain Johann Fehler of the German Navy surrendered his large, Type Xb minelaying submarine to the US Navy at sea in the North Atlantic, off New England. The singularly well-numbered *U234* – uranium has three isotopes, U234, U235, and U238 – had left Kiel in north Germany on 25 March 1945, bound for Japan with Hitler's last technological legacy to his beleaguered Axis ally on the other side of the world. Among the varied cargo, stowed in some of the 1,763-ton boat's thirty mine-tubes, was a total of 560 kilograms (1,232 pounds) of uranium oxide, the compound from which uranium-235, the isotope used in reactors and bombs, is extracted by the process of ion exchange. Nobody seems to have any record of what became of this stockpile, but it is known that US troops captured twelve hundred tons of Belgian Congo ore from central Germany because it was needed for the second US bomb (a huge mass of ore yields only a tiny quantity of uranium).

The first meeting of the American Advisory Committee on Uranium, made up of representatives from the armed forces and the US Bureau of Standards, had taken place in October 1939. In November it recommended to the White House experimental research into the nuclear chain reaction as a possible source of power for submarines. In Britain, where interest in nuclear fission was as strongly encouraged by scientists in exile as in America, the German advance to the Channel prompted the coordination of all nuclear research under the aegis of a body quaintly code-named the 'Tube Alloy Committee.' After plutonium was discovered by a group of American scientists in 1940–41, US interest focused on this man-made element, because it behaves like U235 yet its critical mass is significantly smaller. The British decided in the summer of 1941 to make a bomb. That and the entry into the war of the Soviet Union upon the German invasion in June galvanized Roosevelt and his advisers. On 11 October 1941, Roosevelt proposed collaboration to Churchill, who replied positively in the first week of December, when the US uranium committee decided in principle to change the emphasis from research to development of the American bomb.

Japan's nemesis was mobilized the day before her hubris

led her to attack Pearl Harbor, at a lunch in Washington D.C. It was given by Dr Vannevar Bush, director of the Office of Scientific Research and Development (OSRD), for two scientists. Section I of OSRD and the Corps of Engineers of the US Army worked together on the project, code-named 'Manhattan' from the Engineer District where the Army's role began; Brigadier General Leslie R. Groves led the military, and Dr J. Robert Oppenheimer the scientific side of the project, based at Los Alamos from 1943. At that time, the British, whose suspected lead in the field had helped to mobilize Roosevelt, effectively admitted that they had fallen behind and sent their leading scientists to join the team at Los Alamos.

From August 1944, eighteen B-29 bombers were being modified to carry the new ultimate weapon, long before its operational size, shape or weight was known. The assumption was that it would be very large and very heavy. The 509th Composite Group, 225 officers and 1,052 men, was taken out of training for bombing Germany and assigned to the task of delivering the Bomb. Truman, briefed by Stimson as soon as he took over the White House, was essentially resolved to use the weapon if the trial proved successful. His motive was to shorten the war, to save a million or more American and Allied lives: everyone expected an unparalleled bloodbath as the Japanese suicidally defended their homeland rather than surrender unconditionally, which the Allies persisted in demanding. Warnings from some scientists of the dangers of postblast radiation, then a quantity as unknown as the force of an atomic explosion, and even the possibility of an unstoppable chain reaction in the atmosphere, were disregarded. So was the idea of waiting until the Japanese succumbed to the blockade by sea and air which had so visibly reduced their availability to wage war for the preceding eighteen months – but had not extinguished their resolve.

On 26 July 1945, the thirteen-year-old heavy cruiser *Indianapolis* (9,950 tons, Captain Charles B. McVay, USN), Admiral Spruance's flagship at the Battle of the Philippine Sea, arrived unescorted at Tinian in the Marianas to deliver the detonation device of the first operational bomb. On her way, still unescorted, to Leyte to join Task Force 95 and train for the invasion of Japan, the veteran cruiser was sighted east of the Philippines by submarine *I58* (Commander Hashimoto Mochitsura, IJN) just before midnight on the 29th. Two or possibly three of his salvo of six Long Lances struck home underwater along the starboard side. The cruiser had no sonar, so she had no inkling of the boat's presence; she had been allowed out alone because the Americans had written off the Japanese Navy as a threat in rear areas. The torpedoes set off a series of violent fuel and ammunition explosions which blew off her bow. Her engines drove her under even as she listed at an angle of forty-five degrees.

Because nobody expected American warships to be sunk any more, the 850 men of her crew of 1,199 who got into the water alive had to wait eighty-four hours until they were fortuitously sighted by a Navy patrol plane on 2 August. By then only 316 remained to be rescued. An earlier report by an Army plane of flares fired from the water was ignored. Nobody ashore in the Philippines or in TF 95 noticed that she had not turned up. An Ultra intercept of a message from Hashimoto claiming to have sunk a US battleship was dismissed as the usual Japanese exaggeration. Thus the *Indianapolis* became the last American warship to be sunk, the last submarine-kill and the last victim of the Imperial Japanese Navy in the Second World War. The unforgivable carelessness on the US Navy's part is bad enough; one can only speculate how different history might now look had *I58* caught the unlucky ship on her way *into* Tinian. The US nuclear stockpile at that time consisted of two bombs, with scant prospect of a third for some months.

At Potsdam the Americans were fully aware of the sticking point beyond which the Japanese would not go, even if it meant the invasion of their country and the destruction that would follow. Magic intelligence on and after 11 July from the traffic between Foreign Minister Togo Shigenori in Tokyo and Sato Naotake, his ambassador in Moscow, showed that Tokyo was once again unrealistically trying to arrange Soviet mediation. The intercepts revealed that there would be no surrender unless the Allies guaranteed the survival of the imperial system, the *sine qua non* of Japan's national identity. It was known that the emperor himself backed the Soviet mediation ploy and the choice of Prince Konoye on 12 July as plenipotentiary if Moscow would receive him.

But refusal to give up the monarchy set a condition on the unconditional surrender to which the Allies were irretrievably committed. Allowing the throne to survive, even as a copy of the most constrained European constitutional monarchy, would look like appeasement, which might stiffen the still very powerful Japanese will to resist. It would also upset American and Allied public opinion, which saw Hirohito as a war criminal. It was the throne, and not Hirohito as its occupant, which was the focus of all the main strands of Japanese opinion. The Army wanted to install Hirohito in a specially built, hugely expensive bombproof bunker near Mount Fuji, that other great national symbol, as focus of a last stand. But elements on the Navy staff developed a plot to replace Hirohito with a suitable member of his family. The 'peace party' led by Prince Konoye wanted him to retire as abbot of a Shinto monastery and make way for his small son, Akihito. These details were not known to the Allies, but Stimson was the principal US advocate of leaving the dynasty in place – and of giving the Japanese a clear warning of what would follow rejection of a final demand for surrender.

Oppenheimer was one of many who, before Potsdam, opposed a demonstration of the atomic bomb in such a way that its power would be revealed without significant damage to Japan or its people. The scientist could not think of a credible way of doing it, especially as nobody really knew what would happen when the device was triggered. Other advisers pointed out that a warning of a vast explosion at a fixed time and place would give the Japanese time to move prisoners of war there to prevent the test from taking place. And what if the test went wrong after the president of the United States had issued a solemn warning about the awesome power of the new weapon – what effect would that have on the Japanese will to resist? Oppenheimer accurately estimated that one bomb dropped on a city would have an effect indistinguishable from and no greater than a big B-29 incendiary raid of the kind already in progress, in terms of immediate casualties and total damage.

Having listened to all shades of opinion, Truman appears to have made up his mind on or about 1 June 1945. Unless

Japan gave in, the United States would: (1) drop the bomb on a Japanese industrial target such as a factory surrounded by workers' houses; (2) do it without warning; (3) do it as soon as possible; and (4) not let the Russians in on the secret in advance (the Soviet Union had taken no active interest in nuclear power until 1943, when the German threat was contained, and was therefore far behind). The overriding American concern was to avoid an invasion of Japan if at all possible, even as earnest preparations for the amphibious attack on Kyushu went ahead. There was, however, another way.

Truman went to Potsdam with the intention of holding the Russians to their undertaking to declare war on Japan three months after the surrender of Germany. The administration generally believed that this, not the pie-in-the-sky Bomb, was the last ace which would make even the uniquely intransigent Japanese throw in their hand: the belated appearance of their most dreaded enemy. When Stalin confirmed his intention to go ahead, Truman was hugely relieved and knew that no American invasion of Japan was likely to be needed. But when a full report of the Alamogordo test-firing reached Potsdam, the Bomb was seen as a means of forcing a Japanese surrender without the need for Soviet intervention. Truman, like any politician, had kept his options – Russia or the Bomb to force a capitulation without invasion – open as long as possible. The president's strong and deep-seated distrust of Stalin, whose demands in exchange for intervention were set so high, decided the issue for the Americans: the Bomb was to be dropped before the Russians came in. Ironically, the evidence from Japan indicates that it was not the Bomb, and certainly not the Bomb alone, but the Soviet declaration of war immediately afterward that decided the issue for the Japanese.

The Potsdam Declaration by the Big Three on 26 July, broadcast to Japan in its own language and unloaded all over the landscape by American pamphlet-bombers, promised 'prompt and utter destruction' unless Japan forever renounced militarism and the militarists, gave up its war criminals, withdrew from its overseas conquests since 1895, conceded human rights – and surrendered unconditionally. There was no reference to the throne but only a commitment to the introduction 'of a peacefully inclined and responsible government' based on the free self-determination of the Japanese people. At a press conference on 30 July, the fumbling, confused prime minister, Admiral Suzuki, in line with palace and cabinet policy, which was to play for time and hope for Soviet mediation, used an ambiguous turn of phrase of the kind that makes his language a snare and a delusion for all but the most dedicated foreign student. Apparently he meant, and thought he was saying, that the government would not comment *yet*. The American translators read this as meaning there would be no comment *at all*. Thus the 'wait-and-see' approach was interpreted as a dismissal with contempt; and thus the war, which was partly rooted in verbal misunderstanding about 'sincerity' and Japan's plans in Indochina, came to an unnecessarily apocalyptic climax partly because of another misreading. Seldom can a misconstrued adverbial nuance have had such devastating consequences.

Colonel Paul W. Tibbets, aged thirty, commanding officer of the 509th Composite Bombardment Group, XX Air Force, USAAF, had named his B-29, number 82, *Enola Gay* in honor of his mother. Like any other conscientious field commander, he was not about to ask his men to do something he was not prepared to do himself. He would be in personal charge of the group's special mission. He and his pilot, Captain Robert Lewis, took the plane, painstakingly nurtured to lift seven tons more than the manufacturer recommended, ever so gently into the air from the huge air base at Tinian at 2:45 A.M. on Monday, 6 August 1945. In the Perspex nose with his Norden bombsight was Major Thomas Ferebee, aged twenty-four, the bombardier, who, as ever on a US bombing mission, would be in command of the plane on its final approach and over the target until the moment the bomb-load was released. The bomb bay contained one weapon, grotesquely code-named 'Little Boy' – a dull, gunmetal-black cylinder ten feet six inches long and twenty-nine inches in diameter, weighing ninety-seven hundred pounds. Once airborne at five thousand feet and on their way to Iwo Jima, two weaponeers or armorers completed the assembly of the bomb. From Iwo, two more B-29s, carrying official observers, photographers and scientists, made a rendezvous with the circling *Enola Gay*, and all three flew on to the north-northeast. At 8:15 A.M. Hiroshima time, Ferebee got the bridge he had chosen to aim at in his sights and pressed the button. The bomb took forty-three seconds to fall from thirty-one thousand feet.

A bright light filled the plane [Tibbets told the *Saturday Evening Post*]. The first shock-wave hit us. We were eleven and a half miles slant range from the atomic explosion but the whole airplane cracked and crinkled from the blast . . . We turned back to look at Hiroshima. The city was hidden by that awful cloud . . . mushrooming, terrible and incredibly tall.

The blast, which had been arranged to occur nineteen hundred feet above the ground, was equivalent to 12,500 tons of TNT. The center of Hiroshima was blown away, dematerialized, vaporized, blasted, flattened, charred. So were seventy-eight thousand of its citizens, killed in a split second in the first and last bombing attack on their hometown. Many thousands more died, often lingeringly, later.

I could not understand why our surroundings had changed so greatly in one instant [the Japanese writer Ota Yoko told the American psychiatrist Dr Robert Jay Lifton]. I thought it might have been something which had nothing to do with the war, the collapse of the earth which it was said would take place at the end of the world.

It is not known who first reported the disaster to Tokyo, but the call did not come from Hiroshima, because Hiroshima had no means of telecommunication left. One call came into naval headquarters in Tokyo fifteen minutes after the event, from the Kure naval base on Hiroshima Bay, fifteen miles south of the stricken city, reporting a huge, blinding flash and a vast mushroom cloud. Attempts by the Army to raise anyone at all in Hiroshima by telephone failed. It was only after senior officers had been to the area, threading their way through the detritus left across the country by conventional bombing, which had reduced all communications to chaos, and came back to the battered capital with eyewitness accounts of the unbelievable devastation that the magnitude of the catastrophe was appreciated in Tokyo, 450 miles to the east. This helps to explain the eerily slow reaction of the fragmented Japanese leadership, which went into a state of collective shock. The chronic inability of a divided ruling establishment to make decisions reasserted itself with unprecedented force just when it was least needed.

The Soviet Union duly declared war on the Japanese Empire, as promised, on 8 August. The Supreme War Guidance Council was called together at last for the afternoon of 9 August. By that time, the Red Army had marched into Manchuria – and the southwestern Japanese city of Nagasaki had been demolished by 'Fat Boy,' the second atomic bomb, which killed twenty-four thousand instantaneously and many thousands more slowly – on the same morning. The council split down the middle, symbolizing perfectly the paralyzing indecision which underlay all Japan's troubles and had led to so many rash decisions by so many frustrated individual leaders. The war minister, General Anami Korechika (especially), and the military and naval chiefs of staff, Umezu and Toyoda, were against, and the prime, foreign, and Navy ministers were for accepting Potsdam. Only Hirohito could break the deadlock, and only by direct action, as there was no such thing as an imperial casting vote. A second, wider meeting, an imperial conference, took place that night in the steaming, airless bomb shelter under the palace and produced the same deadlock. Prime Minister Suzuki begged the emperor to intervene. He did. 'The time has come to endure the unendurable,' he said, looking his war minister in the eye. The formulation deliberately echoed his grandfather's sentiments when deferring to Western demands for access to Japan in the 1850s. The junta obeyed.

The Foreign Ministry told its ambassadors in Switzerland and Sweden by radio on the 10th to forward the decision to the Big Four. The Japanese people as yet knew nothing of it.

The Japanese Government is ready to accept the terms listed in the joint declaration issued at Potsdam . . . with the understanding that the said declaration does not include any demand which prejudices the prerogatives of His Majesty as a sovereign ruler.

After frantic consultations, the Allies insisted – in a reply once again distributed by air drop, to the bewilderment of the general population – that:

from the moment of surrender the authority of the Emperor and the Japanese Government to rule the state shall be subject to the Supreme Commander of the Allied Powers, who will take such steps as he deems proper to effectuate the surrender terms.

The last imperial conference of the war took place on the morning of 14 August, Japanese time, in the presence of Hirohito in the stifling shelter. All present wore morning coats or uniform; the haggard emperor wore Army uniform. Amid sweat and tears all round, Hirohito, after listening to the irreconcilable arguments for the last time, told his ministers and officers that it really was all over. What was more, he would go on radio the next morning and tell the people; and he would go to IGHQ as well if need be. Even as some irredeemable Army officers tried to mount a coup, and after long arguments among his military and civilian advisers about the wording, Hirohito cut a recording at the palace with the help of obsequious technicians from NHK. There could of course be no question of a live broadcast by the divine ruler of Japan. At noon on 15 August 1945, astounded Japanese listeners heard a completely unfamiliar, high-pitched voice crackling over their radios, speaking a dialect of Japanese so rarefied as to be incomprehensible to all but the most educated (the court had its own dialect). The linguistic gap was far wider than that between the Queen of England on the one hand and the humblest citizens of Glasgow, New York or Sydney on the other, which is saying a lot. Hirohito intoned, in part:

After pondering deeply the general trends of the world and the actual conditions obtaining in Our Empire today, We have decided to effect a settlement of the present situation by resorting to an extraordinary measure . . .

We declared war on America and Britain out of Our sincere desire to ensure Japan's self-preservation and the stabilization of East Asia, it being far from Our thoughts either to infringe upon the sovereignty of other nations or to embark upon territorial aggrandizement . . .

Despite the best that has been done by everyone . . . the war-situation has developed not necessarily to Japan's advantage, while the general trends of the world have all turned against her interest.

Moreover, the enemy has begun to employ a new and most cruel bomb . . .

We cannot but express the deepest sense of regret to Our allied nations of East Asia, who have consistently cooperated with the Empire toward the emancipation of East Asia . . .

The hardships and sufferings to which Our nation is to be subjected hereafter will certainly be great. We are keenly aware of the inmost feelings of all ye, Our subjects. However, it is according to the dictate of time and fate that We have resolved to pave the way for a grand peace for all the generations to come by enduring the unendurable . . .

Having been able to safeguard and maintain the structure of the Imperial State, We are always with ye, Our good and loyal subjects, relying upon your sincerity and integrity . . .

Unite your total strength, to be devoted to construction for the future. Cultivate the ways of rectitude; foster nobility of spirit; and work with resolution so that ye may enhance the innate glory of the Imperial State and keep pace with the progress of the world.

Those seeking explanations for Japan's failure to face up to its wartime past and for its postwar resurgence as the world's strongest economic power need look no further than here. It has no recorded equal for understatement, ambivalence, distortion or refusal to face reality. Look as hard as you like at the full text and you will not find the word 'defeat' or any of its synonyms in this least submissive and repentant of surrenders.

General Anami had already committed *seppuku*, disemboweling himself with a dagger and then severing a neck artery. The mutineers in the coup plot followed suit, using their samurai swords or their pistols to kill themselves. Many generals and admirals did likewise. Suzuki and his government resigned, to be replaced by Prince Higashikuni, the emperor's younger brother.

But this is a naval history, and we must go back to sea one more time.

As Spruance prepared the invasion of Kyushu in the first days of August, his alter ego, Halsey, used the planes of Task Force 38 to destroy all the remaining assets, afloat or ashore, of the Imperial Japanese Navy, even after the last conventional B-29 raid on the 8th. American and British battleships also shelled industrial and other coastal targets in the first days of the nuclear age. This activity continued until Nimitz ordered the Pacific Fleet to cease fire on 15 August. Ashore, seventy Japanese Army divisions, over a million men, were in position, awaiting an enemy whose eventual arrival they would not be given a chance to dispute. At that time Japan possessed, capable of unaided movement, two aircraft carriers (one damaged) with no planes; three damaged cruisers, forty-one destroyers, most damaged to some degree, and fifty-nine undamaged submarines. There were 829 vessels incapable of movement, some lying on the bottom in shallow water, some floating upside down, some listing, others awash.

The Americans never resolved their wartime problem of unity of command in the Pacific. In April 1945 the Joint Chiefs decided on a reorganization whereby MacArthur would take charge of Army forces all over the ocean and Nimitz all naval, while strategic air forces were to be directed from Washington. This proved academic, because nothing could be done until Okinawa was secure. In May they decided that, for the first stage of the invasion of Japan in the autumn, MacArthur would command on land, including the assault stage, and Nimitz through Spruance at sea. But the Bomb and the Russians put paid to that, leaving the way clear for the apotheosis of five-star General of the Army Douglas MacArthur at the age of sixty-five. Only when Japan surrendered and invasion became superfluous did President Truman, with the consent of the Allies, name him Supreme Commander, Allied Powers. As such, he became the uncrowned king of Japan, over the emperor's head, until 1951, a majestic role which he filled with quite extraordinary magnanimity. There can therefore be no other candidate for the last word in this account of the Pacific Campaign.

First, however, a final balance sheet has to be drawn. Japan and Germany were militarily but not irreversibly crushed: they boasted the world's strongest economies within a generation. The British Empire was on the winning side but dissolved itself in half that time; European colonialism in Asia was destroyed by

Japan. China lost between 2.5 and 13.5 million people – nobody knows the true figure – and, despite postwar revolution and counterrevolution, remains a potential rather than an actual great power, bowed down by the weight of its vast population. The Soviet Union lost twenty million killed but also numbered itself among the victors; yet at time of writing it was on the brink of collapse. The United States, the biggest victor, became the world's greatest power – but also the world's leading debtor-nation in the late 1980s. The principal creditor is of course Japan.

Of the myriads of books about the Second World War, almost none mention its one irreversible result: the number of dead. Over fifty million died worldwide, equivalent to the population of the United Kingdom, France, or Italy today and ten times as many as in 1914–18. Two-thirds died in Europe and one-third in Asia. The majority was civilian. Japan lost 2.25 million killed. The United States lost 290,000 on all fronts, nearly all military and naval, because, alone of all the major belligerents, its territory was virtually untouched by war. The Army lost 234,874, including 52,173 in the Air Corps; the Marines lost more than eighteen thousand, nearly all in the Pacific. The rest belonged to the Navy and the Merchant Marine.

It fell to MacArthur to accept the surrender of Japan on the main deck of the United States battleship *Missouri*, Halsey's new flagship, surrounded by an unprecedented armada of American, British and Allied vessels in Tokyo Bay. The emperor and his new prime minister were not present. Foreign Minister Shigemitsu Mamoru led a delegation of nine, including General Umezu, the only key wartime leader with the face to put in an appearance, up the starboard gangway from the American destroyer which had ferried them from Yokohama. The civilians wore ill-fitting tailcoats, cutaway collars and pinstriped trousers, the military drab uniforms without sidearms. Officers from the Allied powers, including Admiral Fraser of the Royal Navy, Australians and Dutch, stood waiting on deck behind the table bearing the instruments of surrender. After four minutes of tension MacArthur appeared on deck, flanked by Nimitz and Halsey.

The two admirals stepped back to make way for two emaciated lieutenant generals, Percival of Singapore and Wainwright of the Philippines, specially flown in from postliberation convalescence. MacArthur's hands were trembling again, belying the deep and measured tones in which he read a short address. Or perhaps it was only the breeze that moved the paper. Sailors hung silently on MacArthur's words and every possible part of the battleship's towering superstructure. After mellifluously explaining the bureaucratic modalities of the ceremony, where all concerned would sign 'at the places indicated,' the man who won and lost and regained the Philippines and became a legend in his own twilight declared:

It is my earnest hope – indeed the hope of all mankind – that from this solemn occasion a better world shall emerge out of the blood and carnage of the past, a world founded upon faith and understanding, a world dedicated to the dignity of man and the fulfillment of his most cherished wish for freedom, tolerance and justice.

After the laborious process of collecting a multiplicity of signatures on the documents was complete, MacArthur returned briefly to the microphone to say:

Let us pray that peace be now restored to the world, and that God will preserve it always.
  These proceedings are closed.

Acknowledgments and Sources

I should like to thank the following for help with this project.

That we fell out irreconcilably over the treatment of the material which I alone researched does not alter the fact that Edward L. Burlingame, the New York publisher, suggested the basic idea.

My gratitude to Alice Mayhew, vice-president and editorial director of Simon & Schuster, is all the warmer for taking over the project with such enthusiasm. At Hodder & Stoughton, Ion Trewin, editorial director, never wavered in his support and Jane Osborn made valuable suggestions for improvements. My agents, Michael Shaw in London and Peter Ginsberg in New York, both with Curtis Brown, provided moral and material support when it was most needed.

Unstinting in their professional help in Washington were Mr John E. Taylor of the Military Reference branch at the US National Archive; Professor Ronald H. Spector (then director), Dr Dean C. Allard (senior historian), Dr Mark Jacobsen, and (especially) Mrs Kathy Lloyd and staff at the US Naval Historical Center; and in London, Mr G. A. Evans at the Ministry of Defence Library; Mr R. Wythe and staff at the Public Record Office; the staffs of the Imperial War Museum, the London Library and the Twickenham branch of Richmond upon Thames Library.

My wife, as always, gets warmest thanks for tolerating my long absences, either in the United States for research or in my garret-with-a-view while writing, and for disposing of the problems which inevitably arose therefrom.

I should also like to express warm thanks for hospitality, information, encouragement, constructive conversation and ideas to John and Pat Barry, John and Juliet Clibborn, Barber W. Conable, Brian and Virginia Crowe, Stephen Hill, Richard

Norton-Taylor, Peter Preston, Christopher and Valerie Thomas, Frank Vogl, Martin Walker, Michael White.

My apologies to anyone inadvertently left out; nobody mentioned above bears responsibility for any error or omission.

Acknowledgments are due, and gladly made, to the following, their heirs, successors or agents, for quotations from copyright material.

Grafton Books for *FDR: A Biography*, by Ted Morgan; Harper & Row, Inc., for *Silent Victory*, by Clay Blair; Oxford University Press, for *History of the US Naval Operations in World War II*, by Samuel Eliot Morison; Random House, Inc., for *Death in Life*, by Robert Jay Lifton; Simon & Schuster, Inc., for *The Making of the Atomic Bomb*, by Richard Rhodes; the title holders of the *Saturday Evening Post* for 'How to Drop an Atom Bomb,' issue of 8 June 1946 (quoted in Rhodes).

To the best of my knowledge and belief, all the material I have quoted from the United States archives is either US-government copyright and/or free of copyright restriction. If this is not the case, I should like to apologize and, if so advised, to take the opportunity to correct any error or omission for any subsequent edition of this book.

I alone am responsible for conclusions drawn from primary or secondary sources. There is no fiction or 'faction' in this book.

PRIMARY SOURCES

The historical raw material from which this book is largely derived is in the US Naval Historical Center at the Navy Yard and at the US National Archive, both in Washington, D.C. I also learned a lot at the National Air and Space Museum in Washington and aboard USS *Intrepid* in New York Harbor.

At the Naval Historical Center, I made much use of the Second World War histories collection in the USN History Division. These include: the 'Command Summary of Fleet Admiral Chester W. Nimitz, USN' (his 'Gray Book'); the personal papers of Admiral Thomas C. Kinkaid, USN, including his unpublished manuscript account of the war in the Pacific; the Submarine Operational History of World War II; the massive collection of individual 'Interviews and Statements, World War II'; 'Individual Ships' Histories.' I found the following invaluable: 'Records of the Japanese Navy and Related Documents, 1940–1960': Commander South Pacific series (miscellaneous translations and POW interrogations); Commander North Pacific collection of miscellaneous documents of Japanese origin; South Pacific translations from the Combat Intelligence Center 1941–44; Allied Translator and Interpreter Section (ATIS), South-West Pacific, Spot Reports 1941–45; ATIS Current Translations; and ATIS Enemy Publications.

At the National Archive, I drew heavily upon the Strategic Bombing Survey interrogations in Japan (record group 243); War Department General Staff Operations Division papers (record group 165); and the Special Research Histories of intelligence material.

In my text, ellipses in the quotations indicate words omitted by me to avoid repetition or diversion. No other material alterations have been made to direct quotations, except for minor corrections of spelling and grammar.

I found that the following televised material offered useful spurs to an imagination trying to visualize and encompass the great events of half a century ago.

TELEVISION MATERIAL

'America at War,' vol. 3: 'Marine Action,' by US Department of Defense, Stylus Video (UK), 1988.
'Hirohito: Behind the Myth,' by Edward Behr, as shown on BBC1 Television, UK, 24 January 1989.
'Timewatch: Memories of Pearl Harbor,' as shown on BBC2 Television, UK, 5 April 1989.
'The Tokyo Trial,' by Masaki Kobayashi, as shown on Channel 4 TV, UK, 3 and 6 July 1989.

'The World at War,' vol. 3: 'Banzai: Japan Strikes,' by Thames
Television, UK, Thames Video, 1986.

PUBLISHED SOURCES

There is only a handful of previous general accounts of the war
with Japan, the greatest naval conflict in history; most of them
are mentioned below. My labors would have been much harder
but for the trailblazing works of: Samuel Eliot Morison; Messrs
Calvocoressi, Wint, and Pritchard; John Costello; Stephen W.
Roskill; Ronald H. Spector; Christopher Thornc. My gratitude
to them all. There are any number of books dealing with
specialized aspects of this extremely complex conflict, of which
a broadly representative selection is included here.

Agawa, Hiroyuki: *The Reluctant Admiral: Yamamoto and the
    Imperial Navy* (Kodansha International, Tokyo, 1979).
Allard, Dean C; Crawley, Martha L.; and Edmison, Mary
    W.: *US Naval History Sources in the United States* (US
    Naval History Division, Washington, D.C., 1979).
Anonymous: *Japanese Aircraft-Carriers and Destroyers*
    (Japanese author unnamed; Shuppan Kyodosha, Tokyo,
    and Macdonald, London, 1968).
——: *Japanese Battleships and Cruisers* (Japanese author
    unnamed; Shuppan Kyodosha, Tokyo, and Macdonald,
    London, 1968).
Australian War Memorial: *Australia in the War of 1939–1945*
    (21 vols., Canberra, from 1952), especially: *South-West
    Pacific Area, First Year*, by D. McCarthy (1959); *The New
    Guinea Offensives*, by David Dexter (1961); *The Final
    Campaigns*, by Gavin Long (1963).
Bagnasco, Erminio: *Submarines of World War Two* (Arms &
    Armour Press, London, 1977).
Barnhart, Michael A.: *Japan Prepares for Total War: The
    Search for Economic Security, 1919–1941* (Cornell University
    Press, Ithaca, N.Y., 1987).
Beasley, W. G.: *The Modern History of Japan* (Weidenfeld &
    Nicolson, London, third ed., 1981).

Behr, Edward: *Hirohito: Behind the Myth* (Villard Books, New York, 1989).

Blair, Clay: *Silent Victory: The US Submarine War Against Japan* (J. B. Lippincott, Philadelphia/New York, 1975).

Brackman, Arnold C.: *The Other Nuremberg: The Untold Story of the Tokyo War Crime Trials* (Collins, London, 1989).

Breyer, Siegfried: *Battleships of the World* (Conway Maritime, London, 1980).

Brinkley, David: *Washington Goes to War* (Ballantine, New York, 1988).

Buell, Thomas E.: *Master of Seapower: A Biography of Admiral Ernest J. King* (Little, Brown, Boston, 1980).

——: *The Quiet Warrior: A Biography of Admiral Raymond Spruance* (Little, Brown, Boston, 1974).

Calvocoressi, Peter; Wint, Guy; and Pritchard, John: *Total War: The Causes and Courses of the Second World War* (revised ed., Viking, London, 1989).

Campbell, Christy: *World War II Factbook* (Macdonald, London, 1985).

Campbell, John: *Naval Weapons of World War Two* (Conway Maritime, London, 1985).

Conway, Robert H.: *The Navy and the Industrial Mobilization in World War II* (Princeton University Press, Princeton, N.J., 1951).

Costello, John: *The Pacific War* (Collins, London, 1981).

Cray, Ed: *General of the Army: George C. Marshall, Soldier and Statesman* (W. W. Norton, New York, 1990).

Deacon, Richard: *The Silent War: A History of Western Naval Intelligence* (Grafton, London, 1988).

Defense, US Department of: *The 'Magic' Background of Pearl Harbor* (eight vols., Washington, D.C., 1977).

Dixon, Norman F.: *On the Psychology of Military Incompetence* (Jonathan Cape, London, 1976).

Dull, Paul: *A Battle History of the Imperial Japanese Navy* (US Naval Institute Press, Annapolis, Md., 1978).

Ellis, John: *Brute Force: Allied Strategy and Tactics in the Second World War* (André Deutsch, London, 1990).

Fuchida, Mitsuo, and Okumiya, Masatake: *Midway* (US Naval Institute Press, Annapolis, Md., 1955).

Fussell, Paul: *Wartime: Understanding and Behavior in the Second World War* (Oxford University Press, New York, 1989).

Glenton, Bill: *Mutiny in Force X* (Hodder & Stoughton, London, 1986).

Graham, Otis L., and Wander, Meghan Robinson (eds.): *FDR: His Life and Times* (G. K. Hall, Boston, 1985).

Hattori, Takushiro: *The Complete History of the Greater East Asia War* (Masu Publishing Co., Tokyo, 1953; unpublished English translation on microfilm at Library of Congress, Washington D.C., shelf no. 49330).

Hinsley, F. H.: *British Intelligence in the Second World War* (four vols., Her Majesty's Stationery Office, London, from 1979).

Honan, William H.: *Bywater: The Man Who Invented the Pacific War* (Macdonald, London, 1990).

Hopkins, Captain Harold, RN: *Nice to Have You Aboard* (Allen & Unwin, London, 1964).

Hough, Richard: *The Longest Battle: The War at Sea 1939–45* (Weidenfeld & Nicolson, London, 1986).

Humble, Richard: *Aircraft Carriers* (Michael Joseph, London, 1982).

Ienaga, Saburo: *Japan's Last War: World War Two and the Japanese, 1931–1945* (Basil Blackwell, Oxford, 1979).

Ike, Nobutaka (ed.): *Japan's Decision for War: Records of the 1941 Policy Conferences* (Stanford University Press, Stanford, Calif., 1967).

Iriye, Akira: *Power and Culture: The Japanese-American War 1941–1945* (Harvard University Press, Cambridge, Mass., 1981).

James, D. Clayton: *The Years of MacArthur* (two vols., Houghton Mifflin, New York, from 1972).

Keegan, John: *The Price of Admiralty* (Century Hutchinson, London, 1988).

Kemp, Peter: *The Oxford Companion to Ships and the Sea* (Oxford University Press, Oxford, 1976).

Kennedy, Paul: *The Rise and Fall of the Great Powers* (Unwin Hyman, London, 1988).

Kimball, Warren F. (ed.): *Churchill & Roosevelt: The Complete Correspondence* (three vols., Princeton University Press, Princeton, N.J., 1984).

King, Ernest J., with Whitehill, M.: *Fleet Admiral King: A Naval Record* (Eyre & Spottiswoode, London, 1953).

Larrabee, Eric: *Commander in Chief* (Harper & Row, New York, 1987).

Layton, Edwin T.; Pineau, Roger; and Costello, John: *And I Was There: Pearl Harbor and Midway – Breaking the Secrets* (Morrow, New York, 1985).

Leary, William M. (ed.): *We Shall Return! – MacArthur's Commanders and the Defeat of Japan* (University Press of Kentucky, Lexington, 1988).

Lenton, H. T.: *American Battleships, Carriers and Cruisers* (Macdonald, London, 1968).

Lenton, H. T., and Colledge, J. J.: *Warships of World War II* (Ian Allan, London, second ed., 1973).

Lewin, Ronald: *The Other Ultra: Codes, Ciphers and the Defeat of Japan* (Hutchinson, London, 1982).

Liddell Hart, B. H.: *History of the Second World War* (Cassell, London, 1970).

Lifton, Robert Jay: *Death in Life* (Random House, New York, 1967).

Marder, Arthur J.; Jacobsen, Mark; and Horsfield, John: *Old Friends, New Enemies: The Royal Navy and the Imperial Japanese Navy*, vol. II: *The Pacific War 1942–1945* (Oxford University Press, Oxford, 1990).

Masanori, Ito, with Pineau, Roger: *The End of the Imperial Japanese Navy* (W. W. Norton, New York, 1962).

Mason, David: *Who's Who in World War II* (Weidenfeld & Nicolson, London, 1978).

Mason, John T. (ed.): *The Pacific War Remembered: An Oral History Collection* (US Naval Institute Press, Annapolis, Md., 1986).

Middlebrook, Martin, and Mahoney, Patrick: *Battleship: The Loss of the Prince of Wales and the Repulse* (Allen Lane, London, 1977).

Morgan, Ted: *FDR: A Biography* (Grafton, London, 1987).

Morison, Rear Admiral Samuel Eliot, USNR: *History of the US Naval Operations in World War II* (Little, Brown, Boston, from 1947).

——: *The Oxford History of the American People* (Oxford University Press, New York, 1965).

Naito, Hatsuho: *Thunder Gods: The Kamikaze Pilots Tell Their Own Story* (Farrar Strauss & Giroux, New York, 1989).

Okuyima, Masatake; Horikoshi, Jiro; and Caidin, Martin: *Zero: The Story of the Japanese Navy Air Force* (Cassell, London, 1957).

Petillo, Carol Morris: *Douglas MacArthur: The Philippine Years* (Indiana University Press, Bloomington, 1981).

Poolman, Kenneth: *Escort Carrier* (Secker & Warburg, London, 1983).

Popov, Dusko: *Spy/Counterspy* (Weidenfeld & Nicolson, London, 1974).

Potter, E. B.: *Bull Halsey* (US Naval Institute Press, Annapolis, Md., 1986).

——: *Nimitz* (US Naval Institute Press, Annapolis, Md., 1976).

Prange, Gordon W.: *At Dawn We Slept* (McGraw Hill, New York, 1981).

Reischauer, Edwin O.: *The Japanese* (Harvard University Press, Cambridge, Mass., 1977).

Rhodes, Richard: *The Making of the Atomic Bomb* (Simon & Schuster, New York, 1986).

Rohwer, Jürgen: *Axis Submarine Successes 1939–1945* (Patrick Stephens, Cambridge, 1983).

Roskill, Captain S. W., RN: *Naval Policy Between the Wars* (two vols., Collins, London, 1968).

——: *The War at Sea 1939–1945* (three vols., Her Majesty's Stationery Office, London, from 1954).

Rusbridger, James: 'Winds of Warning: Mythology and Fact About Enigma and Pearl Harbor,' *Encounter*, January 1986.

Rusbridger, James, and Nave, Eric: *Betrayal at Pearl Harbor*

(Richard O'Mara Books, London, 1991; Summit Books, New York, 1991).

Schaller, Michael: *Douglas MacArthur: The Far Eastern General* (Oxford University Press, New York, 1989).

Shirer, William L.: *The Rise and Fall of the Third Reich* (Simon & Schuster, New York, 1960; Secker & Warburg, London, 1960).

Simpson, B. Mitchell: *Admiral Harold R. Stark: Architect of Victory 1939–1945* (University of South Carolina Press, 1989).

Sowinski, Larry: *The Pacific War As Seen by USN Photographers during World War II* (Conway Maritime, London, 1981).

Spector, Ronald H.: *Eagle Against the Sun* (Penguin, London, 1987).

Spiller, Roger J. (ed.): *American Military Leaders* (Praeger, New York, 1989).

Taylor, A. J. P.: *English History 1911–1945* (Oxford University Press, Oxford, 1965).

Terkel, Studs: *The Good War* (Penguin, London, 1986).

Thorne, Christopher: *Allies of a Kind: The United States, Britain and the War Against Japan* (Hamish Hamilton, London, 1978).

——: *The Far Eastern War: States and Societies* (Unwin, London, 1986).

Toland, John: *Infamy: Pearl Harbor and Its Aftermath* (Doubleday, Garden City, N.Y., 1982).

Tuchman, Barbara W.: *Stilwell and the American Experience in China 1911–45* (Macmillan, New York, 1971).

US Naval Institute Proceedings: *The Japanese Navy in World War II* (USNI Press, Annapolis, Md., 1969).

Van der Vat, Dan: *The Atlantic Campaign* (Hodder & Stoughton, London; Harper & Row, New York, 1988).

Walton, Francis: *Miracle of World War II: How American Industry Made Victory Possible* (Macmillan, New York, 1956).

Wheal, Elizabeth-Anne; Pope, Stephen; and Taylor, James: *A Dictionary of the Second World War* (Grafton, London, 1989).

Index

ABC-1. *See* 'American-British Conversation Number One'

ABDA (American-British-Dutch-Australian command), 148, 151, 152–3, 154, 158, 169

Abe Hiroaki, Vice-Adm., 228, 282–3

Abe Koki, Rear Adm., 33

Abe Koso, Vice Adm., 260

Abe Noboyuki, Gen., 69

Absolute National Defense Sphere (Japanese), 377, 400, 426

Adachi Hatazo, Lt. Gen., 313, 315, 380

Adak, 234, 235, 333

Adang (Borneo), 152

Addu Atoll, 160

Admiralty Islands, 376, 378, 418, 424

Ainsworth, Rear Adm. Walden L., 339, 340, 387

Ainu, the, 13

Aitape (New Guinea), 380

Akabane Yutaka, 452–3

Akagi (HIJMS), 20, 21–2, 81, 217, 221, 223, 224, 226, 231

Akihito, Emperor, 63, 476

Akiyama Monzo, Rear Adm., 369

Akiyama Teruo, Rear Adm., 339

Alamogordo (N.M.) test-firing, 471, 477

Alaska, 125, 195, 217, 234–5, 331, 332

Albacore (USS), 392

Aleutian Islands, 170, 188, 195, 205, 206, 213, 215, 217, 229, 234–5, 298, 311, 331, 334, 343, 369, 440
  See also Operation AL

Alexander, Gen. Sir Harold, 159, 195

Allen, Maj. Gen. Arthur, 247, 248

Amagiri (HIJMS), 341

Amagi Takahira, Comdr., 227

Ambon, 142, 151

America. *See* United States

America-First Committee. *See* Committee to Defend America First

American Advisory Committee on Uranium, 473

American Black Chamber, The (Yardley), 107

'American-British Conversation Number One' (ABC-1), 94

American-British-Dutch-Australian command. *See* ABDA

Amur River, 66

Anami Korechika, Gen., 480, 482

Andaman Islands, 159, 357

Andreanof Islands, 333

Angaur, 421

Anti-Comintern Pact, 48

ANZAC command, 169, 182

Aoba (HIJMS), 30, 276, 428

Aoki Taijiro, Capt., 226

Arawe (New Britain), 350, 374

'Arcadia' summit conference, 94, 148

Archerfish (USS), 454

Ardennes offensive, of
  Germany, 448

Argonaut (USS), 259

Arima Seiho, Capt., 368

Arizona (USS), 22, 23

Army of Korea (Japanese),
  60, 67, 68

Arnold, Gen. Henry H.
  ('Hap'), 184, 269

Asahi Shimbun, 64

ASBD Number 2 (USS), 323

Aslito airfield (Saipan), 389

Astoria (USS), 229, 258

Atago (HIJMS), 286, 430

Atkins, Lt. Comdr. Barry K.,
  315, 316, 336

Atlanta (USS), 282–3

Atlantic Charter, 357

Atsutasan Maru, 191

Attu, 217, 234, 235, 239, 331,
  332–3, 343, 369

Augusta (USS), 173

Auschwitz, 75

Australasia, 59, 175, 196, 197

Australia, 35, 59, 96, 139, 141,
  148, 151, 153, 158, 169,
  175, 189, 192, 195, 196, 197,
  202–3, 205, 209, 215, 217,
  239, 241, 252, 270, 298, 327,
  385, 403, 404, 418, 447
  Borneo invasion by, 421
  coastwatchers, 255, 340
  and intelligence operations,
    322, 324, 325
  and invasion routes, 359
  mandates of, 345
  military forces of, 136, 137,
    138, 205, 246–50, 295,
    317, 337, 343, 345, 352,
    374, 378, 414, 420, 484
  military vehicles of, 38, 154,
    257, 314, 334–5, 336, 423
  and Operation Cartwheel, 336
  and Papuan campaign, 246–50,
    288, 297
  role of, in Far East, 242–5

'White Australia' policy, 96

Australia (HMAS), 257, 425

Badung Strait, 153

Bailey (USS), 332

Baker Island, 360

Balboa (Panama Canal Zone),
  418

Bali, 142, 149, 152, 153

Balikpapan (Borneo), 150–51

Baltimore (USS), 413

Bandung (Java), 148, 158, 159

Bangkok (Thailand), 450

Banjarmasin (Borneo), 152

Barb (USS), 406

Barber, Lt. Rex, 318

Barbey, Adm. Daniel, 336, 337,
  351, 421, 424, 441

Barton (USS), 284

Basilone, Sgt. John ('Manila
  John'), 279–80

'Bataan Death March,' 165

Bataan Peninsula (Philippines),
  138, 159, 162–5, 245, 261

Batavia (Java), 153, 156,
  157, 158

'Battle of the Atlantic,' 175

Battle of the Bismarck Sea,
  313–15, 316, 337

Battle of Britain, 92, 144

Battle of the Bulge, 120

Battle of Cape Engaño, 437

Battle of Cape Esperance, 331

Battle of Cape St. George, 350

Battle of Chang-ku Feng, 67–9

Battle of the Coral Sea, 206–11,
  212, 213, 215, 216, 229,
  233, 237, 239, 241, 246, 298,
  322, 440

Battle of the East China
  Sea, 463–5

Battle of the Eastern Solomons,
  266–9

Battle of El Alamein,
  288

Battle of Empress Augusta
  Bay, 346–9

Battle of Guadalcanal (naval), 280–88

Battle of the Java Sea, 155–6

Battle of Jutland, 212, 214, 217

Battle of Kolombangara, 340–41

Battle off Kuantan, 39, 40, 134, 160, 161, 315

Battle of Kula Gulf, 339–40

Battle of Leyte Gulf, 427–40, 442

Battle of Midway, 118, 218–34, 239, 298, 317, 320, 322–3, 331, 390, 438, 440, 454
  aftereffects of, 235–8
  foreshadowings of, 212–17

Battle of Midway (film), 220

Battle of North Atlantic, 37

Battle of the North Cape, 419

Battle of the Philippine Sea, 389–95, 399, 400, 428, 474

Battle off Samar, 429, 430, 436, 438, 441, 455

Battle of Santa Cruz, 280–81

Battle of Savo Island, 257–8, 261, 266

Battle of the Sunda Strait, 157–8

Battle of Surigao Strait, 429–33

Battle of Tassafaronga, 290–92

Battle of Tsushima, 22, 55, 81, 82, 115, 319

Battle of Vella Gulf, 342

Battle of Vella Lavella, 344

Belgian Congo, 472, 473

Belgium, 53, 91, 472

Bellinger, Rear Adm. Patrick, 25–6

Bering Sea, 235

Berlin, 471

Betio, 362, 363, 364

Biak island, 380–81, 421

Birmingham (USS), 430

Bismarck, 37

Bismarck archipelago islands, 142, 195, 239, 285, 313, 330, 334, 336, 337, 364–5, 372, 374, 376, 378, 447

Bismarck Sea (USS), 461

Black Chamber (intelligence), 103, 106, 111

Blair, Clay, 193

Blamey, Gen. Sir Thomas Albert, 243–9, 336, 343, 374, 414, 447

Blanchard, Comdr. J. W., 392

Blandy, Rear Adm. William, 382

Bogan, Rear Adm. Gerald F., 430, 431, 435, 436

Boise (USS), 275, 276

Bonin Islands, 342, 414, 420, 460

Borneo, 85, 87, 129, 142, 147, 149, 150, 152, 153, 154, 155, 197, 421, 429, 447
  See also Dutch Borneo; North Borneo

Bose, Subhas Chandra, 356–7

Bougainville, 293, 312, 317, 338, 339, 341, 342, 343, 345, 346, 347, 349, 354, 364, 377, 378, 457

Bradley, Pharmacist's Mate, 2nd Class, John H., 458–9

Brereton, Maj. Gen. Lewis H., 26–7, 28, 154

Brett, Lt. Gen. George H., 245

Brisbane (Australia), 192, 403

Briscoe, Capt. Robert P., 346

Britain, 16, 17, 20, 40, 48, 49, 53, 55, 59, 62, 73, 74, 78, 79, 83, 89, 120, 123, 124, 143, 156, 157, 198, 206, 222, 242, 252, 282, 302, 320, 350, 402, 420, 450, 472, 481
  airpower of, 144, 145
  alliance with U.S., 94, 97, 98, 109, 148, 196, 355, 447
  and atomic bomb, 473
  and Battle of Jutland, 212, 214, 217
  and Cairo Conference, 355
  and Casablanca conference, 307–11, 354
  and German submarines, 173–5, 307, 316, 327, 365
  imperialism of, 357

Britain (*cont.*)
and India, 195
and Indian Ocean, 196, 197,
    205, 308, 329, 416
and intelligence operations,
    104, 108–9, 125, 237, 323,
    324, 326, 440
Japanese rout of, 134–9,
    159–61
and Lend-Lease Act, 93, 98,
    196, 197, 309
in Mediterranean, 35, 92, 117,
    196, 243, 308, 310, 316,
    354, 413
merchant fleet of, 455
Navy of, 35–40, 58, 81, 154,
    155–6, 169, 172, 307,
    315, 379, 407, 426, 455,
    456, 462, 483, 484
in North Africa, 92, 117, 140,
    195, 288, 307
and nuclear weapons tests, 359
participation in Pacific, 416–18
and Pearl Harbor, 110, 112,
    114, 116–17, 119
and Poland, guarantee
    to, 86, 91
possessions of, 85, 121, 141
and 'Quadrant' conference,
    354–5
Singapore naval base of,
    35–9, 111
sound-detection in, 189
submarines of, 328
and Suez Canal, 76
and 'Trident' conference 354
at Washington Conference
    (1921–22), 58, 94–5, 103
See also Battle of Britain;
    England
British North Borneo, 142,
    143, 150
Brooke-Popham, Air Chief
    Marshal Sir Robert, 136
Brookman, Lt. Comdr, W. H.,
    Jr., 228, 230
Brown, Lt. Howard W., 28, 114

Brown, Vice-Adm. Wilson, 181,
    182–3, 205
Brunei, 142
Buckmaster, Capt. Elliot,
    211, 230
Buckner, Lt. Gen. Simon
    Bolivar, 463, 466, 467
Buin (Bougainville), 293, 317,
    318, 320, 321, 341
Buka, 345, 347, 350
Buna (New Guinea), 241–2,
    246, 248–9
Bungo Suido (exit of Inland
    Sea), 464
Bunker Hill (USS), 349
Burke, Capt. Arleigh 350,
    375, 431
Burma, 139, 141, 142, 143, 144,
    148, 149, 159–60, 195, 196,
    287, 308, 309, 311, 313,
    330, 351, 354, 355, 356, 357,
    402, 416
Bush, Dr. Vannevar, 474
Bushido code (Japanese), 51,
    130, 139, 335, 425
Butaritari, 362
    See also Makin Island
Bywater, Hector, 79, 81

Caen (France), 447
Cairo Conference, 355–6, 376
Cairo Declaration, 356
Calhoun, Vice-Adm. William,
    360
California (USS), 23
Callaghan, Rear Adm. Daniel
    J., 282–3, 285, 286, 290
Camranh Bay (Indochina), 154
Canada, 93, 196, 242–3, 333
Canberra (HMAS), 257,
    258, 423
Canton (China), 75
Carlson, Lt. Col. Evans F.,
    259, 359
Caroline Islands, 30, 206, 233,
    251, 266, 281, 298, 299, 342,
    351, 353, 354, 359, 371, 385

Carpender, Vice-Adm. 'Chips,' 336

Cartwheel., *See* Operation Cartwheel

Casablanca conference, 307–11, 354

CAST (Station), 322

Cavalla (USS), 392

Cavite (Philippines), 106

Celebes, 142, 143, 149, 150, 151, 152, 153

Cero (USS), 405

Ceylon, 37, 154, 156, 160, 161

Chamberlain, Neville, 86

Chang Tso-lin, Marshal, 57–8, 61

Chase, Brig. Gen. William, 377

Ch'engtu (China), 386, 450

Chennault, Maj. Gen. Claire L., 67, 413, 423

'Cherry Blossom' secret society (Japan), 61

Chevalier (USS), 344

Chiang Kai-shek, 58, 66, 74, 84, 123, 128, 141, 184, 308, 356, 468, 469

  as Supreme Allied Commander, 196

Chicago (USS), 256, 257, 258, 296

Chicago *Tribune*, 169, 236–8, 322

Chichi Jima, 389, 395

Chikuma (HIJMS), 281

China, 19, 24, 28, 37, 38, 52, 60, 61, 65, 70, 76, 77, 87, 92, 93, 95, 96, 118, 121, 123, 124, 126, 129, 139, 145, 146, 148, 160, 162, 185, 191, 308, 313, 330, 355, 386, 404, 429, 440, 442, 468

  airfields in, 400, 413, 423, 450

  and Big Four, 309

  combat trials in, 145

  deaths in, during war, 484

  intelligence operations vs., 324–5, 326

  Japanese aggression vs., 20, 43, 46, 62, 65–6, 71, 72, 74, 78–9, 84, 86, 143, 450

  Japanese atrocities in, 75, 121, 188

  nationalists in, 58, 66, 74, 84, 128, 141, 184, 196

  and nine-power treaty, 95

  population of, 484

  and 'Sextant' conference, 355–6

  soldiers of, 295

  U.S. 'Open Door' policy in, 17, 76, 95

  Western influence in, 14, 17

  and Yalta Conference, 448, 469–70

  See also China Incident

China Incident, 46, 66, 67, 70, 72–7, 121, 162, 204

Chinese People's Republic. *See* China

Chitose (HIJMS), 267

Chiyoda (HIJMS), 393

Choiseul, 345, 346

Chokai (HIJMS), 256, 257–8

Christie, Rear Adm. Ralph, 404

Churchill, Winston Leonard Spencer, 92, 110, 114, 134, 173, 196, 356, 440, 473

  at 'Arcadia' conference, 148

  at Casablanca conference, 307–8, 310

  letter to Roosevelt, 128

  letters from Roosevelt, 127, 133, 139, 196

  and Pearl Harbor, 117, 133

  at 'Octagon' conference, 416–18

  and 'Trident' conference, 354

  and Yalta Conference, 448, 469

Claggett, Comdr. B. D., 430

Clark, Rear Adm. Joseph J., 394–5

Clarke, Col. Carter, 440

Clausewitz, Karl von, 308, 414

Coastwatchers, 255–6, 257, 318, 340, 345
Coe, Lt. J. W., 300
Cold Bay (Alaska), 332
Colhoun (USS), 268
Colombo (Ceylon), 154, 160
Colorado (USS), 168
Comintern (Communist International), 48
Committee to Defend America First, 97
Conable, 1st Lt. Barber W., 457–8
Congress (U.S.), 30, 93, 94, 96, 109, 133, 134, 201, 238, 470
See also House of Representatives; Senate
Connaway, Comdr. Fred, 405
Connolly, Rear Adm. Richard L., 367, 382, 399
Constitution (U.S.), 357, 470
Coral Sea, 272
See also Battle of the Coral Sea
Corlett, Gen. C. H., 367, 370
Cornwall (HMS), 161
Corregidor, 162, 163, 165, 192, 202, 204, 208, 261, 447
Corvina (USS), 362
Crace, Rear Adm. J. C., 206, 207, 209
Cromwell, Capt. John Philip, 405
Crutchley, Rear Adm. V. A. C., 257
Cunningham, Adm. Sir Andrew, 81, 419
Cunningham, Comdr. Winfield S., 32, 34
Curtin, John, 243, 245, 247, 248, 414
Cushing (USS), 283
Czechoslovakia, 86

Dace (USS), 429–30
Dampier Strait, 313, 314, 316, 350, 364

D'Aquino, Mrs. Iva Ikuko Togori. *See* Tokyo Rose
Darter (USS), 429–30, 438
Darwin (Australia), 55, 153, 192, 195
Daspit, Comdr., 302
Dauntless (HMS), 173–4
Davao (Philippines), 147
Davison, Rear Adm. Ralph, 430
Denmark, 81, 91
Denver (USS), 346, 347
De Ruyter (HNMS), 151, 155, 156
Devereux, Maj. James, 32, 34
Dewey, Thomas E., 439–40, 449
Dickson, Maj. Donald, 289–90
Dobervich, Lt. Michael, 164
Dönitz, Grand-Adm. Karl, 134, 140, 179, 190, 193, 309, 404, 405, 410
Doolittle, Lt. Col. James H., 184, 185, 187–9, 206, 314, 318
Doolittle Raid, 194, 207, 212, 386, 450
Doorman, Rear Adm. Karel, 151, 152, 153, 154–6
Dorsetshire (HMS), 161
Douglas, Capt. A. H., 181
Dow, Comdr. L. J., 264
Dresden (Germany), 450
Driscoll, Mrs. Agnes (née Meyer), 104, 105
Driscoll, Michael B., 105
Drought, Father James M., 123
Duncan (USS), 276
Duncan, Capt. Donald B., 183
Dunkirk (France), 92, 198
Dutch, 14–15, 48, 77, 78, 85, 87, 108, 111, 114, 116, 119, 121, 196, 336, 374, 484
on defensive, 142–4, 148–59, 163
See also Holland; Netherlands
Dutch Borneo, 143
See also Netherlands Borneo
Dutch East Indies, 17, 80, 87,

127, 141, 142, 146, 158, 159,
192, 197, 364, 420, 423
See also Netherlands
East Indies
Dutch Harbor (Alaska), 217,
234, 235, 332
Dutch New Guinea, 250, 386

East China Sea, 394
See also Battle of the East
China Sea
East Indies, 14, 48, 74, 78, 80,
85, 142, 143, 148, 153, 157,
351, 416, 428, 455
See also Dutch East Indies
Eck, Lt. Comdr. Heinz-
Wilhelm, 316
Edson, Col. M. A., 270
Edward VIII, King of England,
54
Efate (New Hebrides), 250
Eichelberger, Gen. Robert L.,
249, 380, 381, 446, 447, 448
Einstein, Albert, 471–2
Eisenhower, Gen. Dwight D.,
175, 309, 355, 470
Ellice Islands, 353, 359, 361, 368
Empress Augusta Bay, 345, 378
See also Battle of Empress
Augusta Bay
Engaño, Cape, 437
Engebi Island, 373
England, 80, 84, 115
See also Britain
Eniwetok Island, 367, 371,
372–4, 377, 384, 387, 394,
422, 441
Enola Gay, 478–9
Enright, Comdr. J. F., 454
Enterprise (USS), 24, 31, 118,
168, 170, 180, 182, 184, 185,
216, 218, 220, 223, 225–6,
228, 231–3, 235, 264, 266–8,
280–82, 285, 286, 290, 393
Esperance, Cape, 275, 282, 287
See also Battle of Cape
Esperance

Espionage Act, 237
Espíritu Santo Island, 250, 264,
268, 269, 272
Essex (USS), 348, 360, 430
Exclusion Act, 96
Exeter (HMS), 155–6, 157, 158
Ezaki Yoshio, Col., 377

Fabyan, Col. George, 101
Falkland Islands, 222
Farenbolt (USS), 276
Farragut (USS), 209
'February Incident' of 1936
(Japanese), 64, 312
Fehler, Capt. Johann, 473
Ferebee, Maj. Thomas, 478–9
Fiji, 169, 175, 182, 195, 197, 205,
217, 236, 251, 253
Fillmore, Millard, 14
Finschhafen, 337, 353
First World War, 16, 35, 57, 93,
94, 101, 104, 137, 190, 192,
212, 294, 301, 307, 327, 328,
359, 404, 406
Fitch, Adm. Aubrey W.,
206–8, 345–6
Fletcher (USS), 284, 291
Fletcher, Adm. Frank Jack, 33,
182, 214, 216, 218, 252, 253,
256, 261
and Battle of the Coral Sea,
206–10
and Battle of the Eastern
Solomons, 266, 268, 269
and Battle of Midway, 218–19,
220, 223, 226, 228, 229
Florida Island, 255, 257
Fluckey, Comdr. Eugene B., 406
'Flying Tigers,' 67
Foote (USS), 348
Ford, Comdr. John, 220
Ford Island, 22, 23, 25
Formidable (HMS), 160
Formosa, 27–8, 119, 128, 146,
298, 356, 359, 394, 407, 408,
409, 413, 420, 422, 423, 432,
441, 451, 456, 463, 468

'Four Freedoms,' 357
Fox, Chief Petty Officer
    Leonard J., 23
France, 17, 36, 49, 53, 75, 91,
    120, 138, 139, 192, 196, 250,
    278, 308, 328, 418, 484
  colonies of, 48, 77, 78, 121
  fall of, 76, 80
  and First World War, 93, 128
  invasion of, planned,
    310, 354–5
  and nuclear weapons tests, 359
  and Poland's borders, 91
  at Washington Conference
    (1921–22), 95
  See also Vichy France
Franklin (USS), 423
Fraser, Adm. Sir Bruce, 419, 484
Fremantle (Australia), 154,
    192, 404
French Frigate Shoal, 214
French Indochina, 38, 62, 80, 84,
    117, 136
French New Caledonia, 197, 250
Friedman, William F., 101–2,
    105, 106–9
Fubuki (HIJMS), 276
Fuchida Mitsuo, Lt. Comdr., 21,
    22, 24, 120
Fukudome Shigeru, Adm., 422,
    427, 430
Fuller, Maj. Gen., 380

Galer, Maj. Robert E., 263
Gambier Bay (USS), 434
Gardner, Capt. M. B., 264
Gatch, Capt. Thomas, 286
Gavutu, 255
Gay, Ens. George, 224, 228
GEA Conference. *See* Greater
    East Asia Conference
Geelvink Bay, 380
Geiger, Maj. Gen. Roy, 345,
    350, 382, 399, 421, 462
Genda Minoru, Capt. ('mad
    Genda'), 21–2, 231,
    411–12, 425–6

George VI, King of England,
    242
German-American Bund, 97
Germany, 36, 66, 69, 70, 74, 80,
    92–4, 97, 122, 194, 198,
    217, 242, 243, 287, 288,
    312, 355, 356, 357, 379, 413,
    416, 468
  and antisubmarine warfare,
    408, 410
  Ardennes offensive of, 448
  and atomic bomb, 471–3
  atrocities of, 316
  Blitzkrieg of, 48, 77, 91
  bombing of, 450, 453
  colonies of, 345
  Czechoslovakia takeover by,
    86
  and 'Enigma' cipher-machine,
    108
  in First World War, 212,
    359, 407
  intelligence operations vs.,
    102–3, 112, 119–20, 125
  vs. Japan, for U.S., 308–9
  under Kaiser, 16, 47, 51
  and *Lebensraum*, 73, 92
  National Socialist party
    (Nazis) in, 17, 43, 47, 48,
    63, 73, 75
  Poland attacked by, 91
  and radar detection, 411
  resistance vs. Allied invaders,
    447
  sigint of, 323
  Soviet Union attacked by, 48,
    67, 85, 91, 92, 124, 125,
    133, 204
  submarines (U-boats) of, 140,
    142, 173–5, 179, 189, 190,
    192, 251, 307, 309, 316,
    328, 365, 404–5, 455, 456
  torpedo problem of, 302
  and Tripartite Pact, 78,
    121, 134
  unconditional surrender of,
    310, 468, 477

vs. U.S., 133–4
See also Nazi-Soviet Pact
'Germany first' policy, of Allies, 94, 133, 175, 251–2, 308
Ghormley, Vice-Adm. Robert L., 242, 250–51, 252, 253, 256, 261, 269, 277
Gilbert Islands, 217, 259, 260, 348, 353, 358, 359–61, 362, 364–5, 367, 368, 370, 376, 385, 405, 412
Giligili (Milne Bay), 247
Glassford, Rear Adm. William A., 150
Gloucester, Cape, 312, 314, 350–51, 374
Goettge, Col. Frank, 262
Goggins, Capt. William, 322
Gona (New Guinea), 246, 248, 249
Gordon, Capt. O. L., 155
Goto Aritomo, Rear Adm., 30, 33, 209, 275, 276
Gray, Lt. James, 225–6
Grayback (USS), 405
Grayling (USS), 167
Great Britain. *See* Britain
Greater East Asia (GEA) Conference, 356–7
Greater East Asia Co-Prosperity Sphere, 143, 356
Gregory, (USS), 270
Grenfell Lt. Comdr. Elton W., 168
Grew, Joseph C., 63, 75, 118, 129
Griswold, Maj. Gen. Oscar W., 340, 350, 378
Groves, Brig, Gen. Leslie R., 474
Guadalcanal, 241, 242, 248, 249, 250, 252, 298, 299, 307, 313, 317, 333, 334, 335, 338, 340, 343, 345, 346, 351, 364, 378, 384, 396, 411, 448
campaign for, 255–6, 261–71, 274–5, 277–80, 288–97

naval actions (major) in struggle for, 290–91
price of, U.S. and Japanese, 297
see also Battle of Guadalcanal
Guam, 30–31, 33, 34, 38, 106, 169, 187, 331, 374, 382, 388, 390, 391, 393, 459
assault of, 399–400
Gudgeon (USS), 168, 191
Gulf of Siam, 38
Gwin (USS), 341

Haguro (Haguro), 155, 392
Hainan, 191
Haines, Comdr. John, 259
Halmahera, 420
Halsey, Adm. William Frederick, Jr. ('Bull'), 24, 297, 311–12, 318, 319, 336, 338, 340, 342–3, 345, 348, 360, 364, 378, 379, 414, 420, 421–2, 451, 467, 482
as American naval hero, 182
background of, 180
and Battle of Leyte Gulf, 429,431–3, 435–6, 437, 439
and Doolittle Raid, 184–6, 188
Guadalcanal messages of, 278, 280
on Japanese propaganda, 423–4
at Japanese surrender ceremonies, 484–5
personality of, 180–81
promotion for, 290
too late for Battle of the Coral Sea, 207, 211, 216
typhoon strikes down, 443–5
Hamaguchi Osachi, 58, 60
Hamasuna, 2nd Lt., 319
Hamlin, Lt. (j.G.), H. S., 157–8
Hammann (USS), 230
Hancock (USS), 465
Hankow (China), 451

Hansell, Brig. Gen. Haywood, 451
Hara Tadaichi, Rear Adm., 208
Hara Takashi, 55
Hara Yosimichi, 88
Harding, Maj. Gen. Forrest, 248–9
Harding, Warren, 94
Harmon, Lt. Gen. Millard, 340
Hart, Adm. Thomas H., 26, 87, 129, 147, 148, 152, 191
Haruna (HIJMS), 222, 277, 394
Hashimoto Mochitsura, Comdr., 475
Hatsukaze (HIJMS), 347–8
Hawaii, 25, 29, 31, 33, 35, 81, 95, 109, 117, 128, 166, 168, 169, 170, 175, 182, 191, 192, 195, 205, 206, 213, 214, 215, 260, 328, 342, 399, 441
 intelligence operations in, 112–14, 322
 strategic position of, 79
 See also Oahu; Pearl Harbor
Hayate (HIJMS), 32
Hearst, William Randolph, 98
Hebern, Edward, 105
Helena (USS), 23, 275–6, 283, 339–40
Helfrich, Vice-Adm. Conrad, 152, 154, 157
Heller, Joseph, 263
Henderson, Maj. Lofton, 222
Hermes (HMS), 160–61
Herring, Lt. Gen. Edmund, 248
Hester, Maj. Gen. John, 339, 340
Heyn, Seaman Allen C., 283–4
Hiei (HIJMS), 283, 285, 286
Higashikuni, Prince, 401, 482
Hill, Rear Adm. Harry, 361, 364, 367, 373
Hinsley, Professor, 110
Hiranuma Kiichiro, Baron, 69
Hirohito, Emperor (formerly Crown Prince), 13, 38, 43, 49, 58, 61, 62, 63, 65, 67, 71, 84, 89, 312, 320, 397, 400, 401, 438, 476
 appeal to, 129
 birth of, 54
 domestic threat to throne of, 64
 early travels of, 53–4
 education of, 54–5
 becomes emperor, 57
 at imperial conferences, 88
 and Japanese surrender, 480–82
 on Japan's early victories, 166
 marriage of, 55, 56–7
Hirose, Rear Adm., 149
Hiroshima, 452, 479
Hiroyuki Agawa, 82
Hiryu (HIJMS), 20, 217, 221, 223, 226, 228, 229, 231
Hitler, Adolf, 17, 43, 46, 48, 66, 70, 97, 100, 133, 310, 311, 328, 330, 448, 453, 470, 473
 in Berlin bunker, 351–2
 Blitzkrieg of, 77, 91
 greatest error of, 93
 and Munich, 86
 and Nazi-Soviet Pact, 104, 124
 Poland attacked by, 69, 91
 rise to power of, 47
 suicide of, 468
Hitokappu Bay, 21
Hitoshi Motoshima, 13
Hiyo (HIJMS), 282, 393
Hoashi Masame, Ens., 39
Hodge, Maj. Gen. John, 462
Hoel (USS), 434
Hokkaido, 13, 469
Holland, 77, 83, 117, 124
 See also Dutch; Netherlands
Hollandia (New Guinea), 378, 379–80, 386, 421, 424
Holmes, Lt. Comdr. Jasper, 216
Holtz Bay, 333
Homma Masaharu, Lt. Gen., 29, 149, 162, 163–4
Hong Kong, 38, 111, 141, 143, 147, 423

Honululu (USS), 291, 292, 341, 425
Honshu, 404, 452
Hood (HMS), 37
Hooper, Capt. Gilbert, 276
Hoover, Herbert, 106, 200, 201
Hoover, J. Edgar, 120
Hoover, Rear Adm. John H., 361, 368
Hopei Province (China), 62
Hopkins, Harry, 130
Horaniu (Vella Lavella), 344
Horii Tomitaro, Maj. Gen., 246, 247, 248
Horikoshi Jiro, 145
Hornet (USS), 169, 184–6, 188, 218, 220, 223, 225, 228, 232, 235, 253, 266, 272, 274, 280–81
Hosogaya Boshiro, Vice-Adm., 234, 235, 331, 332, 333
House of Representatives (U.S.), 80, 133
Houston (USS), 151, 153, 155–7
Houston II (USS), 423
Howe, Quincy, 98
Huggard, 1st Class Fire Controlman Douglas J., 284
Hughes, Charles Evans, 103
Hull (USS), 443
Hull, Cordell, 75, 88, 123–4, 126, 127, 129, 130
Hungnam (Korea), 472
Huon Gulf, 183, 205, 334, 337
Huon Peninsula (New Guinea), 313, 337, 350
Hyakutake Harukichi, Lt. Gen., 265, 268, 274, 278, 378
HYPO (Station), 322–3

Iceland, 93
Ichiki Kiyono, Col., 262, 265
Ie Shima, 466
Iida Shojiro, Lt. Gen., 149
Ijuin Masuji, Rear Adm., 344
Ikawa Tadao, 123

Iki, Lt., 40
Illustrious (HMS), 81
Imamura Hitoshi, Lt. Gen., 149, 157, 158, 292, 321, 353
Imperial Rule Assistance Association (Japanese), 43, 121
Inada Hiroshi, Comdr., 181
Independence, 349, 362
India, 35, 137, 139, 154, 160, 195, 312, 330, 356, 386, 402
See also British India
Indianapolis (USS), 474–5
Indian Ocean, 197, 205, 215, 308, 329, 357, 416
raid, 160–61, 195
Indies. *See* East Indies
Indochina, 74, 78, 80, 84, 86, 87, 124, 125, 127, 128, 129, 136, 141, 154, 222, 325, 351, 451, 478
See also French Indochina
Indomitable (HMS), 36, 160
Indonesia, 87, 158, 159, 356
Inner Mongolia, 62
Inouye Shigeyoshi, Vice-Adm., 30, 206, 209–11, 212, 251
Inukai Tsuyoshi, 62
Iran, 187
See also Persia
Itagaki, Gen., 67, 69
Italy, 46, 53, 78, 81, 92, 94, 108, 121, 134, 308, 310, 312, 330, 351, 354, 355, 370, 456, 484
at Washington Conference (1921–22), 95
Ito Seiichi, Vice Adm., 464
Iwakuro Hideo, Col., 123
Iwo Jima, 389, 394–5, 420, 441, 451, 453, 463, 464, 468, 478
assault of, 457–62
Izaki Shunji, Rear Adm., 340

Jackson, Lt, (j. G.) Leona, 31
Jaluit, 367, 374
Japan
airpower of, 144–7, 352

antisubmarine warfare of, 406–11, 456

Army of, 59, 60, 64, 66, 69, 70, 75, 77, 78, 109, 126, 149, 204–5, 287

atomic bombs dropped on, 478–80

atrocities of, 75, 121, 188, 316

autarky goal for, 71–9

authorianism in, 56–7

bombing of, 183–9, 450–54

and Britain, early rout of, 134–9, 159–62

Cabinet in, 372

carriers of, 368

and China, aggression vs., 20, 43, 46, 62, 65–6, 71–2, 74–5, 78–9, 84, 86, 143, 450

and China Incident, 46

codes and ciphers of, 85, 103–4, 111, 215, 237, 317–18, 322, 440

deaths in Second World War, 484

diplomacy before Pearl Harbor, 120–30

early history of, 13–17

earthquakes in, 56

economy of, 63, 73–4, 121

expansionism of, 15–17, 19–20, 79

extremists in, 57, 78–9

food supply for, 352–3

government organizations in, 73–4

and Greater East Asia Conference, 356–7

Imperial Rule Assistance Association, 43, 121

intelligence apparatus of, 324–7

internationalism in, 63

interservice rivalry in, 411

jungle troops, sufferings of, 292, 353

kamikaze attacks of, 425–7, 435, 441, 442, 443, 445, 453, 461, 463, 464, 465

leadership of, 48–52, 61–5, 72–3, 85, 128, 167–8, 372, 467–8, 480

Long Lance torpedo, 341

MacArthur king of, 483

mandates of, 31, 359

merchant navy of, 455, 456–7

'moral embargo' of, 75

Navy of, 58–60, 64, 70, 77–82, 109, 126, 142, 145–6, 148–51, 194, 204–5, 287–8, 456

'New Economic Order,' 79

overextension of, 194–5

Pearl Harbor attack by, 19, 20–26

population in, 52

and Potsdam Declaration, 477–8, 480–81

propaganda of, 387, 423–4

and radar, 340, 409, 410–11

and Russo-Japanese Neutrality Pact, 70

self-sufficiency, struggle for, 71–9

September 6 resolution of, 88–9

shipbuilding of, 352

shipping, vulnerability in, 296, 365

and Soviet Union, 58, 66–70, 121, 122, 329, 469, 475–6, 480

standard of living in, 63, 452–3

submarines of, 38, 141–2, 168, 170, 181, 189, 230, 269, 272, 284, 328–30, 338, 362, 410–11, 421, 475

Supreme War Guidance Council, 126, 468–9, 480

surrender of, 13, 179, 419, 480–82, 484–5

tankers of, 403, 455

and Tripartite Pact, 92, 108, 121, 124, 351

turning point for, 382, 400
'unconditional surrender'
  demanded of, 310
uranium-ore deposits of, 472
U.S. war declaration
  vs., 133–4
war crimes of, 75
war planning of, 77–8, 142–5,
  204–6, 312–13
at Washington Conference
  (1921–22), 94–6, 103
western civilization in, 53
See also Army of Korea;
  Kwantung Army
Jarman, Maj. Gen. Sanderford,
  396
Jarvis (USS), 258
Java, 111, 142, 143, 146, 147,
  148–9, 150, 152, 153, 154,
  155, 156, 158
Java (HNMS), 155, 156
Java Sea, 149
JCS. *See* Joint Chiefs of Staff
Jehol province (China), 62
Jellicoe, Adm. Lord John, 214
Jintsu (HIJMS), 155, 266,
  268, 340
Johnson, Lt. Col. Chandler
  W., 459
Johnston (USS), 434
Johnston, Stanley, 237
Johnston Island, 298
Johore Strait, 137, 138
Joint Chiefs of Staff (JCS)
  (U.S.), 171, 241, 242, 278,
  354–5, 360, 361, 366, 376,
  379, 413, 447, 451, 460, 483
Jolo, 147, 154
Jones, Brig. Gen. Albert
  M., 162–3
Juneau (USS), 283–4
Junghans, Lt. Comdr. Earl A.,
  25
Junyo (HIJMS), 229, 234, 282
Jutland. *See* Battle of Jutland

Kaga (HIJMS), 20, 217,

223, 225, 226–7,
  228
Kagawa Kiyoto, Capt., 350
Kagoshima Bay, 22
Kajika Sadamichi, Rear
  Adm., 31, 32
Kako (HIJMS), 259
Kakuta Kakuji, Vice-Adm.,
  234, 385
Kaku Tomeo, Capt., 231
Kamide, Capt., 407
Kanemitsu, Serg. Maj., 259–60
Kavieng (New Ireland), 346, 374
Kawaguchi Kiyotake, Maj.
  Gen., 265, 268–9,
  270–71, 278
Kawaguchi Susumi, Comdr., 221
Kawai Chiyoko (Umeryu), 82
Kawasa, Vice Adm., 333, 334
Kempei-tai (Japanese), 453
Kendari (Celebes), 151, 152, 153
Kennedy, Lt. John F., 341–2
Kenney, Lt. Gen. George
  C., 269, 313–14, 345–6,
  348, 374
Kido, Marquis, 400–401,
  468, 469
Kiel (Germany), 473
Kimmel, Adm. Husband E.,
  32–3, 120, 129, 168, 169–70,
  171, 176, 179
Kimmins, Comdr. Anthony,
  370–71, 373
Kimura Masatomi, Rear Adm.,
  313, 314
Kimura Shofuku, Rear Adm.,
  334
Kinashi Takaichi, Lt. Comdr.,
  272
King, Maj. Gen. Edward P., 164
King, Adm. Ernest Joseph 80,
  106, 152, 168, 178, 179,
  180, 183, 184, 219, 242, 253,
  269, 298, 303, 311, 336, 366,
  374, 412
  administrative brilliance
    of, 173–6

King, Adm. Ernest J. (*cont.*)
  background of, 172–3
  and Battle of Midway, 233–4
  and British participation in
    Pacific, 416–18
  at Casablanca conference, 308,
    309, 311
  and intelligence operations,
    214–15, 237–8, 322
  Pacific bias of, 251–2
  personality of, 171, 173,
    177, 244
  and Rabaul operation, 239–41
  and submarine threats,
    140, 142
  whipsaw strategy of, 385, 412
Kinkaid, Vice-Adm. Thomas C.,
  194, 280, 282, 285, 286, 332,
  374–5, 377, 404, 419, 424
  and Battle of Leyte Gulf,
    429–35, 439
Kinugasa (HIJMS), 285
Kiribati, 259
Kirishima (HIJMS), 283, 286
Kiriwina, 336
Kisaragi (HIJMS), 32
Kishi Fukuji, Vice Adm., 408
Kiska, 217, 234, 235, 239, 331,
  332, 333–4, 343, 369
Kita, Consul, 113
Kitamura Sohichi, Lt. Comdr.,
  38
Kitts, Capt. Willard, 292
Knowles, Comdr. Don S., 294
Knox, Frank, 26, 130, 152, 173,
  174, 176, 237, 318
Kobe (Japan), 185, 187, 451
Kodiak, 331
Koga Mineichi, Vice-Adm., 321,
  338, 346, 348, 368, 371–2,
  385, 386
Koiso Kuniaki, Gen., 401, 467–8
Kokoda (New Guinea), 246,
  247, 248
Kolombangara, 339–44
Komai, Comdr., 335–6
Komandorski Islands, 332

Komatsu Sakio, Warrant
  Officer, 391–2
Kondo Nobutake, Vice-Adm.,
  148–9, 153, 160, 217, 229,
  232, 266, 268, 280, 282, 285,
  286, 297
Kongo (HIJMS), 277, 428, 455
Konoye, Prince Fumimaro, 65,
  76, 79, 83, 88, 90, 121, 123,
  124, 126, 401, 468, 469, 476
Korea, 13, 16, 43, 60, 61, 67,
  71, 74, 313, 356, 385, 404,
  442, 472
Koreans, 56, 397
Kossler, Lt. Comdr., 392
Kota Bharu (Malaya), 25,
  37, 136
Kotani, Chief Petty Off., 318
Kreuger, Lt. Gen. Walter, 336,
  380, 381, 424, 425, 440–41,
  446, 448
Kuala Lumpur (Malaya), 136
Kuantan, 38
  See also Battle off Kuantan
Kublai Khan, 425
Kuching (Sarawak), 148
Kure (Japan), 328, 479
Kuribayashi Tadamichi, Lt.
  Gen., 460, 461
Kurile Islands, 21, 331, 406,
  422, 469
Kurita Takeo, Vice Adm.,
  232, 277, 282, 393–4,
  428–31, 433–9
Kurokawa, 1st Lt., 361
Kuroki Toshiro, 1st Lt., 352–3
Kurusu Saburo, 122, 127,
  129, 130
Kusaka Ryunosuke, Rear
  Adm., 79, 224–5, 226, 231,
  321, 422
Kuzume Naoyuki, Lt. Col.,
  380, 381
Kwajalein island, 31, 34, 182,
  260, 366–70
Kwantung Army, 46, 57, 60–62,
  66–8, 69, 70, 74

Kyoto (Japan), 452
Kyushu (Japan), 146, 386, 422,
    452, 477, 482

Lae (New Guinea), 183, 205,
    241, 313, 314, 315, 336–7,
    343, 352
Laffey (USS), 283
Lamon Bay (Philippines),
    147, 163
Langley (USS), 154
Lanphier, Capt. Thomas, 318
La Pérouse Strait, 404
Larson, Capt. Harold, 344
Lasswell, Maj. Alva B., 317
Layton, Comdr. Edwin T.,
    214, 317
Leach, Capt. J. C., 36, 39
League of Nations, 46, 59, 61,
    94, 96, 121
Leander (HMNZS), 340
Leary, Vice-Adm. Herbert F.,
    169, 181, 206, 245
Lee, Vice-Adm. Willis A.,
    282, 286, 389, 390, 394,
    431, 435–6
LeMay, Maj. Gen. Curtis, 423,
    451, 452, 453
Lend-Lease Act, 93, 98, 196, 309
Lengo Channel, 291
Leslie, Lt. Comdr. Maxwell, 226
Lewis, John L., 98
Lewis, Capt. Robert, 478
Lexington (USS), 80, 168, 172,
    181, 182, 206, 207, 210–11,
    212, 237, 360
Lexington II (USS), 366,
    422, 427
Leyte (Philippines), 420, 421,
    424, 427–35, 438, 440–43,
    467, 474
Leyte Gulf, 424, 441
  See also Battle of Leyte Gulf
Lifton, Dr. Robert Jay, 479
Lincoln, Abraham, 55
Lindbergh, Charles A.,
    97

Lindsey, Lt. Comdr. Eugene F.,
    225
Linga Roads, 429
Lingayen Gulf (Philippines), 29,
    147, 148, 162, 191, 445
Liscombe Bay (USS), 362
Little (USS), 270
Liversedge, Col., 339
Lockwood, Vice-Adm. Charles
    A., 300–301, 404, 409
London Naval Conference
    (1930), 46, 60, 63, 81, 189
London naval talks (1934), 63
Long, Capt. Andrew, 104
Long, Huey, 201
Long Island (USS), 261
Los Negros Island, 377
Louisiades, the, 209
Lunga Point (Guadalcanal), 257,
    274, 279
Luzon Island (Philippines), 28,
    29, 147, 162, 388, 413, 414,
    420, 422, 423, 424, 427–31,
    437, 440–49, 463, 467
Lytton, Lord, 61
Lytton Report, 61

MacArthur, Arthur (Douglas'
    father), 198–9
MacArthur, Arthur (Douglas'
    son), 202
MacArthur, Gen. Douglas, 26,
    28, 65, 147, 196, 205, 206,
    207, 249, 251, 252, 253,
    257, 269, 293, 298, 301, 308,
    311–12, 314, 336, 345, 350,
    355, 364, 366, 385, 395, 412,
    434, 462
  and Admiralty Islands, 376–7
  and 'Alamo Force,' 334
  in Australia, 243–5
  background of, 198–9
  and 'Bonus Expeditionary
    Force,' 200–1
  Congressional Medal of
    Honor awarded to,
    164, 198

MacArthur, Gen. (*cont.*)
  intelligence operations of,
    324, 326–7
  as king of Japan, 483
  leadership methods of,
    199–200
  and leapfrog, 343, 386
  and Pearl Harbor, 29–30,
    119, 202
  personality of, 178, 197–8,
    201–2
  and Philippines, 26–7, 87, 112,
    114, 162–4, 197, 200,
    202, 376, 378–81, 413–14,
    418, 419–20, 424–5, 441,
    442, 445–8, 485
  and Port Moresby campaign,
    247–8
  and 'Quadrant' conference,
    374
  and Rabaul, 239–42, 337–8
  and 'Reno IV,' 378
  and Roosevelt, 201–2,
    413, 448–9
  as Supreme Allied
    Commander, South-West
    Pacific, 176, 178, 192,
    197, 203
  and surrender of Japan, 484–5
McCain, Vice Adm. John R.,
    253, 257, 264, 432, 435, 436
McCampbell, Comdr. David, 430
McCawley (USS), 339
McClintock, Comdr. D. H., 430
McClure, Brig. Gen. Robert,
    343
McClusky, Lt. Comdr. Clarence,
    225, 231
McConnell, Comdr. R. P., 154
McCormick, Col. Robert R., 98,
    169, 236–7
McCrane, Water Tender, 2nd
    Class, Joseph C., 444
McKinley, William, 99
McMillin, Capt. G. J., 30
McMorris, Rear Adm. Charles,
    331–2

McVay, Capt. Charles B., 474
Madagascar, 288
Madang (New Guinea), 313, 337
Magic (intelligence material),
    102, 111, 119, 125, 126, 129,
    324, 440, 475
Mahan, Adm. Alfred T., 120
Majuro, 367
Makassar Strait, 149, 151
Makassar Town (Celebes), 152
Make Europe Pay Its War Debts
    Committee (U.S.), 97
Makin Island, 259–61, 299, 359,
    360, 361, 363, 405
Makinami, 350
Malaya, 35, 37, 62, 74, 85, 129,
    136, 137, 142, 143, 146, 149,
    164, 356, 416
  See also British Malaya
Maloelap, 367, 368, 374
Malta, 469
Manchu dynasty (China), 46, 61
Manchukuo, 43, 46, 61, 68, 74,
    121, 357, 442
Manchuria, 52, 60, 69, 70, 76,
    91, 95, 313, 356, 469, 480
  Japanese aggression in, 16,
    43, 46, 57–8, 60–62, 66,
    67, 71, 76
Mangrum, Maj. Richard, 262
Manhattan Project, 471, 474
Manila (Philippines), 143, 146,
    147, 163, 192, 426, 440, 442,
    443, 446–7, 448
Manila Bay, 29, 87, 147, 376
Manus (Admiralty Islands),
    418, 424
Marblehead (USS), 151
Marcus Island, 182, 353,
    360
Mariana Islands, 30, 359–60,
    372, 374, 376, 379, 381, 402,
    410, 412, 413, 426, 432, 451,
    460, 474
  attack on, 382–400
Marshall, Gen. George C., 87,
    88, 117, 164, 174, 175, 180,

202, 203, 239, 244, 308, 366, 413, 440
Marshall Islands, 31, 142, 182, 206, 213, 217, 260, 342, 351, 354, 359, 361–3, 372, 376, 382, 385, 387, 399, 412, 420
conquest of, 366–74
Maruyama Masai, Lt. Gen., 278
Mason, Capt. Charles, 280
Massacre Bay, 333
Massey, Lt. Comdr. Lance, 225
Matsui Iwane, Gen., 75
Matsunaga Sadaichi, Rear Adm., 38
Matsuoka Yosuke, 61, 120–25
Matsuzaki Mitsuo, Lt., 22
May, Andrew Jackson, 408
Maya (HIJMS), 332, 430
'May 15 Incident' (Japanese), 62
Meade (USS), 287
Mediterranean, 35, 92, 117, 196, 243, 308, 310, 316, 354, 413, 419
Meiji, Emperor, 15, 54, 89
Menado Peninsula (Celebes), 150, 152
Merrill, Rear Adm. Stanton, 347–8
Meyer, Agnes. *see* Driscoll, Mrs. Agnes
MI. *See* Operation MI
Micronesia, 359–60, 365, 366, 371
Middle East, 195, 205
Midway Island, 20, 31, 50, 168, 170, 174, 187, 188, 195, 205, 206, 217
See also Battle of Midway
Mikawa Gunichi, Vice Adm., 246, 247, 256, 257–9, 266, 275, 427, 428
Mikuma (HIJMS), 232, 233
Mili, 367, 368, 374
Miller, Lt. Henry L., 184
Milne, James, 39
Milne Bay, 246, 247
Mindanao (Philippines), 147,

202, 266, 376, 379, 381, 420, 423, 429, 448
Mindoro, 441, 443, 455
Minneapolis (USS), 170, 291
Minseito party (Japanese), 53, 58, 60
Miri, 147
Missouri (USS), 484
Mitchell, Maj. John W., 318, 319, 321
Mitchell, Lt. Comdr. Samuel, 223
Mitchell, Brig. Gen. William ('Billy'), 200
Mitscher, Vice Adm. Marc A., 185, 318, 319, 368, 371, 372, 375, 376, 382, 387, 389, 391, 392, 394, 395, 420, 422, 431, 435, 437, 461, 464
Mitsubishi corporation, 53, 145, 147
Mitsui concern, 53
Miwa Shigeyoshi, Vice-Adm., 410–11, 428
Miwa Yoshitake, Comdr., 40
Miyamoto, Comdr., 465
MO. *See* Operation MO
Mogami (HIJMS), 232–3, 432–3
Molotov, Vyacheslav, 356, 469, 471
Moluccas, the, 381, 420
Momsen, Capt. Charles B., 405
Monaghan (USS), 443, 444
Monash, Gen., 244
Mongolia, 68
See also Inner Mongolia; Outer Mongolia
Mongols, 14
Mono, 346
Monssen, 284
Montgomery, Rear Adm. Alfred, 349
Montgomery, Gen. Bernard, 199
Moore, Lt. Comdr. John R., 259
Moosbrugger, Comdr. Frederick, 342
Moresby. *See* Port Moresby

Morgan, Ted, 470
Morimura Tadachi. *See*
	Yoshikawa
	Takeo
Morison, Samuel Eliot, 34, 152,
	255, 290, 315, 319, 395, 444
Morocco, 307, 309
Morotai, 381, 420–21, 460
Morton, Lt. Comdr. Dudley W.
	('Mush'), 403–4
Mott, Comdr., 185–6
Mountbatten, Adm. Lord Louis,
	355, 416
Mount Suribachi (Iwo Jima),
	457, 460
Mount Tapotchau (Saipan), 396
Mukden (Manchuria), 46, 52, 61
Mumma, Comdr. Morton C.,
	Jr., 193, 316
Munda (New Georgia), 293, 338,
	339, 341, 342, 346
Munich, 86
Munson, Lt. Comdr. H. G., 256
Murray, Capt. George D., 24
Musashi (HIJMS), 319, 381, 430
Mussolini, Benito, 47, 76, 134,
	330, 355
Mutsuki (HIJMS), 268
Myitkyina (Burma), 402
Myoko (HIJMS), 347, 430

Nachi (HIJMS), 116, 155, 332,
	428, 433
Nadzab (New Guinea), 337, 352
Nagako, Empress, 55, 63
Nagano, Adm., 90
Nagara (HIJMS), 226, 283
Nagasaki, 13, 14, 452, 480
Nagato (HIJMS), 20, 434
Nagoya (Japan), 185
Nagumo Chuichi, Vice Adm.,
	90, 149, 151, 153, 160–62,
	170, 205, 206, 217, 266, 267,
	280, 384, 385, 397
	and Battle of Midway,
		221–6, 228–9,
		231–2, 234

	and Pearl Harbor, 21, 22, 24,
		33, 112, 114–16, 118
	personality of, 20–21
Naha (Okinawa), 467
Najioka, Adm., 33
Naka (HIJMS), 155
Nakajima Torohiko, Vice-Adm.,
	407
Nakanishi Niichi, Lt. Comdr., 39
Nakaya Kenju, Lt. Comdr., 22
Nampo Shoto, 451
Namur island, 367, 368–9
Nanking (China), 75, 188,
	356, 357
Napoleon, 55
Napoleonic Wars, 407
Nashville (USS), 185, 425, 443
Nassau Bay, 336
Nasu Yumio, 278
National Defense Act (U.S.), 77
National Security Agency.
	See NSA
Nauru island, 206, 211, 214, 353
Nautilus (USS), 222, 228, 259
Nave, Capt. Eric, 116
Nazi-Soviet Pact, 69,
	92, 124
Neches (SS), 182
NEGAT (Station), 322
Nelson, Lt. R. S., 393
Nelson, Adm. Viscount
	Horatio, 81, 98, 115
Neosho (SS), 207, 208
Netherlands, 17, 48, 53, 78,
	80, 89, 91
	See also Dutch; Holland
Netherlands Borneo, 150
	See also Dutch Borneo
Netherlands East Indies, 63,
	87, 121
	See also Dutch East Indies
New Britain, 141, 182, 205, 206,
	239, 312, 313, 314, 337,
	346, 348, 350, 351, 364, 372,
	374, 414
New Caledonia, 195, 205, 206,
	217, 236, 250, 251, 285

See also French New
    Caledonia
New Deal, in U.S., 63, 73,
    98, 202
Newfoundland, 94, 173
New Georgia, 293, 336, 338–9,
    340, 341, 342
New Guinea, 189, 195, 197, 236,
    239, 241, 252, 271, 293, 295,
    296, 311, 313, 316, 317, 330,
    346, 359, 380, 381, 385, 403,
    420, 421, 424, 441, 447, 448
    Allied advances in, 350–51,
        374, 377, 378, 412, 414
    Allied conferences on, 355
    Allied counterattack in, 246–8
    and ANZAC command,
        169, 182
    Japanese advance halted
        in, 205–11
    Japanese on defensive in,
        334, 337–8
    Japanese on offensive in,
        141, 165
    Japanese suffering in, 352
    route, 364, 366
    See also Dutch New Guinea
New Hebrides, 207, 239, 250,·
    268, 272, 285
New Ireland, 346, 374, 375
New Jersey (USS), 371, 422,
    431, 437, 465
New Orleans (USS), 292
Newton, Sir Isaac, 43
New York Daily News, 98, 236
New Zealand, 35, 59, 139, 169,
    196, 197, 242, 250, 252, 262,
    336, 340, 344, 346, 350
Nicobar Island, 357
Niihau, 25
Nimitz, Adm. Chester W., 33,
    167, 169, 175, 182, 197,
    206, 211, 235, 241, 242, 252,
    253, 259, 266, 269, 301, 311,
    312, 319, 336, 345, 348, 349,
    354, 364, 365, 376, 412, 416,
    419, 445

    and Aleutians, 216–17,
        331, 343
    appointments of, 251, 277
    background of, 177–8
    and Battle of Leyte Gulf,
        428, 435–9
    and Battle of Midway, 218,
        220, 221, 230, 234
    and Cairo conference, 355–6
    and Casablanca conference,
        308
    as CINCPAC, 168
    COMSUBPAC under, 298
    and Formosa, 413
    and Gilbert Islands, 348, 359
    and intelligence operations,
        214–15, 237, 317–18,
        322–3
    and Mariana Islands, 378–9,
        382–4
    and Marshall Islands, 366,
        367, 371
    and Okinawa, 462–3, 464
    and Palau Islands, 420, 421–2
    on Pearl Harbor, 179
    personality of, 178, 180
    and submarines, 403, 404
    and surrender of Japan, 484–5
    and Tarawa, 364
Nine Power Treaty, 75
Nishida, Maj. Gen., 373
Nishimura Shoji, Vice Adm.,
    150, 428, 429, 430, 432–3
Nishina Yoshio, 472
Nitto Maru, 185, 186
Nogi Maresuke, Gen., 54
Nomonhan Campaign
    (Japanese), 70
Nomura Kichisaburo, Adm.,
    122, 123–7, 129–30
Normandy, 384, 424, 445, 447
Norris, Maj. Benjamin, 222
North Africa, 92, 117, 140, 195,
    251, 269, 278, 288, 307, 308,
    309, 330, 333, 347
Northampton (USS), 281,
    291, 292

North Borneo, 147, 148, 407
  See also British North Borneo
North Carolina (USS), 251,
  267, 272
Norway, 36, 91, 472
Nouméa (New Caledonia), 250,
  253, 269, 272, 278, 280
NSA (National Security
  Agency), 107
Numfoor, 381
Nuremberg trials, 179

Oahu, 25, 79, 169, 171, 181, 218
O'Bannon (USS), 344
O'Brien (USS), 272
Ocean Island, 206, 211, 214
'Octagon' conference, 416–18,
  420
Office of Naval
  Communications. *See* ONC
Office of Naval Intelligence.
  See ONI
Official Secrets Act, 116
Ofstie, Rear Adm. R. A., 434
Ohara, Comdr., 227
O'Hare, Lt. Edward ('Butch'),
  182
Oigawa Maru, 315
Okada Keisuke, 64
O'Kane, Comdr. Richard
  H., 456
Okinawa, 394, 420, 422, 441,
  460, 462–4, 466–7, 469,
  470, 483
Oklahoma (USS), 23
Oldendorf, Rear Adm. Jesse B.,
  387, 424, 432, 435, 439, 445
Ommaney Bay (USS), 445
Omori Sentaro, Rear Adm.,
  347–8
Onami (HIJMS), 350
ONC (Office of Naval
  Communications), 104,
  106, 321–2
ONI (Office of Naval
  Intelligence), 104,
  106, 321–2

Onishi Takijiro, Adm., 426, 427
Operation A-Go, 385–6, 389
Operation AL, 206, 212, 213,
  217, 233, 234–5, 251
Operation Barbarossa, 92
Operation Cartwheel, 311–12,
  336, 345, 350–51, 354, 365
Operation Catchpole, 373
Operation Flintlock, 367
Operation Forager, 382
Operation F-S, 251
Operation Galvanic, 360, 405
Operation Hurricane, 380
Operation Iceberg, 462
Operation Ichi-Go, 413
Operation I-Go, 317
Operation KA, 266
Operation KE, 296
Operation Kon, 381
Operation MI, 206, 212, 213,
  215, 217, 233, 234, 251
Operation MO, 206, 212, 213,
  215, 234, 251
Operation Overlord, 355, 384
Operation Persecution, 380
Operation Reckless, 379
Operation RO-Go, 346, 349
Operation Sho-Go, 422, 425
Operation SHO-1, 428, 442
Operation Sho-II, 423
Operation Ten-Go, 464
Operation Toenails, 338
Operation Torch, 251, 269, 307
Operation Trinity, 471
Operation Watchtower,
  241–2, 253
Operation X, 317
Operation Y, 317
Operation Z, 80
Opium War of 1839–42, 17
Oppenheimer, Dr. J. Robert,
  474, 476
Op-20-G unit, 106, 214, 322, 323
Ormoc (Philippines), 441
Orwell, George, 57
Osaka (Japan), 185
Osmeña, Sergio, 425

Ota Yoko, 479
Outer Mongolia, 68
Overlord. *See* Operation
    Overlord
Ozawa Jisaburo, Vice Adm.,
    149, 160, 161, 447
  and Battle of Leyte Gulf, 428,
    429, 431, 435, 437–9
  and Mariana Islands,
    385, 387–96

Pacific Military Conference, 311
Pacific War Council, 196
Palembang (Sumatra), 152
Palau Islands, 266, 359, 366, 372,
    376, 379, 385, 395, 407, 420,
    421, 422, 460
Panama, 169, 171
Panama Canal, 141, 168,
    329, 418
Panay (USS), 75
Papua (New Guinea), 141, 183,
    195, 205, 241, 242, 251, 265,
    278, 288, 293, 295, 297, 298,
    313, 314, 315, 317, 336, 378
  campaign for, 245–50
Paramushiro (Kurile Islands),
    331
Parry Island, 373
Patch, Maj. Gen. Alexander M.,
    250, 293, 296–7
Patterson (USS), 258
Patterson, Eleanor ('Cissy'),
    98, 237
Patterson, Capt. Joseph, 98, 237
Patton, Gen. George, 199
Pearl Harbor, 28, 92, 96, 146,
    153, 180, 198, 211, 216, 218,
    259, 268, 317, 362, 366, 368,
    404, 439, 459, 474
  aftereffects of, 140, 167–70,
    191
  conference, 413–14
  confusion or conspiracy, 102,
    105, 109–20, 322–3
  damage at, 23–4, 179, 432
  investigations of, 30

Japanese attack on, 19, 20–26,
    32–3, 37–8, 80, 130, 146,
    149, 190, 212, 426
  Japanese planning for, 30–31,
    50, 78, 143, 165, 320
  and MacArthur, 29, 119, 202
  mock attack on, 172
  retaliation for, 183, 298
  and U.S. war declaration vs.
    Japan, 133
Peking, 106
Peleliu, 421, 460
Peleus (SS), 316
Penang (Malaya), 136
Pensacola (USS), 292
Percival, Lt. Gen. Arthur, 136,
    138–9, 159, 485
Perry, Cdre. Matthew C., 14, 17
Pershing, Gen. John J., 102,
    199, 200
Persia, 356
Persons, Maj. Gen. John, 421
Perth (HMAS), 155, 156, 157
Pétain, Marshal Henri, 86
Petillo, Carol M., 202
Philippines, 63, 65, 77, 78, 79,
    96, 129, 138, 141, 169, 191,
    193, 197, 200, 266, 292, 298,
    320, 356–60, 364, 365, 366,
    376, 385–9, 395, 407, 451,
    475, 485
  and Battle of Leyte Gulf,
    427–30, 435, 436, 439
  decisive battle on, 440–42
  and intelligence operations,
    106, 111, 322
  Japanese air attack on,
    26–9, 31, 35
  Japanese offensive vs., 143,
    147–9, 162–5
  kamikaze attack in, 426
  and MacArthur, 26–7, 87,
    112, 114, 162–4, 197–8,
    200–203, 376, 378–81,
    412, 413–14, 418,
    419–20, 424–5, 441, 442,
    445–8, 485

Philippines (*cont.*)
MacArthur return to, 424–5
and submarines, 300, 409
U.S. invasion of, 424–5,
440–48, 455, 460
U.S. invasion plan for, 413–14,
416–18, 419–20
U.S. preparations for assault,
422–4
Philippine Sea, 431, 435
See also Battle of the
Philippine Sea
Phillips, Sir Thomas, 37, 38, 39
Plan Orange, 376
Plan Z, 386
Plunger (USS), 191
Poland, 46, 69, 86, 91
Pollack (USS), 191
Popov, Dusko, 119
Port Arthur (China), 54
Porter (USS), 281
Portland, 283
Port Moresby (New Guinea),
141, 195, 205, 206, 207, 208,
209, 211, 214, 217, 236, 239,
241–2, 246, 247, 248, 269,
270, 335
Portuguese, 151
Potsdam conference, 471, 476–7
Potsdam Declaration, 477–8, 480
Pownall, Rear Adm. Charles,
366
Princeton (USS), 348, 430
Prince of Wales (HMS),
36–7, 39
Prussia, 16, 49, 50
Purnell, Rear Adm. W. R., 26
Purple (intelligence campaign),
102, 105, 107–9, 112
Putnam, Maj. Paul, 31
Pu-yi, Emperor Henry, 46, 61
Pye, Vice-Adm. William S., 33,
34, 170

'Quadrant' conference, 354–5,
374
Quebec conferences. *See*
'Octagon' conference;
'Quadrant' conference
Queen Elizabeth (RMS), 354
Queensland (Australia), 192,
206, 209
Quezon, Manuel, 65, 164,
202, 449
Quincy (USS), 258

Rabaul (New Britain), 182, 208,
241, 242, 246, 251, 255, 256,
265–6, 293, 296, 313, 317,
318, 319, 321, 338, 355, 364,
372, 385
Allied neutralization of, 351
as Allied objective, 269,
311–12, 338, 345
and Casablanca conference,
311
isolation of, 377
and Japanese on defensive,
343–4, 347–9, 368, 377–8
and Japanese offensive, 141,
205–6, 239–41
and Operation MO, 215
Rangoon (Burma), 159
Rawlings, Vice-Adm. Sir
Bernard, 419
Red Magic (intelligence
material), 104, 105, 107,
109, 112–14
Redman, Comdr. John, 322
Redman, Capt. Joseph, 322
Reich, Comdr. Eli T., 455
Reid, Ens. Jack, 218
Rendova Island, 339
'Reno IV,' 378
Repulse (HMS), 36, 37, 39
Reuben James (USS), 173
Reuters news agency, 186
Rhineland reoccupation, 77
Rhodes, Richard, 471
Richardson, Maj. Gen. Robert,
243–4
Richmond (USS), 331
Rikken facility, 472
Rochefort, Comdr. Joseph J.,

104–5, 178, 206, 214–16,
321, 322, 323
Rockwell, Rear Adm. Francis
W., 147, 332, 333
Rocky Mount (USS), 368
Rodgers, Capt. Bertram, 332
Roi island, 367, 368–9
Rommel, Erwin, 92, 140,
288, 308
Rooks, Capt. A. H., 156
'Room 14.' *See* Black Chamber
Roosevelt, Eleanor, 99
Roosevelt, Franklin Delano, 20,
27, 88, 116, 117, 119, 122,
141, 152, 164, 173, 174, 196,
202, 278, 365, 379, 471, 473
  accusations against, 97–8, 110,
  114, 117
  appeals to Hirohito, 129–30
  and 'Arcadia' conference, 148
  background of, 99–100
  at Cairo conference, 355–6
  and Casablanca conference,
  307–11
  death of, 470
  diplomacy of, 92–3, 125,
  134, 418
  'Four Freedoms' of, 357
  illness of, 469–70
  letters to Churchill, 127, 133,
  139, 196
  and MacArthur, 201–2,
  413, 448–9
  and 'moral embargo' of
  Japan, 75
  Nimitz appointed by, 177
  after Pearl Harbor, 133–4
  proposal on dividing world,
  196
  'quarantine' speech of, 74
  re-elected for fourth term, 439
  on Tokyo raid, 188
  at Yalta Conference, 469–70
Roosevelt, Maj. James, 259
Roosevelt, Sara, 99
Roosevelt, Theodore, 99
Rosenthal, Joseph, 458–9

Rota island, 391
Rowell, Lt. Gen. Sydney, 247
Rupertus, Maj. Gen. William,
255, 421
Rusbridger, James, 115–16
Russell Islands, 338
Russia, 16, 22, 43, 54, 73, 113,
125, 140, 288, 309, 319, 379,
469, 472, 483
  See also Soviet Union
Russo-Japanese Neutrality
Pact, 70
Russo-Japanese War, 101
Ryujo (HIJMS), 152, 229, 234,
266, 267
Ryukyu Islands, 414, 420,
422, 462

Sachs, Alexander, 471
Sadako, Princess, 54
Safford, Lt. Laurence L.,
104, 109
Saidor (New Guinea), 351, 374
Sailfish (USS), 193
Saipan, 372, 382, 384, 387–90,
395–8, 400, 401, 402, 410,
413, 451, 452, 462, 467
Saito Yoshitsugu, Lt. Gen., 384,
387, 397
Sakashima Gunto, 463
Sakhalin, 469
Sakonju Naomasa, Vice
Adm., 428
Salamaua (New Guinea), 183,
205, 241, 313, 334–5, 336–7,
343, 352
Salerno (Italy), 370
Salt Lake City (USS), 276, 331
Samar, 388
  See also Battle off Samar
Samoa, 182, 195, 197, 205, 217,
236, 251
Samuel B. Roberts (USS), 434
Sanananda, 249
San Bernardino Strait,
388, 429, 431, 434,
436, 439

San Francisco (USS), 275, 276, 282, 283
San Francisco Conference, 420
UN, 470
Sansapor, Cape, 380
Santa Cruz Islands, 280
See also Battle of Santa Cruz
Saratoga (USS), 33, 168, 180, 181, 212, 223, 230, 235, 253, 266, 267, 269, 284, 290, 348
Sarawak, 142, 147–8
Sarmi (Wakde), 380
Sasaki Noboru, Maj. Gen., 339, 341
Sato Naotake, 475
Saturday Evening Post, 107, 479
Savo Island, 275, 276, 283, 285, 286, 291
See also Battle of Savo Island
Schmidt, Maj. Gen. Harry, 367, 397, 461
Schreier, 1st Lt. H. G., 458
Scott, Rear Adm. Norman, 275, 276, 282–3
Sculpin (USS), 362, 405
Sealion II (USS), 455
Seawitch (USS), 154–5
Seawolf (USS), 421
Second World War
Anglo-American intelligence collaboration in, 108
atomic bomb in, 167
deaths in, 484
historical background of, 13–18
mobilization for, 72–3
shortening of, 13
start of, 46, 91
winners and losers, 483–4
Seiyukai party (Japan), 53, 58, 62
Selfridge (USS), 344
Senate (U.S.), 87, 100, 133, 470
Sendai (HIJMS), 347
'Sextant' conference, 355, 376
Shad (USS), 405
Shanghai (China), 37, 95, 106

'Shanghai Incident,' 62
Shansi (China), 62
Shark (USS), 147
Shedden, Frederick, 245
Sherman, Capt. Forrest P., 261
Sherman, Rear Adm. Frederick C., 210, 348, 349, 430
Shibasaki Keiji, Rear Adm., 362
Shibata Bunzo, Comdr., 27–8
Shidehara Kijuro, Baron, 58
Shigemitsu Mamoru, 484
Shimada, Adm., 372, 401
Shima Kiyohide, Adm., 428, 433
Shinamo (HIJMS), 454
Shiryu (HIJMS), 217
Shoho (HIJMS), 206, 208, 209, 212
Shokaku (HIJMS), 20, 206, 210, 213, 266, 267, 281, 392
Short, Lt. Wallace, 230
Short, Lt. Gen. Walter C., 120, 170
Shortland Islands, 287, 339, 345
Shoumatoff, Elizabeth, 470
Siam, 62, 136, 141, 357
See also Gulf of Siam
Siberia, 66, 235
Sicily, 308, 310, 355
Signal Intelligence Service.
See SIS
Simard, Capt. Cyril T., 218, 221, 222, 232
Sims (USS), 208
Singapore, 63, 143, 164, 243, 356, 407, 416, 429, 451
British naval base at, 35–9, 111
and Japanese offensive, 136–9, 148, 151, 159
and Washington Conference (1921–22), 95
sio (Huon Peninsula), 337
SIS (Signal Intelligence Service) (U.S.), 106–7
Skipjack (USS), 300–301
Slim, Lt. Gen. Sir William, 196, 402

Smith (USS), 281
Smith, Lt. C.C., 191
Smith, Elizabeth, 101
Smith, Lt. Gen. Holland M.
    ('Howlin' Mad'), 333, 361,
    363, 367, 382, 389, 396–9
Smith, Maj. John, 262
Smith, Maj. Gen. Julian C.,
    361, 363
Smith, Maj. Gen. Ralph C.,
    361, 382, 396
SMR. *See* South Manchurian
    Railway
Society Islands, 418
Solomon Islands, 169, 188, 197,
    205–8, 214, 217, 239, 241,
    246, 247, 272, 298, 318–21,
    330, 341, 351, 366, 381, 385,
    386, 447
  and Absolute National
    Defense
  Sphere, 377
battles in, 265, 269–71, 278,
    280, 282, 285–8, 292, 293,
    296–7, 333–9, 346–7, 350
  Japanese attempt to regain air
    superiority in, 316–17
  Japanese holdouts in, 414
  and Operation Cartwheel, 365
  U.S. invasion of, 242, 250–57,
    259, 311–12, 313
  See also Battle of the Eastern
  Solomons
Somerville, Adm. Sir James,
    160, 161
Sorge, Richard, 67, 68
Soryu (HIJMS), 20, 223, 225,
    226, 227–8, 230
South Dakota (USS), 281,
    286, 391
Southern Manchurian Railway
    (SMR), 46, 58, 61
Soviet Union, 43, 60, 74, 93, 95,
    145, 187, 196, 320, 406, 471,
    473, 483
  Germany attacks, 48, 67, 85,
    91, 92, 124, 125, 133, 204

intelligence gathering on,
    324, 326
  and Japan, 58, 66–70, 121,
    122, 329–30, 469,
    475, 480
  promise to declare war on
    Japan, 356, 469, 477
  and Yalta Conference, 448
  See also Nazi-Soviet Pact;
    Russia
Spence (USS), 348, 443
Sprague, Rear Adm. Clifton A.
    F., 433, 434, 437, 439
Sprague, Rear Adm. Thomas
    L., 424, 433
Spruance, Vice Adm. Raymond
    Ames, 218, 251, 360, 367,
    371, 379, 382, 432, 461, 464,
    467, 474, 482–3
  ability of, 219
  background of, 219
  and Battle of Midway, 174,
    218–20, 223, 226,
    229, 232–4
  and Mariana Islands, 388–99
  personality of, 219
Squalus (USS). *See Sailfish*
    (USS)
Sri Lanka. *See Ceylon*
Stalin, Joseph, 68, 70, 91, 92,
    120, 134, 448, 471, 477
  and Casablanca conference,
    308, 310
  at Tehran Conference, 356
  at Yalta Conference, 469–70
Stalingrad, 288
Stark, Adm. Harold R., 88, 117,
    129, 167–8, 171, 172
Starkenborgh-Stachouwer, A.
    W. L. T. van, 158–9
Stettinius, Edward R., Jr., 471
Stilwell, Gen. Joseph W.
    ('Vinegar Joe'), 196, 402
Stimson, Henry L., 106–7, 117,
    326, 471, 474, 476
St. Lô, 435
St. Louis (USS), 341

Strong (USS), 339
Stump, Adm. Felix B., 433–4
Sturgeon (USS), 150
Sudetenland (Czechoslovakia), 86
Suez Canal, 35, 76
Sugawara, Lt. Gen., 136
Sugiyama Haijime, Gen., 90, 402
Sulu archipelago, 154, 387
Sumatra, 142, 143, 149, 152, 153, 197, 416
Sumiyoshi Tadashi, Maj. Gen., 278
Sunda Strait, 152, 158
  See also Battle of the Sunda Strait
Supreme Court (U.S.), 96
Supreme War Guidance Council (Japanese), 126, 468–9, 480
Surabaya (Java), 151, 155, 157, 158, 192
Surigao Strait, 388
  See also Battle of Surigao Strait
Sutherland, Lt. Gen., 26–7, 28, 245, 269, 311, 376
Suzuki Kantaro, Adm., 64, 468, 478, 480, 482
Suzuki Sasaki, Lt. Gen., 440
Suzuki Yoshio, Vice Adm., 428, 434
Sweeney, Lt. Walter C., 222
Swinburne, Capt. Edward R., 455
Swordfish (USS), 191
Sydney (Australia), 195
'Symbol.' *See* Casablanca conference

Tabata Sunao, Lt. Comdr., 362
Tagami Hachiro, Lt. Gen., 380
Taiho (HIJMS), 391–2
Taisho, Emperor, 16, 54, 57
Taivu Point (Guadalcanal), 268
Taiwan, 27

Takagi Takeo, Vice-Adm., 155–6, 208, 389
Takahashi, Vice Adm., 149, 150
Takanami (HIJMS), 291, 292
Takao (Formosa), 408
Takao (HIJMS), 286, 430
Takashina, Lt. Gen., 399
Takeda, Lt. Col., 399
Talbot, Rear Adm. A. G., 418
Tama (HIJMS), 438
Tambor (USS), 232–3
Tanabe, Lt. Comdr., 230
Tanaka Giichi, Gen., 58
Tanaka Raizo, Rear Adm., 268, 285–7, 288, 290–92
Tanambogo, 255
Tang (USS), 456
Tanimoto, Petty Off. 1st Class, 318
Tarakan, 149, 150
Tarawa, 360, 361, 367, 369, 405
  invasion of, 362–4
Tautog (USS), 191
Tawitawi (Sulu archipelago), 387, 389
Tehran Conference, 356, 469
Tennant, Capt. W. G., 36
Terauchi Hisaichi, Field Marshal Count, 149, 159, 163, 427
'Terminal.' *See* Potsdam conference
ter Poorten, Gen. Hein, 158
Thach, Lt. Comdr. John S., 228
Thailand, 37, 62, 129, 143
Theobald, Rear Adm. Robert A., 217, 235, 331
Thomas, Platoon Serg. Ernest I., 459
Thomas, Comdr. W. D., 434
Tibbets, Col. Paul W., 478
Timor, 142, 151
Tinian, 382, 387, 398, 400, 474, 478
Tinosa (USS), 302
Tisdale, Rear Adm. Mahlon, 291
Titanic (SS), 231
Tjilatjap (Java), 151, 158

Tobruk, 288

Togami, Lt. Comdr., 181–2

Togo Hcihachiro, Adm., 22, 55, 115, 319

Togo Shigenori, 50, 475

Tojo Hideki, Gen., 43, 90, 121, 123, 126, 127, 128, 397, 467
  and Burma front, 287
  new responsibilities for, 313, 372, 400
  out of office, 401
  as scapegoat, 401
  at Tokyo war crimes trial, 50, 401

Tokugawa dynasty (Japanese), 14

Tokyo
  bombing of, 183–8, 314, 452, 461
  earthquake in region, 56
  population of, 453
  war-crimes trial, 50, 401

Tokyo Express, 262, 268, 269, 270, 274, 276, 277, 282, 288, 291, 297, 339, 340, 342, 381, 407

Tokyo Rose (Mrs. Iva Ikuko Togori D'Aquino), 387, 435, 458, 470

Toland, John, 115

Tomonaga, Lt., 221, 222

Torch. See Operation Torch

Torokina, Cape, 345, 347, 349

Toyoda Soemu, Adm., 385, 394, 395, 422, 428, 480

Toyoda Teijiro, Adm., 124, 126

Treasury Islands, 345, 346

Treaty of Commerce and Navigation, 76

Treaty of Kanagawa, 15

Treaty of Versailles, 47
  See also Versailles peace conference

Treaty of Washington of 1922, 37, 53, 63
  See also Washington Conference

'Trident' conference, 354

Trincomalee (Ceylon), 160

Tripartite Pact, 78, 92, 108, 121, 124, 134, 351

Trobriand group, 336

Truk (Caroline Islands), 30, 206, 233, 251, 266, 281, 299, 302, 342, 348, 353, 362, 371, 372, 376, 379, 385, 395, 405

Truman, Harry S., 242, 476–7, 483
  and atomic bomb, 474
  becomes president of U.S., 470–71

Tsushima, See Battle of Tsushima

Tulagi, 205, 206, 207, 208, 215, 239, 241, 242, 251, 255, 270, 274, 287, 292, 293, 317, 340, 342, 364

Turnage, Maj. Gen. A. H., 347

Turner, Vice Adm. Richmond Kelly, 256, 269, 272–7, 282, 319, 336–8, 343, 361–4, 367, 369, 382, 389, 395, 398, 461, 462
  background of, 252–3
  and Battle of Savo Island, 257–8, 261
  personality of, 252–3

Twitty, Brig. Gen. Joseph J., 323

Uchimura, 1st Lt., 352

Ueda Kenkichi, Lt. Gen., 68

Ugaki Matome, Vice-Adm., 318–19

Ulithi, 422, 432, 435, 441, 444

Ultra (intelligence campaign), 111, 112, 119, 235, 237, 241, 321, 322, 324, 326, 405, 440, 475

Umeryu. See Kawai Chiyoko

Umezu Yoshijiro, Gen., 402, 480, 484

Umnak Island, 235

UN. See United Nations

United Kingdom. See Britain

United Nations (UN), 180, 308,
    309, 330, 354, 356, 470
United Nations Declaration, 358
United States
  Air Corps, 87, 484
  airpower of, 145, 146, 183–6,
      218–19, 277, 292,
      296, 352
  alliance with Britain, 94,
      97–8, 109, 148, 196–7,
      354–5, 447
  Army, 87, 106, 109, 112, 117,
      171, 184–5, 198, 251, 484
  and atomic bomb, 13, 471–80
  Boeing B-29 Superfortress,
      386, 450–52, 453
  bombing policy of, 450–54
  and Cairo conference, 356, 376
  carriers of, 368
  and Casablanca conference,
      307–11, 354
  commitments of, to two
      transoceanic wars, 308–9
  deaths in Second World
      War, 484
  diplomacy before Pearl
      Harbor, 120–30
  entry into war of, 91
  and 'Germany first' policy,
      133–4
  industrial capacity of, 133
  intelligence apparatus of, 19,
      100–109, 214–16, 321–4
  interservice rivalry in, 411–12
  isolationism in, 97–8
  and July 1941 intercepted
      message from Japan, 19,
      84, 110
  as leading debtor-nation, 484
  as leading world power,
      307, 484
  leapfrog strategy of, 342
  and Lend-Lease Act, 93, 98,
      197, 309
  'manifest destiny' of, 14
  manufacturing productivity
      of, 384

  Marines, 34, 274, 278–9,
      289–90, 457–62, 484
  medical expertise of, 294–5
  merchant shipping, 141–2, 484
  military might of, 384
  'moral embargo' of Japan, 75
  Navy, 79, 80, 103–4, 109, 112,
      117–18, 140, 144, 167,
      171–81, 183–5, 192, 194,
      198, 231, 252, 258, 276,
      287, 298–303, 456, 484
  neutrality of, 92–3
  New Deal in, 63, 73, 98, 202
  and 'Octagon' conferences,
      416–18, 420
  on offensive, 298
  'Open Door' policy of,
      17, 76, 95
  and Pearl Harbor attack,
      20–26
  and Pearl Harbor conference,
      413–14
  political leadership of, 469–71
  and Potsdam Conference,
      471, 475–7
  and 'Quadrant' conference,
      354–5, 374
  and San Francisco conference,
      420
  Seabees, 250, 467
  shipbuilding of, 455
  Signal Corps, 102, 106
  State Department, 75, 106, 129
  submarines of, 167, 189–93,
      218, 256, 259, 298–303,
      382, 402–10, 454–6, 473
  and surrender of Japan, 484
  Tokyo bombing by, 183–8
  torpedoes, 299–303
  and 'Trident' conference, 354
  troops of, 294–5
  unity of command issue
      for, 239–40, 364–6,
      369–71, 483
  war declaration vs. Japan,
      133
  at Washington Conference

(1921–22), 58, 59,
94–6, 103
and Yalta Conference, 448,
469–70
See also Congress;
Constitution; Joint Chiefs
of Staff; Supreme Court
Ushijima Mitsuru, Lt. Gen.,
463–4, 467
Ushi Point air base (Tinian), 398

Vandegrift, Maj. Gen.
Alexander Archer, 242,
253, 257, 262, 272, 274, 280,
289, 290, 293, 345
Vanuatu, 250
Vasey, Maj. Gen. George,
248, 249
Vatican, 53
Vella island, 347
Vella Lavella 340, 343, 344
See also Battle of Vella
Lavella
Versailles peace conference, 16,
59, 94, 102
See also Treaty of Versailles
Vian, Rear Adm. Philip, 419
Vichy France, 84, 86
Victorious (HMS), 416
Vila airfields (New Georgia),
338, 339
Vincennes (USS), 258
Vogelkop peninsula (New
Guinea), 380, 381

Wahoo (USS), 403–4
Wainwright, Gen. Jonathan
Mayhew, 162, 164, 165, 485
Wakatsuki Reiijiro, 60–61
Wakde, 380
Wake Island, 24, 30, 31–4, 40,
142, 144, 146, 170, 181, 216,
217, 233, 353, 360
Waldron, Lt. Comdr. John,
223, 224
Walker, Capt. Frank C.,
123, 344

Wallace, Henry, 470
Waller, Capt. H. M. L., 157
Walsh, Bishop James E.,
123, 127
Wanawana, 341
Wang-Ching-wei, 356
Ward (USS), 25
Warspite (HMS), 160
Wasatch (USS), 431
Washington (USS), 272, 275,
280, 286
Washington Conference
(1921–22), 55, 58, 94–6, 103
See also Treaty of Washington
of 1922
Washington Times-Herald, 237
Wasp (USS), 169, 251, 261, 266,
268, 272, 275
Watanabe, Capt., 319, 321
Watson, Brig. Gen. T. E., 373
Wavell, Gen. Sir Archibald, 138,
148, 154, 159, 195
Weimar Republic, 73
Welles, Orson 98
Wenneker, Vice-Adm. Paul,
328, 402
Westinghouse Corporation,
300, 303
West Virginia (USS), 23, 119
Wewak (New Guinea), 313, 337,
380, 403
Weyler, Rear Adm. G. L., 432
Wheeler, Burton K., 97–8
Wilhelm II, Kaiser of Germany,
47
Wilkinson, Rear Adm.
Theodore S., 343, 344, 345,
421, 424
Willkie, Wendell, 440
Wilson, Woodrow, 93, 94
Winters, Capt. Thomas H.,
Jr., 427
Woodlark island, 336
World War I. *See* First World
War
World War II. *See* Second
World War

Wotje, 366, 367, 368, 374
Wright, Rear Adm. Carleton,
    290–92
Wymer, Lt. 'Tex,' 28

Yabagi (HIJMS), 464, 465
Yalta Conference, 448, 469–70
Yamaguchi Tamon, Rear Adm.,
    229, 231
Yamagumo (HIJMS), 362, 405
Yamaji (medical officer), 265
Yamamoto Isoroku, Fleet
    Adm., 38, 75, 90, 119, 126,
    172, 194, 205, 258, 266, 280,
    385, 411
  background of, 81–2
  and Battle of the Coral Sea,
    211, 212
  and Battle of Midway, 229–34
  on danger of overoptimism,
    166
  as gambler, 40, 64, 78, 82–3
  importance of, 320–21
  killing of, 318–19
  and London Naval Conference
    of 1930, 60
  and London naval talks
    (1934), 63 .
  messages from, to Nagumo,
    114–15, 116
  and Operation AL, 213, 234–5
  and Operation I-Go, 317
  and Operation MI, 213
  and Operation MO, 206, 213
  overall plan of, 217
  and Pearl Harbor attack,
    20–21, 24, 78, 80–81
  personality of, 82–3
  physical appearance of, 82
  and submarine aircraft-
    carriers, 411

  and submarines, 329
  and Tokyo air defenses, 186
  on war with U.S., 83–4
Yamashita Tomoyuki, Lt. Gen.,
    138, 149, 164, 427, 440
  jungle warfare of, 136
  and Luzon holding operation,
    443, 446, 447
Yamato (HIJMS), 40, 217, 231,
    266, 381, 428, 430, 434,
    454, 464–5
Yamazaki, Col., 333
Yamishiro (HIJMS), 428
Yap Island, 395, 420
Yardley, Herbert O. 101–3,
    105, 106
Yasuda Takeo, Lt. Gen., 472
Yellow Sea, 404
Yokohama (Japan), 56
Yokoi Shoichi, Corp., 399
Yokosuka (Japan), 408, 454
Yokota, Comdr., 269, 284
Yonai Mitsumasa, 401
Yorktown (USS), 168, 182, 206,
    207, 210, 211, 213, 216,
    218, 220, 223, 226, 227, 228,
    229–30, 360
Yoshihito, Crown Prince, 54
Yoshikawa Takeo, Ens. (alias
    Morimura Tadachi), 113
Yubari (HIJMS), 32, 33
Yugiri (HIJMS), 350
Yugumo (HIJMS), 344

Zaibatsu cartel (Japanese), 53,
    63
Zero fighter plane, 145–7, 297
Zhukov, Gen. Georgi, 69, 70, 92
Zuiho (HIJMS), 235, 281
Zuikaku (HIJMS), 20, 206, 235,
    266, 393, 428, 438